ANALYSIS AND SYNTHESIS OF SINGULAR SYSTEMS

ANALYSIS AND SYNTHESIS OF SINGULAR SYSTEMS

ZHIGUANG FENG
JIANGRONG LI
PENG SHI
HAIPING DU
ZHENGYI JIANG

Series Editor

QUAN MIN ZHU

Academic Press is an imprint of Elsevier
125 London Wall, London EC2Y 5AS, United Kingdom
525 B Street, Suite 1650, San Diego, CA 92101, United States
50 Hampshire Street, 5th Floor, Cambridge, MA 02139, United States
The Boulevard, Langford Lane, Kidlington, Oxford OX5 1GB, United Kingdom

Library of Congress Cataloging-in-Publication Data
A catalog record for this book is available from the Library of Congress

British Library Cataloguing-in-Publication Data
A catalogue record for this book is available from the British Library

ISBN: 978-0-12-823739-7

For information on all Academic Press publications
visit our website at https://www.elsevier.com/books-and-journals

Publisher: Mara Conner
Acquisitions Editor: Sonnini R. Yura
Editorial Project Manager: Rafael G. Trombaco
Production Project Manager: Sojan P. Pazhayattil
Designer: Victoria Pearson

Typeset by VTeX

To our families

Contents

Preface

Singular systems, also called descriptor systems, and generalized state-space systems, frequently appear in various engineering systems, such as vehicle suspension systems, flexible robots, large-scale electric networks, chemical engineering systems, and complex ecosystems. Such systems provide a more natural description of dynamic systems than the standard state-space systems, due to the fact that singular systems can preserve the structure of physical systems more than accurately by including nondynamic constraints and impulsive elements. On the other hand, singular systems are, in essence, differential equations coupled with functional equations, and thus the stability problem for singular systems is much more complicated than that for standard state-space systems, because it requires considering not only stability, but also regularity and absence of impulses (for continuous-time singular systems) and causality (for discrete-time singular systems). For these reasons, singular systems not only have great practical significance, but also are of theoretical interest.

Different analysis and synthesis problems, including admissibility and admissibilization, state-feedback control, static output feedback control, bounded real lemma and H_∞ control, dissipative control and filtering, reliable control and filtering, and sliding mode control for linear singular systems and nonlinear singular systems are all thoroughly studied. Less conservative techniques, such as the slack matrix method, the Wirtinger-based inequality, the reciprocally convex combination approach and two equivalent sets, combined with the linear matrix inequality (LMI) technique, are applied to singular systems. This book includes eight chapters. The part of Introduction is given in Chapter 1. Dissipative control and filtering for discrete-time linear singular systems are considered in Chapter 2. For nonzero initial conditions, the H_∞ control with transients problem is solved in Chapter 3. Considering the time delay, the problems of delay-dependent H_∞ control and dissipative synthesis for singular delay systems are stated in Chapters 4 and 5, respectively. For singular Markovian systems, by applying equivalent sets technique, some new formulation of dissipativity conditions are obtained in Chapter 6. Chapter 7 carries out sliding mode control (SMC) problem for singular stochastic Markov systems (SSMSs). In Chapter 8, for nonlinear singular systems, by using Takagi–Sugeno (T-S) fuzzy model to describe, the issues of admissibility analysis and controller design for T-S fuzzy singular systems are investigated.

In sum, the book provides a systematic theory about analysis and synthesis of singular systems by introducing recent theoretical findings. The purpose of this book is to propose a base for further theoretical research or guidance of engineering applications; it can serve as a reference for undergraduate and postgraduate students who are interested in singular systems.

Acknowledgments

We are deeply indebted to our colleagues Prof. James Lam, and Prof. Wei Xing Zheng for their great contributions regarding the contents of the book.

The financial support of the National Natural Science Foundation of China under Grants 61741305, 61763045, Natural Science Foundation of Heilongjiang Province of China under Grant YQ2019F004, the Fundamental Research Funds for the Central Universities under Grant 3072020CFJ0409, the China Postdoctoral Science Foundation under Grant 2018M63034, and Grant 2018T110275, the Fundamental Research Project of Natural Science Foundation of Shaanxi Province under Grant 2020JM-552 are gratefully acknowledged.

Harbin, China — *Zhiguang Feng*
Yan'an, China — *Jiangrong Li*
Adelaide, Australia — *Peng Shi*
Wollongong, Australia — *Haiping Du*
Wollongong, Australia — *Zhengyi Jiang*

May 2020

Acronyms and symbols

ADE	Algebraic and differential equation
BRL	Bounded real lemma
DOF	Dynamic output-feedback
ESPR	Extended strictly positive realness
FMB	Fuzzy-model-based
GARI	Generalized algebraic Riccati inequality
GLE	Generalized Lyapunov equation
KYPL	Kalman–Yakbovich–Popov Lemma
LMI	Linear matrix inequality
IT2	Interval type-2
ODE	Ordinary differential equation
PRL	Positive real lemma
SMJS	Singular Markovian jump system
SOF	Static output-feedback
SSMS	Singular stochastic Markov system
SVD	Singular value decomposition
$\mathbb{R}$	set of real numbers
$\mathbb{R}^+$	set of nonnegative real numbers
$\mathbb{R}^n$	n-dimensional Euclidean space
$\mathbb{R}^{n\times m}$	set of $n \times m$ real matrices
l_2	space of infinite summable vector sequences
$L_2[0,\infty)$	space of square integrable functions
$\in$	belong to
$\triangleq$	defined as
$\square$	end of proof
$\lvert x\rvert$	absolute value of the number x, or Euclidean norm of the vector x
$\lVert f\rVert_2$	$\sqrt{\sum_{k=0}^{\infty}\lvert f(k)\rvert^2}$ for $f \in l_2$, or $\sqrt{\int_0^{\infty}\lvert f(t)\rvert^2 dt}$ for $f \in L_2[0,\infty)$
$\langle x, y\rangle_\tau$	$\int_0^\tau x(t)^T y(t)dt$ for $x, y \in L_2[0,\infty)$, or $\sum_{k=0}^{\tau} x^T(k)y(k)$ for any vector $x, y \in l_2$
$\lVert \mathcal{G}\rVert_\infty$	H_∞ norm of the operator $\mathcal{G}$
I	identity matrix
A^T	transpose of the matrix A
A^{-1}	inverse of the matrix A
$\mathrm{sym}(A)$	$A + A^T$
$\mathrm{diag}(A_1,\ldots,A_n)$	block diagonal matrix with $A_1,\ldots,A_n$ on the diagonal
$A > B$	$A - B$ is positive definite
$A \geq B$	$A - B$ is positive semidefinite
$\begin{bmatrix} A & B \\ \star & C \end{bmatrix}$	$\begin{bmatrix} A & B \\ B^T & C \end{bmatrix}$
$\left[\begin{array}{c\|c} A & B \\ \hline C & D \end{array}\right]$	explicitly partitioned matrix of $\begin{bmatrix} A & B \\ C & D \end{bmatrix}$

CHAPTER 1

Introduction

1.1 Background

The model of the standard state-space system described by ordinary differential equations (ODEs) is widely used in linear control theory, because the state space approach does not only reveal various properties of the system, but also offers us effective system analysis and synthesis methods. In many practical systems, however, some algebraic constrained laws are imposed on the state components, which lead to the singular system description. A singular system essentially composes of a set of algebraic and differential equations (ADEs), containing information of the static and dynamic constraints of a plant.

Singular systems, also called descriptor systems, semistate space systems and generalized state-space systems, can provide a more natural description of dynamic systems than the standard state-space systems, due to the fact that singular systems can preserve the structure of physical systems more accurately by including nondynamic constraints and impulsive elements. Moreover, because the standard state-space system is a special case of the singular system, the singular system form can describe more practical systems than the standard state-space form. Singular systems have strong application background and are frequently employed to model circuit systems [25,137], economic systems [120], constrained mechanical systems [34,67], aircraft control systems [159] and chemical processes [82]. In this book, linear singular systems and nonlinear singular systems are studied.

Apart from the different form compared with the standard state-space systems, singular systems have fundamental differences as pointed out as follows [34,209]:

- A singular system may not have a solution. If a solution exists, there may be more than one solutions. This point is different with standard state-space systems, which have a unique solution for any initial conditions.
- For any initial conditions, the response of a singular system may contain impulse terms and the derivatives of these impulses or noncausal behaviors. Standard state-space systems do not have impulsive or noncausal behaviors.

Analysis and Synthesis of Singular Systems
https://doi.org/10.1016/B978-0-12-823739-7.00008-2

- A singular system usually contains three dynamic models: finite dynamic models, infinite models (which lead to the undesired impulsive behaviors), infinite static models. However, a standard state-space system only contains finite dynamic models.
- The transfer function of a singular system may contain a polynomial matrix, which is not strictly proper, whereas that of a standard state-space system is strictly proper.

Therefore the analysis and the synthesis problems for singular systems are more complicated than those of standard state-space systems, because it is required to consider not only the stability, but also the regularity and nonimpulsiveness (for continuous-time singular systems), or causality (for discrete-time singular systems) characteristics. In sum, investigating singular systems is significant both in practice and theory. The core of studying singular systems basically consists of analysis and synthesis problems such that the singular systems have a desired or satisfactory property, which mainly contains admissibility, performance, and robustness.

The analysis problem in the control-theoretic context is to establish conditions under which a system is guaranteed to have these properties. Admissibility of singular systems is a property as important as the stability of standard state-space systems. A singular system is said to be admissible if it is asymptotically stable, regular, and impulse-free (for continuous-time singular systems) or causal (for discrete-time singular systems). The existence and uniqueness of solution to a singular system can be guaranteed by the regularity. Nonimpulsiveness for a continuous-time singular system means there is not impulsive behavior with the consistent initial conditions, whereas causality for a discrete-time singular system means that the states of a system in the past do not depend on the state in the future. Asymptotic stability guarantees that the state of a singular system approaches the equilibrium when time goes to infinity.

The performance of a dynamic system is usually characterized by an input-output relationship. Among various performance specifications, bounded real (H_∞ performance) and positive real (passivity performance) are two of the most popular choices. The H_∞ performance represents the maximum gain of the system, which characterizes the worst-case norm of the regulated outputs over all exogenous inputs with bounded energy. By designing the loop gain of the system to be less than unity, the closed-loop system is guaranteed to be asymptotically stable. Positive real property is widely used in adaptive control, absolute stability, and robust stable analysis. The stability of a closed-loop system can be realized by passivity control such that the phase lag of the system is less than 180 degrees [201]. However, H_∞ control based on the small gain theorem, and positive real control based on the positive real lemma, both consider the gain and phase performance separately. This may lead to conservative results when used in applications. The dissipativity theory introduced in [180] not only unifies the H_∞ and positive real control theory, but also provides a more flexible and less conservative robust control design as it allows a better trade-off between the gain and phase performances. Based on an input-output energy-related consideration, dissipativity theory has generalized many independent theorems or lemmas, for example, the passivity theorem, bounded real lemma (BRL), Kalman–Yakbovich–Popov lemma (KYPL) and the circle criterion, and provides a unified framework for the analysis and design of control systems [114]. Dissipative systems are very useful for a wide range of fields, such as system, circuit, network, and control theory [12], [147]. It gives strong links amongst physics, systems theory, and control engineering. Since its introduction, the theory of dissipativity has attracted extensive attention in system control, such as for nonlinear systems [242] and for linear systems [130].

Time delays are sources of instability and poor performance of a dynamical system. They always exist in many dynamical systems [89]. Consequently, many stability results and controller design approaches of delay systems have been reported in the literature. Singular time-delay systems are in essence delay differential equations coupled with functional equations, and thus the robust stability problem for singular systems is much more complicated than that for state-space systems, because it requires to consider not only stability robustness, but also regularity and causality (absence of impulses), which may affect the stability of the system. The problems arising from singular time-delay systems are significant both in theory and in practice. A considerable number of studies have been devoted to singular time-delay systems, such as the results on continuous-time systems [156,226], discrete-time cases [88,233], and the references therein. It is worth noticing that various methods were developed to obtain less conservative results. The free-weighting matrices approach [53,231], Jensen inequality method [89], the reciprocally convex combination approach [48], Wirtinger inequality [149], and delay partitioning method [4] are used in many papers. In this paper, the delay partitioning and the reciprocally convex combination approaches have been used to reduce conservative of results for bilinear system with time-varying delay.

On the other hand, with respect to the uncertainties, the method of sliding mode control (SMC) has fast response and good robustness as competitive advantages. Sliding mode control, which is based on the theory of variable structure systems, has been widely applied to robust control of nonlinear systems. The sliding mode control employs a discontinuous control law to drive the state trajectory toward a specified sliding surface and maintain its motion along the sliding surface in the state space [232]. The dynamic performance of the sliding mode control system has been confirmed as an effective robust control approach with respect to system uncertainties and unknown disturbance when the system trajectories belong to predetermined sliding surface.

The synthesis problem in the control-theoretic context is to design a controller or a filter such that the closed-loop system has a desired or satisfactory behavior. Generally speaking, an effective way to synthesize controllers/filters investigation is often based on some performance-based criteria, under which the controlled/filtered system has the desired properties. Then, a controller or a filter will be designed to guarantee the closed-loop system or the filtering error system to satisfy the criterion. For filter design problem, among various filter design methods, Kalman filtering approach is one of the most popular way, which estimates the state vector by optimizing the covariance of the estimation error [2]. However, to use the Kalman filtering approach, the exact statistical properties of the external noises are required, but it is not always satisfied in practical applications. Therefore an alternative approach called H_∞ filtering, which does not need the exact information of the external noises, has received much attention. Since dissipativity theory provides a better trade-off between gain and phase performances, dissipative filtering, which is more general and unifies the H_∞ and passive filtering, has also been an effective approach. Designed filters may be classified into full-order filters and reduced-order filters. When the order of the filter equals the order of the plant, it is called full-order filter; when it is lower, it is called a reduced-order filter. When considering some large-scale systems, a full-order filter may be not suitable to apply, and the reduced-order filter is a better choice.

1.2 Research problems

This book studies the analysis and synthesis problems of linear singular systems and nonlinear singular systems. Some widely encountered factors are taken into consideration, they are time-delay, disturbance, uncertainties,

Markovian, and nonlinear characteristics. Studying singular systems will be more difficult than their standard state-space counterparts, because the regularity and nonimpulsiveness (for continuous case) or causality (for discrete case) characteristics need to be considered along with the stability. The approaches employed to study standard state-space systems cannot be applied directly, and improved methods or new approaches will be derived. The detailed problems being dealt with are listed as follows:

- How to establish conditions under which the singular systems are admissible or/and dissipative for a singular system with or without time-delay, or uncertainty or Markovian or nonlinear characteristic?
- Based on the analysis results, how to design a state-feedback controller or a static output-feedback controller such that the closed-loop singular systems are admissible or/and dissipative? For filter design problem, how to design a full-order or reduced-order filter such that the filtering error system is admissible or/and dissipative?

1.3 Literature review

1.3.1 Singular systems

Formally speaking, singular systems are the generalization of the standard state-space systems. A great many fundamental concepts and results of singular systems have been obtained by extending the concepts and methods utilized in standard state-space systems. Originally, two frequency domain methods, including the geometric approach [91] and the polynomial matrix method [139,168], are used to study control problems of singular systems. Many problems are solved by using the two methods, such as solvability, controllability, and observability [24], disturbance decoupling [49], pole assignment [23], and observer design [139]. However, the mathematical tools used in geometric approaches are very abstract, which makes the computation difficult. Polynomial matrix methods require the same stability margin of all the designed controllers. With the development of the state-space method for standard state-space systems, more and more researchers have extended the method to singular systems. Some fundamental definitions and the control system design results of singular systems are provided systematically by using state-space method in [25]. Until now, a great number of results on singular system about various topics have been obtained, such as minimal realization [61,68], linear-quadratic optimal control [83], model reduction [207], H_2 control [71]. Among the above topics, the following three aspects for singular systems will be reviewed:

Admissibility and admissibilization

As mentioned above, admissibility contains regularity, nonimpulsiveness or causality, and stability, and hence is an important concept for a singular system. Admissibilization problem is to design a feedback controller such that the closed-loop system is admissible. The definition of admissibility is proposed in [25], which contains a basic criterion for checking a singular system whether it is admissibility by calculating the degree of the characteristic polynomial and the roots of the characteristic equation of a singular system. However, the criterion is not used directly, especially for a system of high dimensions. It is well known that the Lyapunov equation approach is a powerful tool in studying standard state-space systems. A Lyapunov theorem of singular systems is given in terms of Riccati equation in [90] under the assumption of regularity. With the assumption, the state feedback control problem is solved in [25]. However, the assumption is very restrictive, because singular systems are not always regular in practice and the regularity of a singular system may be destroyed by feedback control. Without the regularity assumption, Takaba et al. in [161] propose a necessary and sufficient condition in terms of a generalized Lyapunov equation (GLE) for characterizing the admissibility of singular systems. The admissibilization method is proposed in [166], which involves decomposing system matrices. Although the Riccati equation approach plays an important role in the early period of the state-space theory development, the computational difficulties of some kinds of Riccati equation have limited its application. On the other hand, the way of decomposing system matrices makes the design procedure indirect and cumbersome. By using linear matrix inequalities (LMIs), which can be solved via efficient Matlab® LMI toolbox, Sedumi, or Yalmip [9], a necessary and sufficient admissibility condition is established in [129]. However, the condition in [129] contains an equality constraint and a nonstrict inequality, which makes the condition difficult to apply. By proving the equivalence of two sets, the admissibility condition in terms of strict LMIs is provided in [71]. The same strict LMIs are obtained in [209] by using different proof procedure and the admissibilization problem is solved, which can be designed by solving LMIs. Moreover, another necessary and sufficient admissibility condition for continuous-time singular systems is proposed with strict LMIs in [14]. For discrete-time singular system, the admissibility conditions are given in terms of GLE in [70]. Moreover, the assumptions of regularity and causality are needed. Also, under the regularity assumption, a necessary and sufficient admissibility condition is provided in terms of non-strict LMIs in [144], and the controller designed method needs two steps. Without the regularity assumption, the direct state-feedback controller design method is established in [217]. However, the admissibility condition also contains a nonstrict LMI. To obtain tractable and reliable conditions, an equivalent admissibility condition in terms of strict LMIs is proposed in [208].

H_∞ *control and dissipative control*

H_∞ optimal control to suppress disturbances is introduced when there are external disturbances with unknown statistical properties [178]. Very recently, H_∞ control results for singular systems have been addressed in the literature, which guarantee the closed-loop system to be regular, impulse-free, and delay-dependent robustly stable, and meet an H_∞ norm bound constraint on disturbance attenuation. For instance, a generalized algebraic Riccati equation (GARE) solution and LMI approaches are used to the H_∞ control, and dissipative control problem are established in [144], [173], respectively. Due to the effectiveness of solving LMIs and wide applicability of the LMI approach, many efforts are devoted to LMIs conditions. Without the jw-axis zeros condition and rank conditions on the plants, a necessary and sufficient condition for H_∞ control problem is reduced to solving certain generalized algebraic Riccati inequalities (GARIs) or equivalent LMIs [129]. However, to obtain a dynamic output-feedback controller, some equality conditions and nonstrict LMIs should be solved firstly, which leads to computational complexity. To overcome the problem, strict LMIs conditions for H_∞ state-feedback control problem are presented in [165]. For discrete-time singular systems, a BRL in terms of strict LMIs is given in [229], which makes the condition more tractable. However, the method presented in [229] cannot be easily used to solve the controller design problem. A new BRL and H_∞ control problem are solved in [15] by using the augmentation system approach, which can also be employed to study static output-feedback control problem. However, the H_∞ control problem mentioned above are standard H_∞ optimization problem, in which it is always assumed that the initial condition of the system is zero. Actually, the initial states are often uncertain and might be nonzero, which may affect the performance level of disturbance attenuation of the H_∞ control considerably. The case of H_∞ control with transients for singular systems remains an unsolved, yet important, problem with many technical issues to be addressed.

On the other hand, the BRL only uses the gain information, and positive real lemma (PRL) only considers the phase information, which may lead to conservative results in many applications. Dissipativity property provides a unified framework for considering the gain and phase information simultaneously, which have played an important role in control theory and applications [180]. For continuous-time singular systems, without the constraint on the choice of the system realization, a necessary and sufficient dissipativity condition is established and the state-feedback control problem is solved in [127]. The corresponding output-feedback controller design method is provided in [128]. However, the results in [127] and [128] are nonstrict LMIs, which often give rise some computational difficulties. The new KYP lemma for the dissipativity of singular system is characterized in terms of strict LMI in [39]. For discrete-time singular systems, there exists only one necessary and sufficient dissipativity condition reported up to now. The dissipativity condition in terms of nonstrict LMIs is proposed in [29].

For the motivations given earlier, it is natural to pose a research problem as follows:

Problem 1.1. How to study the H_∞ control problem of singular systems when the initial condition is nonzero? How to establish a necessary and sufficient dissipativity condition in terms of strict LMIs to check the admissibility and dissipativity of discrete-time singular systems effectively? How to develop a SOF controller design method such that the closed-loop system is admissible and dissipative?

Filtering

The filtering problem for singular system has received great attention due to its practical significance in the field of signal processing and communication, control applications. A classical method is Kalman filtering, which is developed for singular systems in [138]. However, the requirements of knowing exact information on both the external noises and the model of system are not satisfied in practice. An alternative method called H_∞ filtering has received much attention, because the class of the noise signals only needs to be bounded energy. The H_∞ filtering problem for singular system is firstly investigated in [206], where a necessary and sufficient condition for solving the H_∞ filtering problem is proposed in terms of LMIs. The reduced-order filter is designed in [210] such that the filtering error system is admissible with a prescribed H_∞ performance. By using a similar method, the reduced-order energy-to-peak filtering problems for continuous-time and discrete-time singular systems are tackled in [251] and [250], respectively. It should be pointed out that the conditions obtained in [206], [210],

[251], and [250] involve rank and equality constraints, which introduce numerical difficulties. Moreover, to obtain the desired filter parameters, the complicated matrix structure is needed in them. A reduced-order l_2-l_∞ filter design method for discrete-time singular systems is given in terms of strict LMI in [124]. To the best knowledge of the author, no reduced-order dissipative filter design method for discrete-time singular systems has been reported to date.

Hence, a natural research problem is given as follows:

Problem 1.2. How to design a reduced-order filter in terms of strict LMSs, which can be obtained by solving the LMIs directly, such that the filtering error singular system is admissible and dissipative?

1.3.2 Singular systems with time-delay

Time-delay, often indicating a source of instability and oscillation, is unavoidable in many practical systems, such as chemical plants, neural networks, and networked control systems [179]. Increasing attention has been focused on the delay-dependent analysis results for standard time-delay systems, and various new approaches have been proposed to reduce their conservatism. Many effective methods established in standard state-space time-delay systems have been successfully extended to singular time-delay systems. In this subsection, a number of effective methods for reducing conservatism of the results of time-delay systems are firstly introduced. In what follows, the feedback control and filter synthesis of time-delay singular systems are reviewed.

Techniques to reducing conservatism of results

It is well known that delay-dependent stability conditions are generally less conservative than delay-independent ones, especially in the case when the time-delay is small. The aim of reducing the conservatism of these delay-dependent stability criteria is to establish new stability criteria to provide a maximal allowable delay as large as possible. Among different techniques, the reduced conservatism is mainly obtained from constructing an improved Lyapunov function and employing a better bound on some weighted cross products. For continuous-time systems, by utilizing Park's and Moon's inequalities, delay-dependent stability conditions are proposed in [141] and [133] based on Lyapunov theory, respectively. The descriptor model transformation method and the corresponding Lyapunov function are proposed in [51] to improve the criterion in [141]. Less conservative results generated by introducing the free-weighting matrices method are presented in [66] to investigate the system with time-varying delay in an interval. By using a convex combination approach to estimate an upper bound of the derivative of Lyapunov functional accurately, improved stability criteria are proposed in [151]. However, the difference between delay bounds has to be approximated, because of the inversely weighted nature of the coefficients caused by the Jensen inequality. The reciprocally convex approach, which can deal with the inversely weighted convex combination of the quadratic integral terms directly is provided in [140]. The results in [140] are less conservative than those in [151], and have less decision variables. To further reduce the conservatism, a new Lyapunov functional, which involves some triple integral terms, is constructed in [160]. Another effective and popular method is the delay-partitioning method, which can improve the stability criteria significantly. It is first proposed for systems with constant time-delay in [60]. Then the method is extended to the time-varying case in [134]. Recently, some new and effective methods are proposed to study the stability analysis problem for discrete-time systems with time-varying delay. By constructing a novel Lyapunov functional and using the bounding inequalities for certain cross products, less conservative results are obtained in [53] compared with those in [55]. Furthermore, some useful terms ignored in [53] are introduced in the new Lyapunov functional defined in [229]. By combining a new Lyapunov functional with the free weighting matrix technique, the results in [53] are improved by the result in [229]. To reduce the conservatism of the results in [53], the convex combination approach proposed in [151] is extended to the discrete-time delay systems. Two new and less conservative stability criteria are given in [150]. Moreover, a delay-partitioning method is firstly utilized to deal with the stability analysis problem for discrete-time system with time-varying delay in [132], which have significantly reduced the conservatism compared with existing results. The reciprocally convex approach is also extended to discrete-time systems with time-varying delay in [104], which improves the results in [150]. In sum, two classes approaches are established to make the LMIs hold more easily for reducing the conservatism: one is to establish novel Lyapunov functional and the other is to establish new integral inequality. Up to now, the delay-partitioning method is the most effective way to establish the Lyapunov functional, and the reciprocally convex approach is the best integral inequality.

H_∞ *control and dissipative control*

The robust H_∞ control problem for uncertain singular time-delay systems is investigated in [72,81,213]. However, the sufficient conditions in these three references are delay-independent, and thus conservative. To reduce the conservatism, an equivalent model transformation approach is provided in [50], and the delay-dependent conditions for H_∞ control problem of singular systems with time-delay are obtained in terms of LMIs. The singular system concerned should be transformed to an augmented system to using the results in [50], which makes the analysis and synthesis procedure relatively intricate. Free-weighting matrix methods, which are employed for the standard state-space system are extended to singular systems with time-delay in [196,214] by using neither system transformation nor bounding technique. Moreover, state-feedback controllers are designed in [196,214] such that the closed-loop systems are admissible, while satisfying a prescribed H_∞ performance level. By using the delay-partitioning method and introducing a three-integral term to construct a new Lyapunov functional, an improved delay-dependent BRL is presented in terms of LMIs, and corresponding feedback control problem is solved in [135]. The above-mentioned results are concerned with continuous-time singular systems. In the discrete-time cases, the robust H_∞ control problem for time-delay discrete-time singular systems with parameter uncertainties is solved in [212] based on the delay-independent matrix inequality condition. The delay-dependent case is considered in [122] by using the restricted system equivalent transformation. Although this condition plays a key role in solving the H_∞ control problem, it involves semidefinite and nonlinear problems, and is thus difficult to implement numerically. By applying the delay-partitioning method, strict LMI conditions are proposed to solve the H_∞ control problem in [45].

For dissipative control of state-space systems with time-varying delays, some results have been reported. Different from using the Lyapunov method for analyzing the stability of time-delay system as in [54], dissipative theory is employed in [21]. A considerable number of studies have been devoted to dissipative control for singular systems. Dissipativity theory and KYP conditions are generalized to nonlinear and linear continuous singular systems in [74]. When time-delay appears, there are few results for such systems. By using an LMI approach, sufficient delay-dependent conditions for dissipative control are established in [101]. Based on the extended Itô stochastic differential formula, the problem of dissipative control for stochastic singular time-delay systems is tackled in [117]. Robust dissipative control conditions are presented in [28,143] for time-delay singular systems, but the conditions are delay-independent. The delay-dependent conditions in terms of LMIs for singular time-delay systems are shown in [123], which reduces the conservatism of the results obtained in [143]. It should be pointed out that conservatism still remains in these results. Moreover, a parameter which denotes the level of dissipativeness is often not considered when the problem of dissipative control is investigated.

Hence, a meaningful research problem is raised naturally as follows:

Problem 1.3. How to establish a new admissibility/H_∞ analysis/dissipativity condition in terms of LMIs for singular time-delay systems to further reduce the conservatism of existing results?

Filtering

Considerable attention has been devoted to the H_∞ filtering problem due to its major theoretical significance in the past decade [56,183], and great many applications in the aerospace industry [145], TV tracking systems [205], and data segmentation [186]. Moreover, the H_∞ filtering results based on the theory of state-space systems have been successfully extended to singular time-delay systems. For continuous-time singular time-delay systems, the robust H_∞ filtering problem is dealt with in [230] such that the filtering error singular system is admissible with a prescribed H_∞ level. When both discrete and distributed delay appear, the robust H_∞ filtering problem for singular systems with norm bounded uncertainty are solved in [222]. However, the results in [230] and [222] are delay-independent conditions, which have higher conservatism. By system transformation and decomposing system matrices, a delay-dependent condition of H_∞ filtering for singular time-delay systems is given in [50]. To overcome the complexity introduced by the system transformation and the decomposition of system matrices, an LMI-based delay-dependent condition for the existence of H_∞ filter is proposed in [195]. For discrete-time cases, robust H_∞ filter design method for singular systems with polytopic uncertainty is considered in [78] without system matrix decomposition, or requiring additional assumption on the systems. By using the reciprocally convex combination method, the reliable dissipative filtering problem for singular systems is studied in [41]. Although the results in [78] improve the results in [56,241] when recovering the state-space case, there is still room for improvement.

Recently, passive filtering problem has also received increasing attention. The delay-dependent passive filtering problem for discrete-time singular systems with time-varying delay is investigated in [80], where a finite sum inequality proposed in [241] is employed. The existence condition of a filter is obtained in [115] such that the singular error system with Markovian jump parameters is admissible and satisfies the proposed passivity performance. The mixed H_∞ and passivity performance as a special case of dissipativity is first given [130]. Based on this definition, the mixed H_∞ and passive filtering problem for continuous-time singular systems with time-delay is considered in [187], where the time-delay is constant and the uncertainty is not considered.

In actual implementation, conventional filters for a multi-input-multi-output plant may lead to unsatisfactory performance, due to the temporary failures of sensors, which results in incomplete signal delivered to actuators. Therefore the reliable filter design problem has attracted increasing research attention. By employing adaptive method, adaptive reliable H_∞ filters are designed to compensate the sensor failure effects on systems in [219]. Benefiting from the delay-partitioning method, the problems of reliable H_∞ filtering for discrete time-delay systems with randomly occurring nonlinearities and Markovian jump systems with partially unknown transition probabilities are solved in [110,111], respectively. However, there are few results on dissipative filtering, which is more general and unifies the H_∞ and passive filtering. By using sector-nonlinearity modeling techniques, a (Q, S, R)-dissipative fuzzy filter is designed for a class of nonlinear systems rewritten by a T-S fuzzy model in [112]. A sufficient condition for dissipative filtering problem of linear discrete systems is proposed in terms of LMI in [92]. A nonfragile dissipative filtering problem for a class of nonlinear discrete-time systems with sector bounded nonlinearities is investigated in [220]. However, the important problem of reliable dissipative filter for discrete singular systems with time-varying delay and uncertainties remains to be considered.

Motivated by the discussion, a natural research problem is given as follows:

Problem 1.4. How to investigate the reliable dissipative filtering problem for singular systems both with time-varying delay and uncertainty?

1.3.3 Singular Markovian jump systems (SMJSs)

Practically not all the systems can be appropriately described by the linear time-invariant model, due to the abrupt changes in their structures and parameters caused by some discrete events, such as failures, repairs, and disconnection of some components, or sudden environmental disturbances. As a special class of stochastic hybrid systems, Markovian jump systems have been widely employed to model the systems with sudden variation in their dynamic characterization, see, for example, manufacturing systems in [126], networked control systems in [125], and fault-tolerant control systems in [163]. Therefore a great deal of attention has been devoted to the study of the Markovian jump systems over the past decades, such as stabilization in [37,204,236], H_2 or H_∞ control in [26,155], passivity control and reliable mixed passivity control in [154] and [153], robust extended dissipative control in [152].

H_∞ control and dissipative control

When abrupt changes appear in singular systems, it is reasonable to model them with SMJSs and a considerable number of studies have been devoted to the control of SMJSs. For both continuous-time and discrete-time Markovian singular systems, necessary and sufficient conditions guaranteeing the systems to be admissible and stochastically stable are proposed in [209]. Moreover, the corresponding H_∞ control problem is also solved in the book. However, the results of H_∞ control include an upper bound constraint, which makes the condition highly conservative. By using some inequalities, a more tractable result without the upper bound constraints is obtained in [7]. Unfortunately, the results are obtained in terms of nonstrict LMIs, which make it difficult to solve. Based on the technique proposed in [165], the conditions in terms of strict LMIs of H_∞ control problem for singular Markovian system are given in [235]. For SMJSs with time-delay, delay-dependent H_∞ control problem is investigated in [193] but the results in [193] are applicable to SMJSs with singular value decomposition (SVD) form.

Therefore the following problem will be investigated:

Problem 1.5. How to design a H_∞ controller such that the closed-loop SMJS is admissible and strictly (Q, S, R)-dissipative?

Reliable control

The results of H_∞ control problem are obtained under a full reliability assumption that all sensors, control components, and actuators of the systems are in a good working condition. However, in practical engineering systems, it is unavoidable to encounter the failures of actuators or sensors, which may lead to intolerable performance of the closed-loop systems

[164,225]. Therefore it is necessary to design a reliable controller that can tolerate the actuator failures and guarantee the required performance of the closed-loop system. More and more attention has been paid to the study of reliable control for dynamic systems. By modeling sensor failures and actuator failures as a scaling factor and a disturbance, a reliable H_∞ controller is designed for linear systems such that the closed-loop system is asymptotically stable and satisfies the H_∞ performance in [224]. For singular system, the problem of reliable H_∞ control for uncertain systems with actuator failures and multiple time delays is investigated in [225]. For Markovian jump system, the robust reliable H_∞ control problem for discrete-time systems is solved in [19]. For reliable dissipative control problem, the state feedback controllers and impulsive controllers are designed in [234] for Makovian jump systems with actuator failures and impulsive effects, such that the closed-loop system is robustly stable and strictly (Q, S, R)-dissipative. Reliable dissipative control for singular Markovian jump systems has never been tackled.

Therefore the following problem will be investigated:

Problem 1.6. How to design a reliable state-feedback controller such that the closed-loop SMJS with actuator failures is admissible and strictly (Q, S, R)-dissipative?

1.3.4 T-S fuzzy singular systems

Additionally, fuzzy logic plays a significant role in system modeling and data mining with characterising uncertainty [8,95,203]. The Takagi–Sugeno (T-S) fuzzy rule model is quite a efficient method to approximate the complicated nonlinear systems, which combines the mathematical theory and the fuzzy logic theory [93,176,244]. Based on "IF-THEN" rules, by "blending" every local linear system, the T-S fuzzy-model-based approach is able to handle nonlinearities existing in the application systems [98,119]. A lot of studies have been done on the research of T-S fuzzy systems [99,100,218]. On account of the interval T-S fuzzy approach, the issue of fuzzy control for nonlinear networked control systems with parameter uncertainties and packet dropouts is investigated in [97]. By utilizing delay partitioning approach, the dissipativity analysis is addressed for T-S fuzzy singular systems with constant time delay in [197]. For the interval time-varying delay, a variable delay composition approach is used to solve the issue of the stability of Takagi–Sugeno fuzzy systems in [198]. In [248], the issue of delay-dependent dissipative control of T-S fuzzy singular model with uncertainties and time delay is investigated. However, there exists ample room to improve the proposed results. A new integral inequality is devised to reduce the conservatism of linear systems with a discrete distributed delay in [228], which is much tighter than other existing integral inequalities. This greatly motivates us to extend this new integral inequality technique to singular systems and reduce the conservatism of results about T-S fuzzy singular systems. Therefore the following problem will be investigated:

Problem 1.7. How to obtain less conservative conditions such that the T-S fuzzy singular systems with time delays are admissible?

1.3.5 Type-2 fuzzy singular systems

The aforementioned T-S fuzzy systems are all about type-1 fuzzy systems. When uncertainties are involved in nonlinear plant, type-2 fuzzy sets are presented in [227] to capture the uncertainties effectively, and can provide a better performance than type-1 fuzzy sets [102]. A number of applications have been reported in autonomous mobile robots in [63], and extended Kalman filtering in [77]. Since 1971, many different kinds of type-2 fuzzy sets are proposed, such as interval-valued fuzzy sets [59], interval type-2 (IT2) fuzzy sets [131], and so on. Two excellent references have revealed the history of the development of fuzzy sets and the relationships between them in detail in [10], [11]. However, these mentioned type-2 fuzzy set theory is not mainly for fuzzy-model-based (FMB) control problem, and few results study the type-2 FMB control problems. Recently, IT2 fuzzy systems can provide better performance than that of type-1 fuzzy systems [148]. On the other hand, IT2 fuzzy model has outstanding feature on describing the nonlinear plant subject to parameter uncertainties, which can be captured by the lower and upper membership functions [87]. Therefore there has been a growing interest in studying FMB control of IT2 fuzzy systems. By using the information of footprint of uncertainties and the lower and upper membership functions, a new IT2 controller is proposed in [85] to guarantee the closed-loop IT2 fuzzy system to be stable. The extended dissipative control problem by state feedback and dynamic output feedback for IT2 fuzzy systems is investigated in [96]. By using Lyapunov theory, a type of IT2 filter is designed for IT2 fuzzy systems, such that the filtering error system is D stable and satisfies some required performance in [94]. In [13], the stability of an IT2 fuzzy model is investigated, and the result is applied on a half-vehicle active suspension model. The nonlinear networked control system with parameter uncertainty is modeled by IT2 fuzzy system, and the IT2 fuzzy predictive controller is designed to stabilize the plant in [116]. The state-feedback control problem is solved by designing IT2 fuzzy controller sharing the same lower and upper membership functions with the considered IT2 fuzzy systems in [87]. For continuous-time IT2 fuzzy systems, an IT2 fuzzy controller with different low and upper membership functions is synthesized to guarantee the closed-loop systems to be stable in [86]. For discrete-time IT2 fuzzy systems, the IT2 fuzzy SOF controller is designed to study the reliable stability with mixed H_2/H_∞ performance in [57]. By constructing a fuzzy Lyapunov function, the IT2 fuzzy static output feedback controller is designed for IT2 fuzzy systems in [243]. However, the equality constraint $MK = KB$ in [243] makes the result difficult to solve. Moreover, it should be pointed out that no results about static output feedback control of singular IT2 fuzzy systems have been reported, which is another motivation of this study.

Therefore the following problem will be investigated:

Problem 1.8. How to design a state-feedback controller and static output-feedback controller such that the closed-loop IT2 fuzzy singular systems are admissible?

1.4 Book outline

- **Chapter 2** investigates the dissipative control and filtering of delay-free singular systems. Firstly, the dissipative control problem is considered for continuous-time singular systems. By using two equivalent sets, a novel dissipativity analysis condition is proposed in terms of strict LMIs. Based on this criterion, the state-feedback controller is designed. Secondly, the system augmentation approach is extended to the dissipative control problem of discrete-time singular systems. By giving an equivalent representation of the solution set, a necessary and sufficient dissipativity condition is proposed in terms of strict LMI, such that the singular system is admissible and dissipative. Then by using the system augmentation method, the state-feedback control and SOF control problems are dealt with. The reduced-order filtering problem is transformed to a SOF control problem, and a reduced-order filter is obtained directly by solving LMIs, such that the filtering error system is admissible and dissipative. The effectiveness and applicability of the results are demonstrated by numerical and simulation examples.
- **Chapter 3** addresses H_∞ control with transients for singular systems. For singular system with nonzero initial condition, a new performance

measure is defined firstly. Then a necessary and sufficient condition is given to guarantee the singular system is admissible with an H_∞ performance. Based on this, the state feedback control problem is solved in terms of LMIs. Moreover, the relationship between the new H_∞ performance and standard H_∞ performance is revealed.

- **Chapter 4** explores the delay-dependent admissibility and H_∞ control for discrete-time singular delay systems. Firstly, a triple-summation term is introduced into the Lyapunov function, and the improved reciprocally convex combination approach is utilized to bound the forward difference of the double summation term. The new admissibility criterion, given in terms of LMIs, ensures the regularity, causality, and stability of the considered system. Then the result is extended to singular systems with polytopic uncertainty and disturbances. Delay partitioning technique has been introduced to derive improved results for robust stability and stabilization problems of linear uncertain discrete-time singular systems with state delay, which guarantees the closed-loop system is admissible. Moreover, the proposed new results have been utilized to investigate robust H_∞ control problem, which assures the resulting closed-loop system is admissible with an H_∞ disturbance attenuation. Less conservative and easily verifiable conditions have been formulated in terms of strict LMIs, involving no decomposition of the system matrices. It is also proved that the conservatism of the results is nonincreasing with the reduction of the partition size.
- **Chapter 5** studies the dissipativity analysis and dissipative synthesis of singular systems with time-delay. Firstly, by utilizing the improved reciprocally convex approach and integral inequality, the α-dissipativity analysis condition for discrete-time singular systems with time-varying delay is proposed. Secondly, by employing delay-partitioning method, the less conservative dissipativity conditions are obtained for continuous-time singular systems with time delay. Based on the criteria, the state-feedback controller design method is provided such that the closed-loop system is admissible and strictly α-dissipativity. Finally, the reliable dissipative filter design method is presented for singular systems with time delay and sensor failures such that the filtering error system subject to possible sensor failures is admissible and strictly (Q, S, R)-dissipative.
- **Chapter 6** considers the admissibilization, H_∞ control and reliable dissipative control problems for SMJSs by state-feedback control. Firstly, by proposing two equivalent sets, the necessary and sufficient admissibi-

lization criterion of SMJSs is given in terms of strict LMIs. Secondly, for time delay SMJSs, a less conservative H_∞ control method is presented benefitting from the equivalent sets. Secondly, a sufficient condition in terms of strict LMIs is derived to guarantee that the unforced SMJSs with actuator failure is stochastically admissible and strictly (Q, S, R)-dissipative. Based on the proposed dissipativity analysis results, the condition for the existence of the state-feedback controller is such that the closed-loop system is stochastically admissible and strictly (Q, S, R)-dissipative.

- **Chapter 7** investigates the SMC problem for SSMSs. Firstly, a novel mean square admissibility condition is given in terms of strict LMIs by using replacement of matrix variables. Based on the criterion, the desired state-feedback controller is designed such that the closed-loop system is mean square admissible. Then, the method is applied to solve SMC problem of SSMSs. Numerical examples are provided to demonstrate the applicability of the theoretic results developed.
- **Chapter 8** studies the admissibility analysis for Takagi–Sugeno (T-S) fuzzy singular system with time delay and admissibilization of IT2 fuzzy singular system, respectively. Firstly, a novel tighter integral inequality is utilized to derive a sufficient delay-dependent criterion such that the considered system is admissible. Secondly, based on Lyapunov stability theory, state feedback control criterion and SOF control method are proposed to guarantee the closed-loop system to be admissible. To obtain less conservative results, the information of mismatched membership functions is employed. Numerical examples are given to illustrate the effectiveness of the proposed techniques.

CHAPTER 2

Dissipative control and filtering of singular systems

In this chapter, the problems of dissipative control and filtering of singular systems are investigated. Based on two equivalent sets and parameterizing the solutions of the constraint set, dissipative control for continuous-time singular system and discrete-time singular system have been studied, respectively. Necessary and sufficient conditions are established in terms of strict LMI, which makes the conditions more tractable. By using the system augmentation approach, state feedback controller and static output feedback controller design methods are proposed to guarantee that the closed-loop discrete-time singular systems are admissible and strictly (Q, S, R)-dissipative. Then, the results are applied to tackle the reduced-order filtering problem. The effectiveness of the obtained results in this section are illustrated by numerical examples.

2.1 Dissipative control of continuous-time singular systems

Dissipativity property provides a unified framework considering the gain and phase information simultaneously, which has played an important role in control theory and applications [180]. Many basic tools, such as passivity theorem, bounded real lemma, Kalman–Yakubovic–Popov (KYP) lemma and circle criterion are special cases under dissipativity theory that have attracted considerable attention [29], [40], [189]. In this section, the problems of dissipativity analysis and dissipative control of continuous-time singular systems are addressed by employing the two equivalent sets.

2.1.1 Problem formulation

Consider a class of linear continuous singular systems described by

$$\begin{cases} E\dot{x}(t) = Ax(t) + B_w w(t), \quad x(0) = x_0, \\ z(t) = Cx(t) + D_w w(t), \end{cases} \tag{2.1}$$

where $x(t) \in \mathbb{R}^n$ is the state vector; x_0 is the initial condition; $w(t) \in \mathbb{R}^p$ represents the exogenous input, which includes disturbances to be rejected,

Analysis and Synthesis of Singular Systems
https://doi.org/10.1016/B978-0-12-823739-7.00009-4

and $z(t) \in \mathbb{R}^q$ is the controlled output; A, B_w, C, and D_w are constant matrices with appropriate dimensions. In contrast with standard linear systems with $E = I$, the matrix $E \in \mathbb{R}^{n\times n}$ has $0 < \text{rank}(E) = r < n$. First, we give some definitions and lemmas based on unforced system (2.1):

Definition 2.1. [25]
1. The pair (E, A) is regular if $\det(sE - A)$ is not identically zero.
2. The pair (E, A) is impulse-free if $\deg\{\det(sE - A)\} = \text{rank}(E)$.

In view of this, we introduce the following definition for singular system (2.1):

Definition 2.2.
1. The singular system in (2.1) is said to be regular and impulse-free if the pair (E, A) is regular and impulse-free.
2. The singular system in (2.1) is said to be asymptotically stable if all the finite roots of $\det(sE - A) = 0$ have negative real parts.
3. The singular system in (2.1) or the pair (E, A) is said to be admissible if the system is regular, impulse-free and asymptotically stable.

For a supply rate $s(w, z) = \begin{bmatrix} w \\ z \end{bmatrix}^T S \begin{bmatrix} w \\ z \end{bmatrix}$ with $S \in \mathbb{R}^{(p+q)\times(p+q)}$, the definition of dissipativity is given as follows:

Definition 2.3. The singular system (2.1) is said to be strictly dissipative with respective to the supply rate $s(w, z)$, under zero initial condition, if for any $t_1 \geq 0$ and for any $w \in L_2[0, t_1]$, the following inequality holds:

$$\int_0^{t_1} s(w(t), z(t))dt < 0.$$

Denote $S = \begin{bmatrix} S_{11} & S_{12} \\ \star & S_{22} \end{bmatrix}$, $M = \begin{bmatrix} 0 & I \\ C & D \end{bmatrix}^T S \begin{bmatrix} 0 & I \\ C & D \end{bmatrix}$ as that in [127]; the following lemma gives a necessary and sufficient dissipativity condition for singular system (2.1):

Lemma 2.4. [165,199] *For matrix $E \in \mathbb{R}^{n\times n}$ with $\text{rank}(E) = r \leq n$, denote E_L and E_R are full column rank with $E = E_L E_R^T$, $\text{rank}(E_L) = \text{rank}(E_R) = r$, and let $P = P^T$ such that $E_L^T P E_L > 0$ and Q is nonsingular. U with full row rank and Λ with full column rank are the left and right null matrices of matrix E, respectively,* *that is, $UE = 0$ and $E\Lambda = 0$. Then, $PE + U^TQ\Lambda^T$ is nonsingular, and its inverse is expressed as*

$$(PE + U^TQ\Lambda^T)^{-1} = \bar{P}E^T + \Lambda\bar{Q}U, \tag{2.2}$$

where $\bar{P} = \bar{P}^T$ and $\bar{Q}$ is nonsingular such that

$$E_R^T\bar{P}E_R = (E_L^TPE_L)^{-1}, \ \bar{Q} = (\Lambda^T\Lambda)^{-1}Q^{-1}(UU^T)^{-1}. \tag{2.3}$$

Lemma 2.5. *The following sets are equivalent:*

$$\mathbb{A} = \left\{X \in \mathbb{R}^{n\times n} : E^TX = X^TE \geq 0, \ X \text{ is nonsingular}\right\},$$
$$\mathbb{B} = \left\{X = PE + U^T\Phi\Lambda^T : P = P^T \in \mathbb{R}^{n\times n}, \ E_L^TPE_L > 0, \ \Phi \in \mathbb{R}^{(n-r)\times(n-r)}\right\},$$

$\mathbb{B}$ is nonsingular, where E_L, E_R, $U^T \in \mathbb{R}^{n\times(n-r)}$ and $\Lambda \in \mathbb{R}^{n\times(n-r)}$ are defined in Lemma 2.4, respectively.

Proof. (Sufficiency) Let $X = PE + U^T\Phi\Lambda^T$, then we have $E^TX = X^TE = E_RE_L^TPE_LE_R^T \geq 0$, and X is nonsingular based on Lemma 2.4.

(Necessity) Without loss of generality, denote $X = \begin{bmatrix} X_{11} & X_{12} \\ X_{21} & X_{22} \end{bmatrix}$, $E = \begin{bmatrix} I_r & 0 \\ 0 & 0 \end{bmatrix}$, where $X_{11} \in \mathbb{R}^{r\times r}$ and $X_{22} \in \mathbb{R}^{(n-r)\times(n-r)}$, then we have $E_L = E_R = \begin{bmatrix} I_r \\ 0 \end{bmatrix}$ and $U^T = \Lambda = \begin{bmatrix} 0 \\ I_{n-r} \end{bmatrix}$. By using $E^TX = X^TE$, we have $X_{12} = 0$, $X_{11} \geq 0$. Due to $\text{rank}(E^TX) = \text{rank}(X_{11}) = r$, we arrive at $X_{11} > 0$. Recalling X is nonsingular, it yields that X_{22} is nonsingular. Then, we can find a matrix $P = P^T = \begin{bmatrix} X_{11} & X_{21}^T \\ X_{21} & I_{n-r} \end{bmatrix}$ and $\Phi = X_{22}$ such that $E_L^TPE_L = X_{11} > 0$ and $X = PE + U^T\Phi\Lambda^T$. □

Remark 2.6. It should be noted that Lemma 2.5 is different from that in [42], because the set X_1 in [42] and the set $\mathbb{A}$ in this section are different. On the other hand, the matrices E_L and E_R satisfying $E = E_L^TE_R$ are not used in [42]. The system considered in [42] is discrete-time singular system, and the synthesis results in [42] are just sufficient conditions, whereas the system considered in this section is continuous-time singular systems, and the synthesis results in terms of strict LMIs are necessary and sufficient conditions for delay free singular systems.

Lemma 2.7. [127] *Consider partition of M as $M = \begin{bmatrix} M_{11} & M_{12} \\ \star & M_{22} \end{bmatrix}$, $M_{11} \in \mathbb{R}^{n\times n}$, and suppose that $M_{11} \geq 0$. Then, the singular system in (2.1) is admissible and strictly dissipative if and only if there exists a matrix X such that the following matrix equality and inequalities hold:*

$$E^T X = X^T E \geq 0, \tag{2.4}$$

$$\begin{bmatrix} \mathrm{sym}(A^T X) + M_{11} & X^T B_w + M_{12} + A^T W \\ \star & \mathrm{sym}(W^T B_w) + M_{22} \end{bmatrix} < 0, \tag{2.5}$$

where $W \in \mathbb{R}^{n\times p}$ is a matrix satisfying $E^T W = 0$.

2.1.2 Main results

Based on the above, we present a new necessary and sufficient dissipativity condition below.

Theorem 2.8. *Suppose that $M_{11} \geq 0$. Then, the singular system (2.1) is admissible and strictly dissipative if and only if there exist a symmetric matrix P and a nonsingular matrix Φ such that the following LMIs hold:*

$$E_L^T P E_L > 0, \tag{2.6}$$

$$\begin{bmatrix} \mathrm{sym}(A^T X) + M_{11} & X^T B_w + M_{12} + A^T W \\ \star & \mathrm{sym}(W^T B_w) + M_{22} \end{bmatrix} < 0, \tag{2.7}$$

where

$$X = PE + U^T \Phi \Lambda^T,\ M_{11} = C^T S_{22} C,\ M_{12} = C^T S_{12}^T + C^T S_{22} D_w,$$
$$M_{22} = S_{11} + \mathrm{sym}(S_{12} D_w) + D_w^T S_{22} D_w.$$

E_L, U, and Λ are defined in Lemma 2.5, and $W \in \mathbb{R}^{n\times p}$ is a matrix satisfying $E^T W = 0$.

Proof. From the inequality in (2.5), we can see that $\mathrm{sym}(A^T X) + M_{11} < 0$. Then, it yields $\mathrm{sym}(A^T X) < 0$ from $M_{11} \geq 0$, which implies that matrix X is nonsingular. Therefore together with the condition in (2.4), the matrix X in Lemma 2.7 satisfies the set $\mathbb{A}$ in Lemma 2.5. By the two equivalent sets in Lemma 2.5, we have the desired result. □

Remark 2.9. The result in Theorem 2.8 is more general, because it covers serval important performance analysis criteria as special cases. When $S =$

$\begin{bmatrix} -\mathcal{R} & -\mathcal{S} \\ \star & -\mathcal{Q} \end{bmatrix}$, the result in Theorem 2.8 changes to $(\mathcal{Q}, \mathcal{S}, \mathcal{R})$-dissipativity criterion ([130], [162], [201]). When $S = \begin{bmatrix} -\gamma^2 I & 0 \\ \star & I \end{bmatrix}$, the condition in Theorem 2.8 will be a new bounded real lemma ([129], [173]). A new formulation of positive real lemma can be obtained by setting $S = \begin{bmatrix} 0 & -I \\ \star & 0 \end{bmatrix}$ ([237], [247]).

Remark 2.10. It can be seen that the equality constrain $E^T X = X^T E$ and nonstrict LMI $E^T X \geq 0$ in Lemma 2.7 are removed in the dissipativity criterion in Theorem 2.8, which makes the condition easier to check.

Next we will study the dissipative control problem by using state feedback. The aim is to design the following controller:

$$u(t) = Kx(t) + Jw(t) \tag{2.8}$$

for the open-loop system

$$\begin{cases} E\dot{x}(t) = Ax(t) + Bu(t) + B_w w(t), \quad x(0) = x_0, \\ z(t) = Cx(t) + Du(t) + D_w w(t) \end{cases} \tag{2.9}$$

with $u(t) \in \mathbb{R}^m$, B, and D denoting constant matrices with appropriate dimensions, such that the closed-loop system

$$\begin{cases} E\dot{x}(t) = (A + BK)x(t) + (B_w + BJ)w(t), \\ z(t) = (C + DK)x(t) + (D_w + DJ)w(t) \end{cases} \tag{2.10}$$

is admissible and strictly dissipative. A necessary and sufficient dissipative control method is given in the following theorem:

Theorem 2.11. *Suppose that $S_{22} = N^T N \geq 0$. There exists a state feedback controller in form of (2.8) such that the closed-loop system in (2.10) is admissible and strictly dissipative if and only if there exist a symmetric matrix $\bar{P}$, a nonsingular matrix $\bar{\Phi}$, matrices H and G, such that the following LMIs hold:*

$$E_R^T \bar{P} E_R > 0, \tag{2.11}$$

$$\begin{bmatrix} \mathrm{sym}(AY^T + BH) & \Xi_{12} & (YC^T + H^T D^T)N^T \\ \star & \Xi_{22} & (VC^T + D_w^T + G^T D^T)N^T \\ \star & \star & -I \end{bmatrix} < 0, \tag{2.12}$$

where

$$\begin{aligned} Y &= (\bar{P}E^T + \Lambda\bar{\Phi}U)^T; \\ \Xi_{12} &= AV^T + BG + B_w + YC^T S_{12}^T + H^T D^T S_{12}^T; \\ \Xi_{22} &= \mathrm{sym}(S_{12}CV^T + S_{12}D_w + S_{12}DG) + S_{11}. \end{aligned}$$

E_R, U, and Λ are defined in Lemma 2.5, and $V \in \mathbb{R}^{p\times n}$ is a matrix satisfying $EV^T = 0$. Then, a desired controller can be obtained by $K = HY^{-T}$, $J = G - KV^T$.

Proof. Based on Theorem 2.8, the closed-loop system is admissible and strictly dissipative if and only if the following inequalities hold:

$$E_L^T P E_L > 0, \tag{2.13}$$

$$\Pi = \begin{bmatrix} \mathrm{sym}(A_c^T X) + C_c^T S_{22} C_c & X^T B_{wc} + C_c^T (S_{12}^T + S_{22} D_{wc}) + A_c^T W \\ \star & \mathrm{sym}(W^T B_{wc}) + S_{11} + \mathrm{sym}(S_{12} D_{wc}) + D_{wc}^T S_{22} D_{wc} \end{bmatrix} < 0, \tag{2.14}$$

where

$$A_c = A + BK,\ B_{wc} = B_w + BJ,\ C_c = C + DK,\ D_{wc} = D_w + DJ;$$

$$\Pi = \mathrm{sym}\left(\begin{bmatrix} X^T & 0 \\ W^T & I \end{bmatrix} \begin{bmatrix} A_c & B_{wc} \\ 0 & 0 \end{bmatrix} + \begin{bmatrix} 0 \\ S_{12} \end{bmatrix} \begin{bmatrix} C_c & D_{wc} \end{bmatrix} \right) + \begin{bmatrix} 0 & 0 \\ 0 & S_{11} \end{bmatrix} + \begin{bmatrix} C_c^T \\ D_{wc}^T \end{bmatrix} S_{22} \begin{bmatrix} C_c & D_{wc} \end{bmatrix}.$$

Considering $S_{22} = N^T N \geq 0$ and employing Schur complement equivalence, (2.14) is equivalent to

$$\begin{bmatrix} \mathrm{sym}\left(\begin{bmatrix} X^T & 0 \\ W^T & I \end{bmatrix} \begin{bmatrix} A_c & B_{wc} \\ 0 & 0 \end{bmatrix} + \begin{bmatrix} 0 \\ S_{12} \end{bmatrix} \begin{bmatrix} C_c & D_{wc} \end{bmatrix} \right) + \begin{bmatrix} 0 & 0 \\ 0 & S_{11} \end{bmatrix} & \begin{bmatrix} C_c^T \\ D_{wc}^T \end{bmatrix} N^T \\ \star & -I \end{bmatrix} < 0. \tag{2.15}$$

Performing the congruence transformation to (2.15) by $\mathrm{diag}\left(\begin{bmatrix} Y^T & V^T \\ 0 & I \end{bmatrix}, I\right)$, with $Y = (PE + U^T \Phi \Lambda^T)^{-T}$, $V = -W^T X^{-T}$, and noting that $X =$

$PE + U^T \Phi \Lambda^T$, $YX^T = I$, $VX^T + W^T = 0$, we have

$$\begin{bmatrix} \text{sym}\left(\begin{bmatrix} A_c & B_{wc} \\ S_{12}C_c & S_{12}D_{wc} \end{bmatrix}\begin{bmatrix} Y^T & V^T \\ 0 & I \end{bmatrix}\right) + \begin{bmatrix} 0 & 0 \\ 0 & S_{11} \end{bmatrix} & \begin{bmatrix} Y & 0 \\ V & I \end{bmatrix}\begin{bmatrix} C_c^T \\ D_{wc}^T \end{bmatrix} N^T \\ \star & -I \end{bmatrix} < 0. \tag{2.16}$$

Then setting $H = KY^T$, $G = KV^T + J$ in (2.16), we obtain (2.12). Noting that $E_L^T PE_L = (E_R^T \bar{P} E_R)^{-1}$, (2.14) is equivalent to (2.11). □

Remark 2.12. It should be remarked that a necessary and sufficient dissipative control criterion is proposed in Theorem 2.11. Compared with the controller design method in [127], the equality and nonstrict LMI constraints $EY^T = YE^T \geq 0$ have been moved, which makes the numerical computations more tractable and reliable.

2.1.3 Examples

In this section, numerical examples are provided to show the advantages on numerical computations and the state feedback control of the equivalent sets approach.

Example 2.1. Consider the following singular system:

$$\begin{aligned} \begin{bmatrix} 1 & 0 & 0 \\ 0 & 1 & 0 \\ 0 & 0 & 0 \end{bmatrix} \dot{x}(t) &= \begin{bmatrix} 0 & 1 & 0 \\ -3 & -2 & 1 \\ 0 & 0 & -1 \end{bmatrix} x(t) + \begin{bmatrix} 0 \\ 0 \\ 1 \end{bmatrix} w(t), \\ z(t) &= \begin{bmatrix} 2 & 5 & -2 \end{bmatrix} x(t) + 3w(t). \end{aligned}$$

We choose $E_L = E_R = \begin{bmatrix} 1 & 0 \\ 0 & 1 \\ 0 & 0 \end{bmatrix}$, $U^T = \Lambda = \begin{bmatrix} 0 \\ 0 \\ 1 \end{bmatrix}$. Based on Theorem 2.8, the following three performances are discussed respectively:

- H_∞ case: $S_{11} = -\gamma^2 I$, $S_{12} = 0$, $S_{22} = I$. By solving the LMIs in (2.6)–(2.7), the minimal value of γ is $\gamma_{min} = 3.5551$ and the corresponding decision variables P and Q are computed to be

$$P = \begin{bmatrix} 40.8318 & 0.7772 & -6.0000 \\ 0.7772 & 18.2773 & -14.9999 \\ -6.0000 & -14.9999 & 0 \end{bmatrix}, \quad Q = 10.6386.$$

To verify the effectiveness of the result, the maximal singular values (MSVs) of the considered system are depicted in Fig. 2.1. From the figure, we can see that the value of γ_{min} is very near the supremum of MSVs, which implies that the system satisfies the prescribed H_∞ performance.

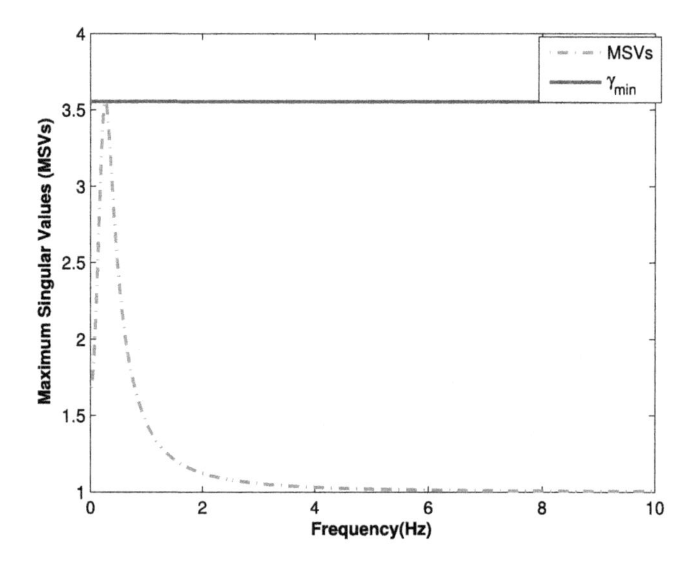

Figure 2.1 Maximal singular values and γ_{min}.

- General dissipative case: $S_{11} = -1.6$, $S_{12} = -0.5$, $S_{22} = 0.4$. Similarly, to verify the dissipativity of the considered system, we can verify whether the LMIs in Theorem 2.8 are feasible. By solving the LMIs, feasible matrices P and Q are computed to be

$$P = \begin{bmatrix} 10.8800 & 0.2980 & -1.4001 \\ 0.2980 & 4.8022 & -3.4971 \\ -1.4001 & -3.4971 & 0 \end{bmatrix}, \quad Q = 3.3810.$$

According to the definition of $(\mathcal{Q}, \mathcal{N}, \mathcal{R})$-dissipativity in [201], we can see the considered system is $(-0.4, 0.5, 1.6)$-dissipativity.

- Positive realness case: $S_{11} = 0$, $S_{12} = -I$, $S_{22} = 0$. By solving the LMIs in Theorem 2.8, the conditions in (2.6) and (2.7) are satisfied with the following solution:

$$P=\begin{bmatrix}14.5494 & 1.8136 & 1.9068\\ 1.8136 & 4.0456 & 4.5228\\ 1.9068 & 4.5228 & 0\end{bmatrix},\quad Q=3.0400.$$

Therefore the system is positive real.

Example 2.2. Consider a singular system with the following parameters:

$$E=\begin{bmatrix}1 & 0 & 0\\ 0 & 1 & 0\\ 0 & 0 & 0\end{bmatrix},\ A=\begin{bmatrix}0 & 1 & 0\\ -3 & -2 & 1\\ 0 & 0 & -1\end{bmatrix},\ B_w=\begin{bmatrix}0\\ 0\\ 1\end{bmatrix},$$

$$C=\begin{bmatrix}1 & 3 & -1\\ 0 & 0 & 0\end{bmatrix},\ D_w=\begin{bmatrix}2\\ 0\end{bmatrix},\ D=\begin{bmatrix}0\\ 3\end{bmatrix},\ B=\begin{bmatrix}0\\ 1\\ 0\end{bmatrix}.$$

Choose $E_L=E_R=\begin{bmatrix}1 & 0\\ 0 & 1\\ 0 & 0\end{bmatrix}$, $U^T=\Lambda=\begin{bmatrix}0\\ 0\\ 1\end{bmatrix}$. To illustrate the effectiveness of Theorem 2.11, we consider the H_∞ case, that is $S_{11}=-\gamma^2 I$, $S_{12}=0$, $S_{22}=I$. By solving the LMIs in Theorem 2.11, the minimal value of γ is obtained as $\gamma_{min}=2.2472$, and the corresponding solutions of decision variables are

$$\bar{P}=\begin{bmatrix}0.0459 & -0.0011 & 0.0425\\ -0.0011 & 0.1079 & -0.2776\\ 0.0425 & -0.2776 & 0\end{bmatrix},\ \bar{\Pi}=0.8001,$$

$$H=\begin{bmatrix}0.0000 & -0.1099 & -0.0007\end{bmatrix},\ G=-0.0030.$$

Then, the feedback controller is $K=\begin{bmatrix}-0.0244 & -1.0215 & -0.0009\end{bmatrix}$, $J=-0.0021$.

2.1.4 Conclusion

In this section, a new equivalent sets approach is proposed to investigate the problems of admissibilization and dissipative control of singular systems. By employing an equivalent parametrization of the constrained sets, a novel necessary and sufficient dissipativity condition of singular systems is presented without the equality constraint. Based on the criterion, a necessary and sufficient condition for the existence of a state feedback controller is established to render the closed-loop system to be admissible and dissipative. Two examples are given to demonstrate the effectiveness of the obtained results.

2.2 Dissipative control of discrete-time singular systems

The problem of dissipative control of discrete-time singular systems is investigated in this section. Based on parametrizing the solutions of the constraint set, a necessary and sufficient dissipativity condition is established in terms of strict LMI, which makes the condition more tractable. By using the system augmentation approach, an SOF controller design method is proposed to guarantee that the closed-loop system is admissible and strictly (Q, S, R)-dissipative.

2.2.1 Problem formulation

Consider a class of linear discrete-time singular systems described by

$$\left\{\begin{array}{rcl} Ex(k+1) & = & Ax(k)+Bu(k)+B_w w(k), \ x_0=x(0), \\ z(k) & = & Cx(k)+Du(k)+D_w w(k), \\ y(k) & = & C_y x(k)+D_y w(k), \end{array}\right. \tag{2.17}$$

where $x(k) \in \mathbb{R}^n$ is the state vector; $u(k) \in \mathbb{R}^m$ is the control input; $w(k) \in \mathbb{R}^l$ represents a disturbance, which belongs to l_2; $z(k) \in \mathbb{R}^q$ is the controlled output; $y(k) \in \mathbb{R}^g$ is the measurement output; matrices E, A, B, B_w, C, D, D_w, C_y, and D_y are constant matrices with appropriate dimensions and $\operatorname{rank}(E) = r \leq n$.

Before moving on, some definitions and lemmas are given, which will be used in deriving the main results.

Definition 2.13. [209]

1. The singular system in (2.17) is said to be regular if $\det(zE - A)$ is not identically zero.
2. The singular system in (2.17) is said to be causal if $\deg\{\det(zE - A)\} = \operatorname{rank}(E)$.
3. The singular system in (2.17) is said to be stable if the moduli of the roots of $\det(zE - A) = 0$ are less than 1.
4. The singular system in (2.17) is said to be admissible if it is regular, causal, and stable.

Definition 2.14. [29] The system in (2.17) ($u(k) = 0$) is said to be strictly (Q, S, R)-dissipative if there exists a scalar $\alpha > 0$, and under zero initial state $x_0 = 0$, the following inequality holds:

$$\langle z, Qz\rangle_\tau + 2\langle z, Sw\rangle_\tau + \langle w, Rw\rangle_\tau \geq \alpha\langle w, w\rangle_\tau, \ \forall\tau \geq 0. \tag{2.18}$$

As in [29], $Q \leq 0$ is assumed. Consequently, there exists a matrix $Q_-^{\frac{1}{2}} = (-Q)^{\frac{1}{2}} \geq 0$ satisfying $-Q = (Q_-^{\frac{1}{2}})^2$.

Lemma 2.15. [29] *Let the matrices Q, S, and R be given with Q and R real symmetric. Then the system in (2.17) is admissible (when $u(k) = 0$, $w(k) = 0$) and strictly (Q, S, R)-dissipative, if and only if there exists a real symmetric and invertible matrix X such that*

$$E^T XE \geq 0, \tag{2.19}$$

$$\begin{bmatrix} A^T XA - E^T XE & A^T XB_w - C^T S & C^T Q_-^{\frac{1}{2}} \\ \star & B_w^T XB_w - D_w^T S - S^T D_w - R & D_w^T Q_-^{\frac{1}{2}} \\ \star & \star & -I \end{bmatrix} < 0. \tag{2.20}$$

Lemma 2.16. *The following two sets are equivalent:*

$$\begin{aligned} \mathcal{X}_1 &= \{X \in \mathbb{R}^{n\times n} : E^T XE \geq 0, \ \mathrm{rank}(E^T XE) = r, \ X = X^T\}; \\ \mathcal{X}_2 &= \{X = P - E_0^T UE_0 : P > 0, \ E_0 E = 0, \ E_0 E_0^T > 0, \ E_0 \in \mathbb{R}^{(n-r)\times n}, \\ &\quad U = U^T\}. \end{aligned}$$

Proof. (Sufficiency) When $X \in \mathcal{X}_2$, one has $E^T XE = E^T PE \geq 0$ and $\mathrm{rank}(E^T XE) = r$, which imply $X \in \mathcal{X}_1$.

(Necessity) Without loss of generality, set $E = \begin{bmatrix} I_r & 0 \\ 0 & 0 \end{bmatrix}$ and $X = \begin{bmatrix} X_1 & X_2 \\ X_2^T & X_3 \end{bmatrix}$, where $X_1 = X_1^T \in \mathbb{R}^{r\times r}$ and $X_3 = X_3^T \in \mathbb{R}^{(n-r)\times(n-r)}$. Then one has $E_0 = \begin{bmatrix} 0 & I_{n-r} \end{bmatrix}$, and it yields from $E^T XE \geq 0$ that $X_1 \geq 0$. Combining with $\mathrm{rank}(E^T XE) = \mathrm{rank}(X_1) = r$, one has $X_1 > 0$. By constructing $P = \begin{bmatrix} X_1 & X_2 \\ X_2^T & X_2^T X_1^{-1} X_2 + \epsilon I \end{bmatrix} > 0$ with $\epsilon > 0$ and $U = U^T = X_2^T X_1^{-1} X_2 + \epsilon I - X_3$, $X = P - E_0^T UE_0$ is obtained. □

2.2.2 Dissipative control

The aim of this section is to design an SOF controller in the form of $u(k) = Ky(k)$ such that the closed-loop system

$$\begin{cases} Ex(k+1) &= (A+BKC_y)x(k) + (B_w + BKD_y)w(k), \\ z(k) &= (C+DKC_y)x(k) + (D_w + DKD_y)w(k) \end{cases} \tag{2.21}$$

is admissible and strictly (Q, S, R)-dissipative.

Firstly, the new dissipativity analysis conditions of system (2.17) (when $u(k) = 0$), in terms of strict LMIs, are proposed in following theorem:

Theorem 2.17. *Let the matrices Q, S, and R be given with Q and R real symmetric and $Q \leq 0$. The following statements are equivalent:*

(i) *System (2.17) is admissible and strictly (Q, S, R)-dissipative.*

(ii) *There exist matrices $P > 0$, and $U = U^T$ such that the following LMI holds:*

$$\begin{bmatrix} -E^TPE + A^TVA & A^TVB_w - C^TS & C^TQ_-^{\frac{1}{2}} \\ \star & B_w^TVB_w - D_w^TS - S^TD_w - R & D_w^TQ_-^{\frac{1}{2}} \\ \star & \star & -I \end{bmatrix} < 0, \tag{2.22}$$

where $V = P - E_0^TUE_0$.

(iii) *There exist matrices $P > 0$, $U = U^T$, $\mathcal{F}$, and $\mathcal{G}$ such that the following LMI holds:*

$$\begin{bmatrix} -\mathcal{E}^T\mathcal{P}\mathcal{E} + \mathrm{sym}(\mathcal{L}^T\mathcal{S} + \mathcal{F}\mathcal{A}) & -\mathcal{F} + \mathcal{A}^T\mathcal{G}^T \\ \star & \mathcal{V} - \mathcal{G}^T - \mathcal{G} \end{bmatrix} < 0, \tag{2.23}$$

where

$$\mathcal{E} = \begin{bmatrix} E & 0 \\ 0 & I \end{bmatrix}, \quad \mathcal{P} = \begin{bmatrix} P & 0 \\ 0 & R \end{bmatrix}, \quad \mathcal{L} = \begin{bmatrix} C & D_w \end{bmatrix},$$

$$\mathcal{A} = \begin{bmatrix} A & B_w \\ Q_-^{\frac{1}{2}}C & Q_-^{\frac{1}{2}}D_w \end{bmatrix}, \quad \mathcal{V} = \begin{bmatrix} V & 0 \\ 0 & I \end{bmatrix}, \quad \mathcal{S} = \begin{bmatrix} 0 & -S \end{bmatrix}.$$

Proof. (i)$\Longleftrightarrow$(ii): The equivalence between item (i) and item (ii) are obtained by using Lemma 2.15 and Lemma 2.16.

(iii)⇒(ii): The following LMI can be derived by pre- and post-multiplying (2.23) with $\begin{bmatrix} I & \mathcal{A}^T \end{bmatrix}$ and $\begin{bmatrix} I & \mathcal{A}^T \end{bmatrix}^T$:

$$\bar{V} = \begin{bmatrix} -E^T PE + A^T VA & A^T VB_w - C^T S \\ \star & B_w^T VB_w - D_w^T S - S^T D_w - R \end{bmatrix} - \begin{bmatrix} C^T QC & C^T QD_w \\ \star & D_w^T QD_w \end{bmatrix} < 0,$$

which is equivalent to (2.22) by utilizing Schur complement equivalence.

(ii)⇒(iii): By employing Schur complement equivalence, condition (2.22) is equivalent to

$$\begin{aligned} & \begin{bmatrix} -E^T PE + A^T VA - C^T QC & A^T VB_w - C^T S - C^T QD_w \\ \star & B_w^T VB_w - D_w^T S - S^T D_w - R - D_w^T QD_w \end{bmatrix} \\ = \; & -\mathcal{E}^T \mathcal{P} \mathcal{E} + \mathrm{sym}(\mathcal{L}^T \mathcal{S}) + \mathcal{A}^T \mathcal{V} \mathcal{A} < 0, \end{aligned}$$

where $\mathcal{P}$, $\mathcal{A}$, and $\mathcal{V}$ are defined in (2.23). On the other hand, there always exists a matrix $\mathcal{G}$ such that $\mathcal{V} - \mathcal{G}^T - \mathcal{G} < 0$ and

$$\begin{bmatrix} -\mathcal{E}^T \mathcal{P} \mathcal{E} + \mathrm{sym}(\mathcal{L}^T \mathcal{S}) + \mathcal{A}^T \mathcal{V} \mathcal{A} & 0 \\ 0 & \mathcal{V} - \mathcal{G}^T - \mathcal{G} \end{bmatrix} < 0. \tag{2.24}$$

By pre- and post-multiplying (2.24) by $\begin{bmatrix} I & -\mathcal{A}^T \\ 0 & I \end{bmatrix}$ and $\begin{bmatrix} I & -\mathcal{A}^T \\ 0 & I \end{bmatrix}^T$, it yields that

$$\begin{bmatrix} -\mathcal{E}^T \mathcal{P} \mathcal{E} + \mathrm{sym}(\mathcal{L}^T \mathcal{S}) + \mathcal{A}^T \mathrm{sym}(\mathcal{V} - \mathcal{G}) \mathcal{A} & \mathcal{A}^T (-\mathcal{V} + \mathcal{G} + \mathcal{G}^T) \\ \star & \mathcal{V} - \mathcal{G} - \mathcal{G}^T \end{bmatrix} < 0. \tag{2.25}$$

By setting $\mathcal{F} = \mathcal{A}^T (\mathcal{V} - \mathcal{G})$, one gets inequality (2.23). □

Remark 2.18. The advantage of Item (iii) of Theorem 2.17 lies in separating the Lyapunov matrix P and the system matrices A and C, which is utilized widely [142]. However, if one uses Item (iii) of Theorem 2.17 to design the SOF controller, the term $F_2 Q_-^{\frac{1}{2}} (C + DK)$ will appear with $\mathcal{F} = \begin{bmatrix} F_1 & F_2 \\ F_3 & F_4 \end{bmatrix}$, which makes the condition in (2.23) difficult to solve. Therefore the separation of the controller K and system matrices D will be helpful for solving the controller design problem.

In the following, the SOF controller will be designed such that the closed-loop system in (2.21) is admissible and strictly (Q, S, R)-dissipative. To this end, one augments the closed-loop singular system in (2.21), as follows, by considering $u(k)$ as a state component and choosing $\bar{x}(k) = \begin{bmatrix} x^T(k) & u^T(k) \end{bmatrix}^T$ as the new system state:

$$\begin{cases} \bar{E}\bar{x}(k+1) & = & \bar{A}\bar{x}(k) + \bar{B}_w w(k), \\ z(k) & = & \bar{C}x(k) + D_w w(k), \end{cases} \tag{2.26}$$

where

$$\bar{E} = \begin{bmatrix} E & 0 \\ 0 & 0 \end{bmatrix}, \ \bar{A} = \begin{bmatrix} A & B \\ KC_y & -I \end{bmatrix}, \ \bar{B}_w = \begin{bmatrix} B_w \\ KD_y \end{bmatrix}, \ \bar{C} = \begin{bmatrix} C & D \end{bmatrix}.$$

Before giving the main result, the equivalence of admissibility and dissipativity between the systems in (2.21) and (2.26) is proved firstly. The following two equations are true:

$$z\bar{E} - \bar{A} = \begin{bmatrix} zE - A & -B \\ -KC_y & I \end{bmatrix} = \begin{bmatrix} I & -B \\ 0 & I \end{bmatrix} \begin{bmatrix} zE - A - BKC_y & 0 \\ 0 & I \end{bmatrix} \begin{bmatrix} I & 0 \\ -KC_y & I \end{bmatrix}$$

and

$$\begin{aligned} & \bar{C}(z\bar{E} - \bar{A})^{-1}\bar{B}_w + D_w \\ & = \begin{bmatrix} C & D \end{bmatrix} \begin{bmatrix} I & 0 \\ KC_y & I \end{bmatrix} \begin{bmatrix} (zE - A - BKC_y)^{-1} & 0 \\ 0 & I \end{bmatrix} \begin{bmatrix} I & B \\ 0 & I \end{bmatrix} \begin{bmatrix} B_w \\ KD_y \end{bmatrix} + D_w \\ & = (C + DKC_y)(zE - A - BKC_y)^{-1}(B_w + BKD_y) + D_w + DKD_y, \end{aligned}$$

which derive that the determinants of $z\bar{E} - \bar{A}$ and $zE - A - BKC_y$ are the same, and the transfer functions of the systems in (2.21) and (2.26) are equal, respectively. By using Definitions 2.13 and 2.14, the admissibility and dissipativity of system in (2.21) are equivalent to those in (2.26).

The following corollary proposes SOF controller design method by utilizing the system augmentation approach:

Corollary 2.19. *The system in (2.21) is admissible and strictly (Q, S, R)-dissipative if and only if there exist matrices $P > 0$, $U = U^T$, $\mathcal{F}$, and $\mathcal{G}$ such that the following inequality holds:*

$$\begin{bmatrix} -\bar{\mathcal{E}}^T \mathcal{P} \bar{\mathcal{E}} + \text{sym}(\bar{\mathcal{L}}^T \mathcal{S} + \mathcal{F}\bar{\mathcal{A}}) & -\mathcal{F} + \bar{\mathcal{A}}^T \mathcal{G}^T \\ \star & \bar{\mathcal{V}} - \mathcal{G}^T - \mathcal{G} \end{bmatrix} < 0, \tag{2.27}$$

where

$$\bar{\mathcal{E}} = \begin{bmatrix} \bar{E} & 0 \\ 0 & I \end{bmatrix}, \ \mathcal{P} = \begin{bmatrix} P & 0 \\ 0 & R \end{bmatrix}, \ \bar{\mathcal{L}} = \begin{bmatrix} \bar{C} & D_w \end{bmatrix}, \ \mathcal{S} = \begin{bmatrix} 0 & -S \end{bmatrix},$$

$$\bar{\mathcal{A}} = \begin{bmatrix} \bar{A} & \bar{B}_w \\ Q_-^{\frac{1}{2}}\bar{C} & Q_-^{\frac{1}{2}}D_w \end{bmatrix}, \ \bar{\mathcal{V}} = \begin{bmatrix} \bar{V} & 0 \\ 0 & I \end{bmatrix}, \ \bar{V} = P - \bar{E}_0^T U \bar{E}_0.$$

Proof. By applying the Items (i) and (iii) of Theorem 2.17 to the augmented singular system in (2.26), and then using the equivalence of the admissibility and dissipativity between the system in (2.21) and the system in (2.26), the results follow. □

Remark 2.20. It can be seen that the inequality in (2.27) is in terms of bilinear matrix inequality, which can be solved by utilizing the existing numerical method [113]. Moreover, the H_∞ control problem and the passivity control problem can also be addressed by setting $-Q = I$, $S = 0$, $R = \gamma^2$ and $-Q = 0$, $S = I$, $R = 0$ in (2.27), respectively.

To obtain a more tractable SOF control condition for system (2.21), the following result is given:

Theorem 2.21. *There exists an SOF controller such that the closed-loop system in (2.21) is admissible and strictly (Q, S, R)-dissipative if there exist matrices* $P = \begin{bmatrix} P_{11} & P_{12} \\ \star & P_{22} \end{bmatrix} > 0$, $U = U^T$, F_{11}, F_{12}, F_{13}, F_{21}, F_{22}, F_3, G_{11}, G_{13}, G_{21}, G_{22}, G_3, *and M such that the following LMI holds:*

$$\Lambda = \left[\begin{array}{cc|c|cc|c} \Lambda_{11} & \Lambda_{12} & \Lambda_{13} & \Lambda_{14} & \Lambda_{15} & -F_{21} + C^T Q_-^{\frac{1}{2}} G_3^T \\ \star & \Lambda_{22} & \Lambda_{23} & \Lambda_{24} & \Lambda_{25} & -F_{22} + D^T Q_-^{\frac{1}{2}} G_3^T \\ \hline \star & \star & \Lambda_{33} & \Lambda_{34} & \Lambda_{35} & -F_3 + D_w^T Q_-^{\frac{1}{2}} G_3^T \\ \hline \star & \star & \star & \Lambda_{44} & \Lambda_{45} & -G_{21} \\ \star & \star & \star & \star & \Lambda_{55} & -G_{22} \\ \hline \star & \star & \star & \star & \star & I - G_3 - G_3^T \end{array}\right] < 0, \quad (2.28)$$

where

$$\begin{aligned} \Lambda_{11} &= -E^T P_{11} E + \mathrm{sym}(F_{11}A + LMC_y + F_{21} Q_-^{\frac{1}{2}} C); \\ \Lambda_{12} &= F_{11}B - LF_{12} + F_{21} Q_-^{\frac{1}{2}} D + A^T F_{13}^T + C_y^T M^T + C^T Q_-^{\frac{1}{2}} F_{22}^T; \\ \Lambda_{13} &= -C^T S + F_{11}B_w + LMD_y + F_{21} Q_-^{\frac{1}{2}} D_w + C^T Q_-^{\frac{1}{2}} F_3^T; \end{aligned}$$

$$
\begin{aligned}
\Lambda_{14} &= -F_{11} + A^T G_{11}^T + C_y^T M^T L^T + C^T Q_-^{\frac{1}{2}} G_{21}^T; \\
\Lambda_{15} &= -LF_{12} + A^T G_{13}^T + C_y^T M^T + C^T Q_-^{\frac{1}{2}} G_{22}^T; \\
\Lambda_{22} &= \mathrm{sym}(F_{13}B - F_{12} + F_{22} Q_-^{\frac{1}{2}} D); \\
\Lambda_{23} &= -D^T S + F_{13} B_w + M D_y + F_{22} Q_-^{\frac{1}{2}} D_w + D^T Q_-^{\frac{1}{2}} F_3^T; \\
\Lambda_{24} &= -F_{13} + B^T G_{11}^T - F_{12}^T L^T + D^T Q_-^{\frac{1}{2}} G_{21}^T; \\
\Lambda_{25} &= -F_{12} + B^T G_{13}^T - F_{12}^T + D^T Q_-^{\frac{1}{2}} G_{22}^T; \\
\Lambda_{33} &= -R + \mathrm{sym}(F_3 Q_-^{\frac{1}{2}} D_w - D_w^T S); \\
\Lambda_{34} &= B_w^T G_{11}^T + D_y^T M^T L^T + D_w^T Q_-^{\frac{1}{2}} G_{21}^T; \\
\Lambda_{35} &= B_w^T G_{13}^T + D_y^T M^T + D_w^T Q_-^{\frac{1}{2}} G_{22}^T; \\
\Lambda_{44} &= P_{11} - \bar{E}_{01}^T U \bar{E}_{01} - G_{11} - G_{11}^T; \\
\Lambda_{45} &= P_{12} - \bar{E}_{01}^T U \bar{E}_{02} - LF_{12} - G_{13}^T; \\
\Lambda_{55} &= P_{22} - \bar{E}_{02}^T U \bar{E}_{02} - F_{12} - F_{12}^T,
\end{aligned}
$$

and $\bar{E}_0 = \begin{bmatrix} \bar{E}_{01} & \bar{E}_{02} \end{bmatrix}$ *with* $\bar{E}_0 \bar{E} = 0$, $\bar{E}_0 \bar{E}_0^T > 0$, $\bar{E}_0 \in \mathbb{R}^{(n+m-r)\times(n+m)}$, $\bar{E}_{01} \in \mathbb{R}^{(n+m-r)\times n}$, $\bar{E}_{02} \in \mathbb{R}^{(n+m-r)\times m}$, $L^T = \begin{bmatrix} I_m & 0_{m\times(n-m)} \end{bmatrix}$. *Then the SOF controller can be obtained by* $K = F_{12}^{-1} M$.

Proof. Let the matrices $\mathcal{F}$ and $\mathcal{G}$ in (2.27) be of the following forms:

$$
\mathcal{F} = \begin{bmatrix} F_1 & F_2 \\ 0 & F_3 \end{bmatrix}, \quad \mathcal{G} = \begin{bmatrix} G_1 & G_2 \\ 0 & G_3 \end{bmatrix}
$$

with

$$
F_1 = \begin{bmatrix} F_{11} & LF_{12} \\ F_{13} & F_{12} \end{bmatrix}, \quad F_2 = \begin{bmatrix} F_{21} \\ F_{22} \end{bmatrix}, \quad G_1 = \begin{bmatrix} G_{11} & LF_{12} \\ G_{13} & F_{12} \end{bmatrix}, \quad G_2 = \begin{bmatrix} G_{21} \\ G_{22} \end{bmatrix}.
$$

By setting $M = F_{12} K$, the inequality in (2.27) is obtained. □

Remark 2.22. The nonsingularity of the matrix F_{12} in Theorem 2.21 is satisfied without loss of generality. If it is not the case, then one can choose a sufficient small scalar θ such that $\bar{F}_{12} = F_{12} + \theta I$, satisfying the inequality in (2.28). Then the matrix K can be replaced with $\bar{F}_{12}^{-1} M$.

Remark 2.23. The SOF controller design method is given in terms of strict LMIs in Theorem 2.21, which can be easily solved by standard software. Although some SOF control problem for singular systems have been reported such as [22], [169], the controller design method contains equality constrain, which makes the computation difficult. The result in Theorem 2.21 can also be used to deal with state-feedback dissipative control problem for singular systems. For the computation complexity, the total number of scalar decision variables of Theorem 2.21 is $0.5(n+m)(n+m+1)+0.5(n+m-r)(n+m-r+1)+2n^2+m(2n+m+g)+q(2n+2m+l+q)$.

2.2.3 Illustrative example

In this section, one example is provided to illustrate the effectiveness of the proposed approach. Theorem 2.17, which provides a necessary and sufficient dissipativity condition, will be used to check the applicability of the static output feedback controller design.

Example 2.3. Consider a singular system (2.17) with following parameters:

$$\begin{aligned} E &= \begin{bmatrix} 1 & 0 \\ 0 & 0 \end{bmatrix}, \ A = \begin{bmatrix} 1.6 & 0.8 \\ 0.6 & 1.2 \end{bmatrix}, \ B = \begin{bmatrix} 1.8 \\ 1 \end{bmatrix}, \ B_w = \begin{bmatrix} 0.1 \\ 0.4 \end{bmatrix}, \\ C &= \begin{bmatrix} 0.5 & 0.6 \end{bmatrix}, \ D = 0.6, \ D_w = 0.8, \ C_y = \begin{bmatrix} 1 & 0 \end{bmatrix}, \ D_y = -0.5. \end{aligned}$$

Given

$$Q = 1.2, \ S = 0.8, \ R = 1.6$$

and to test the admissibility and dissipativity of this open-loop system, one chooses $E_0 = \begin{bmatrix} 0 & 1 \end{bmatrix}$. Solving the LMI in (2.22), no feasible solutions can be found, which means this open-loop system is not admissible and strictly (Q, S, R)-dissipative. To make the closed-loop system in (2.21) admissible and strictly (Q, S, R)-dissipative, the SOF controller design method in Theorem 2.21 is employed. By solving the LMI in (2.28), one has

$$M = -0.5634, \ F_{12} = 0.5520,$$

and $K = F_{12}^{-1}M = -1.0207$. Then the closed-loop system becomes

$$\begin{cases} \begin{bmatrix} 1 & 0 \\ 0 & 0 \end{bmatrix} x(k+1) &= \begin{bmatrix} -0.2372 & 0.8000 \\ -0.4207 & 1.2000 \end{bmatrix} x(k) + \begin{bmatrix} 1.0186 \\ 0.9103 \end{bmatrix} w(k), \\ z(k) &= \begin{bmatrix} -0.1124 & 0.6000 \end{bmatrix} x(k) + 1.1062 w(k). \end{cases} \tag{2.29}$$

Utilizing Theorem 2.17 to system (2.29), one can find that the LMI in (2.22) is feasible, and the closed-loop system in (2.29) is admissible and strictly $(1.2, 0.8, 1.6)$ dissipative.

2.2.4 Conclusion

This subchapter studied the problem of dissipative control for discrete-time singular system. By using the system augmentation approach, a static output feedback controller design method is proposed to guarantee that the closed-loop singular system is admissible and strictly (Q, S, R)-dissipative. An example is given to demonstrate the effectiveness of the results.

2.3 Dissipative filtering of singular systems

In this section, the SOF controller designed approach will be applied to design the dissipative reduced-order filtering of singular systems.

2.3.1 Reduced-order dissipative filtering

Consider a class of discrete-time singular system:

$$\left\{\begin{array}{rcl} Ex(k+1) &=& Ax(k) + B_w w(k),\ x_0 = x(0), \\ z(k) &=& Cx(k) + D_w w(k), \\ y(k) &=& C_y x(k) + D_y w(k), \end{array}\right. \tag{2.30}$$

where $x(k) \in \mathbb{R}^n$ is the state vector; $w(k) \in \mathbb{R}^l$ represents a disturbance, which belongs to l_2; $z(k) \in \mathbb{R}^q$ is the controlled output; $y(k) \in \mathbb{R}^g$ is the measurement output; matrices E, A, B_w, C, D_w, C_y, and D_y are constant matrices with appropriate dimensions. To estimate controlled output $z(k)$, the following reduced-order filtering is constructed:

$$\left\{\begin{array}{rcl} \hat{x}(k+1) &=& A_f\hat{x}(k) + B_f y(k),\ \hat{x}(0) = 0, \\ \hat{z}(k) &=& C_f\hat{x}(k) + D_f y(k), \end{array}\right. \tag{2.31}$$

where $\hat{x}(k) \in \mathbb{R}^m$ $(0 < m \le n)$ is the state vector of the filter; $\hat{z}(k) \in \mathbb{R}^q$ is the estimation of $z(k)$; matrices A_f, B_f, C_f, and D_f are filter parameters to be determined.

Denote $\breve{x}(k) = \begin{bmatrix} x^T(k) & \hat{x}^T(k) \end{bmatrix}^T$ and the estimation error $\breve{z}(k) = z(k) - \hat{z}(k)$, then the filtering error singular system derived from the singular system in (2.30) and the filter in (2.31) is

$$\begin{cases} \breve{E}\breve{x}(k+1) &= \breve{A}\breve{x}(k)+\breve{B}_w w(k), \\ \breve{z}(k) &= \breve{C}\breve{x}(k)+\breve{D}_w w(k), \end{cases} \tag{2.32}$$

where

$$\begin{aligned} \breve{E} &= \begin{bmatrix} E & 0 \\ 0 & I \end{bmatrix}, \ \breve{A}=\begin{bmatrix} A & 0 \\ B_f C_y & A_f \end{bmatrix}, \ \breve{B}_w=\begin{bmatrix} B_w \\ B_f D_y \end{bmatrix}, \\ \breve{C} &= \begin{bmatrix} C-D_f C_y & -C_f \end{bmatrix}, \ \breve{D}_w=D_w-D_f D_y. \end{aligned}$$

The aim is to design a filter in (2.31) such that the filtering error system in (2.32) is admissible and strictly (Q, S, R)-dissipative.

By carrying out simple manipulation, the following equations hold:

$$\begin{aligned} \breve{A} &= \tilde{A}+HK\tilde{C}_y, \ \breve{B}_w=\tilde{B}+HK\tilde{D}_y, \\ \breve{C} &= \tilde{C}+JK\tilde{C}_y, \ \breve{D}_w=D_w+JK\tilde{D}_y, \end{aligned} \tag{2.33}$$

where

$$\begin{aligned} \tilde{A} &= \begin{bmatrix} A & 0 \\ 0 & 0 \end{bmatrix}, \ \tilde{B}=\begin{bmatrix} B_w \\ 0 \end{bmatrix}, \ \tilde{C}=\begin{bmatrix} C & 0 \end{bmatrix}, \ \tilde{C}_y=\begin{bmatrix} 0 & I \\ C_y & 0 \end{bmatrix}, \\ \tilde{D}_y &= \begin{bmatrix} 0 \\ D_y \end{bmatrix}, \ H=\begin{bmatrix} 0 & 0 \\ I & 0 \end{bmatrix}, \ J=\begin{bmatrix} 0 & -I \end{bmatrix}, \ K=\begin{bmatrix} A_f & B_f \\ C_f & D_f \end{bmatrix}. \end{aligned}$$

Then the system in (2.32) can be rewritten as

$$\begin{cases} \breve{E}\breve{x}(k+1) &= \tilde{A}\breve{x}(k)+H\breve{u}(k)+\tilde{B}w(k), \\ \breve{z}(k) &= \tilde{C}\breve{x}(k)+J\breve{u}(k)+D_w w(k), \\ \breve{y}(k) &= \tilde{C}_y\breve{x}(k)+\tilde{D}_y w(k), \end{cases} \tag{2.34}$$

with $\breve{u}(k)=K\breve{y}(k)$. By now, one can see that the filter design problem is transferred to an SOF control problem, that is, to design a controller K guaranteeing the system in (2.32) to be admissible and strictly (Q, S, R)-dissipative. Define $\bar{x}(k)=\begin{bmatrix} \breve{x}^T(k) & \breve{u}^T(k) \end{bmatrix}^T$ as a new state variable, and the system in (2.34) is equivalent to the following augmented one:

$$\begin{cases} \bar{E}\bar{x}(k+1) &= \bar{A}\bar{x}(k)+\bar{B}_w w(k), \\ \breve{z}(k) &= \bar{C}\bar{x}(k)+\bar{D}_w w(k), \end{cases} \tag{2.35}$$

where

$$\bar{E} = \begin{bmatrix} \breve{E} & 0 \\ 0 & 0 \end{bmatrix}, \; \bar{A} = \begin{bmatrix} \tilde{A} & H \\ K\tilde{C}_y & -I \end{bmatrix}, \; \bar{B}_w = \begin{bmatrix} \tilde{B} \\ K\tilde{D}_y \end{bmatrix},$$

$$\bar{C} = \begin{bmatrix} \tilde{C} & J \end{bmatrix}, \; \bar{D}_w = D_w.$$

Therefore by substituting the matrices E, A, B, C_y, B_w, D_y, C, and D in Theorem 2.21 with $\breve{E}$, $\tilde{A}$, H, $\tilde{C}_y$, $\tilde{B}$, $\tilde{D}_y$, $\tilde{C}$, and J in (2.33), respectively, one can get the following reduced-order dissipative filter design method:

Theorem 2.24. *There exists a filter in (2.31) such that the filtering error system in (2.32) is admissible and strictly* (Q, S, R)*-dissipative if there exist matrices* $P = \begin{bmatrix} P_{11} & P_{12} \\ \star & P_{22} \end{bmatrix} > 0$, $U = U^T$, F_{11}, F_{12}, F_{13}, F_{21}, F_{22}, F_3, G_{11}, G_{13}, G_{21}, G_{22}, G_3, *and* M *such that the following LMI holds:*

$$\Theta = \begin{bmatrix} \Theta_{11} & \Theta_{12} & \Theta_{13} & \Theta_{14} & \Theta_{15} & -F_{21} + \tilde{C}^T Q_-^{\frac{1}{2}} G_3^T \\ \star & \Theta_{22} & \Theta_{23} & \Theta_{24} & \Theta_{25} & -F_{22} + J^T Q_-^{\frac{1}{2}} G_3^T \\ \star & \star & \Theta_{33} & \Theta_{34} & \Theta_{35} & -F_3 + D_w^T Q_-^{\frac{1}{2}} G_3^T \\ \star & \star & \star & \Theta_{44} & \Theta_{45} & -G_{21} \\ \star & \star & \star & \star & \Theta_{55} & -G_{22} \\ \star & \star & \star & \star & \star & I - G_3 - G_3^T \end{bmatrix} < 0, \tag{2.36}$$

where

$$\begin{aligned}
\Theta_{11} &= -\breve{E}^T P_{11} \breve{E} + \mathrm{sym}(F_{11}\tilde{A} + LM\tilde{C}_y + F_{21} Q_-^{\frac{1}{2}} \tilde{C}); \\
\Theta_{12} &= F_{11}H - LF_{12} + F_{21} Q_-^{\frac{1}{2}} J + \tilde{A}^T F_{13}^T + \tilde{C}_y^T M^T + \tilde{C}^T Q_-^{\frac{1}{2}} F_{22}^T; \\
\Theta_{13} &= -\tilde{C}^T S + F_{11}\tilde{B} + LM\tilde{D}_y + F_{21} Q_-^{\frac{1}{2}} D_w + \tilde{C}^T Q_-^{\frac{1}{2}} F_3^T; \\
\Theta_{14} &= -F_{11} + \tilde{A}^T G_{11}^T + \tilde{C}_y^T M^T L^T + \tilde{C}^T Q_-^{\frac{1}{2}} G_{21}^T; \\
\Theta_{15} &= -LF_{12} + \tilde{A}^T G_{13}^T + \tilde{C}_y^T M^T + \tilde{C}^T Q_-^{\frac{1}{2}} G_{22}^T; \\
\Theta_{22} &= \mathrm{sym}(F_{13}H - F_{12} + F_{22} Q_-^{\frac{1}{2}} J); \\
\Theta_{23} &= -J^T S + F_{13}\tilde{B} + M\tilde{D}_y + F_{22} Q_-^{\frac{1}{2}} D_w + \tilde{D}^T Q_-^{\frac{1}{2}} F_3^T; \\
\Theta_{24} &= -F_{13} + H^T G_{11}^T - F_{12}^T L^T + J^T Q_-^{\frac{1}{2}} G_{21}^T; \\
\Theta_{25} &= -F_{12} + H^T G_{13}^T - F_{12}^T + J^T Q_-^{\frac{1}{2}} G_{22}^T; \\
\Theta_{33} &= -R + \mathrm{sym}(F_3 Q_-^{\frac{1}{2}} D_w - D_w^T S);
\end{aligned}$$

$$\begin{aligned}\Theta_{34} &= \tilde{B}^T G_{11}^T + \tilde{D}_y^T M^T L^T + D_w^T Q_-^{\frac{1}{2}} G_{21}^T;\\ \Theta_{35} &= \tilde{B}^T G_{13}^T + \tilde{D}_y^T M^T + D_w^T Q_-^{\frac{1}{2}} G_{22}^T;\\ \Theta_{44} &= P_{11} - \bar{E}_{01}^T U \bar{E}_{01} - G_{11} - G_{11}^T;\\ \Theta_{45} &= P_{12} - \bar{E}_{01}^T U \bar{E}_{02} - LF_{12} - G_{13}^T;\\ \Theta_{55} &= P_{22} - \bar{E}_{02}^T U \bar{E}_{02} - F_{12} - F_{12}^T;\end{aligned}$$

and $\bar{E}_0 = \begin{bmatrix} \bar{E}_{01} & \bar{E}_{02} \end{bmatrix}$ *with* $\bar{E}_0 \bar{E} = 0,\ \bar{E}_0 \bar{E}_0^T > 0,\ \bar{E}_0 \in \mathbb{R}^{(n+m+q-r)\times(n+2m+q)}$, $\bar{E}_{01} \in \mathbb{R}^{(n+m+q-r)\times(n+m)}$, $\bar{E}_{02} \in \mathbb{R}^{(n+m+q-r)\times(m+q)}$, $L^T = \begin{bmatrix} I_{m+q} & 0_{(m+q)\times(n-q)} \end{bmatrix}$.
Then the desired filter can be obtained by $K = F_{12}^{-1} M = \left[\begin{array}{c|c} A_f & B_f \\ \hline C_f & D_f \end{array}\right]$.

Proof. Set

$$F_1 = \begin{bmatrix} F_{11} & LF_{12} \\ F_{13} & F_{12} \end{bmatrix},\ F_2 = \begin{bmatrix} F_{21} \\ F_{22} \end{bmatrix},\ G_1 = \begin{bmatrix} G_{11} & LF_{12} \\ G_{13} & F_{12} \end{bmatrix},\ G_2 = \begin{bmatrix} G_{21} \\ G_{22} \end{bmatrix},$$

and let the matrices $\mathcal{F}$ and $\mathcal{G}$ be the following forms:

$$\mathcal{F} = \begin{bmatrix} F_1 & F_2 \\ 0 & F_3 \end{bmatrix},\ \mathcal{G} = \begin{bmatrix} G_1 & G_2 \\ 0 & G_3 \end{bmatrix}.$$

Then noting that $K = F_{12}^{-1} M$, one can obtain the inequality in (2.27) from the inequality in (2.36) by straightforward manipulation. Therefore the admissibility and dissipativity of system (2.35), which is equivalent to those of system (2.32) are proved. □

Remark 2.25. The reduced-order filter design method is given in terms of strict LMIs in Theorem 2.24, and the total number of scalar decision variables of Theorem 2.24 is $0.5(n+2m+q)(n+2m+q+1) + 0.5(n+m+q-r)(n+m+q-r+1) + 2(m+n)(n+m+q) + (m+q)(2n+4m+g+3q) + (q+l)q$. The reduced-order filtering problem is also investigated in [79], [118] and [210], respectively. However, to obtain the desired filter parameters, a complicated matrix structure is needed, and the rank of the difference of two decision variables should be less than the order of the filter in [210]. For the method developed in this section, the filter parameters can be obtained directly by solving the LMI in (2.36), which avoids considering the rank constraint, non-strict LMIs in [210], [118], or constructing some complicated matrices in [79].

2.3.2 Illustrative example

In this subsection, an example is provided to illustrate the effectiveness of the proposed approach and filter design method.

Example 2.4. In this example, a first-order filter in the form of (2.31) will be designed for the following discrete-time singular systems:

$$\left\{\begin{array}{rcl}\begin{bmatrix}1 & 1 & 0\\ 1 & -1 & 1\\ 2 & 0 & 1\end{bmatrix}x(k+1) & = & \begin{bmatrix}-1 & 0.5 & 1\\ -1 & -0.3 & 1\\ 0.5 & 0 & 1\end{bmatrix}x(k)+\begin{bmatrix}-0.1\\ 0\\ 0.1\end{bmatrix}w(k),\\ z(k) & = & \begin{bmatrix}-3.2 & 0 & 3.2\\ 3.2 & 0 & 1.6\\ 0 & 0 & 3.2\end{bmatrix}x(k)+\begin{bmatrix}-0.1\\ 0.5\\ 0.1\end{bmatrix}w(k),\\ y(k) & = & \begin{bmatrix}1 & 1 & 0\\ 1 & 1 & 0\\ 0 & 0 & 1\end{bmatrix}x(k)+\begin{bmatrix}0.1\\ 0\\ 0.1\end{bmatrix}w(k).\end{array}\right. \tag{2.37}$$

From the value of $\bar{E}$, one gets

$$\bar{E}_{01}=\begin{bmatrix}1 & 1 & -1 & 0\\ 0 & 0 & 0 & 0\\ 0 & 0 & 0 & 0\\ 0 & 0 & 0 & 0\\ 0 & 0 & 0 & 0\end{bmatrix},\quad \bar{E}_{02}=\begin{bmatrix}0 & 0 & 0 & 0\\ 1 & 0 & 0 & 0\\ 0 & 1 & 0 & 0\\ 0 & 0 & 1 & 0\\ 0 & 0 & 0 & 1\end{bmatrix}.$$

By setting

$$Q=-\begin{bmatrix}0.6200 & -0.8000 & 0.1600\\ -0.8000 & 1.2500 & -0.0500\\ 0.1600 & -0.0500 & 1.0100\end{bmatrix},\quad S=\begin{bmatrix}-0.1\\ 0.5\\ 0.2\end{bmatrix},\quad R=1.5,$$

and solving the LMI in (2.36), the matrix

$$K=\left[\begin{array}{c|ccc}0.1256 & 1.3812 & -1.7594 & -0.7707\\ \hline -0.0637 & 0.0899 & -0.5044 & -0.8806\\ 0.1001 & 1.8047 & -1.3853 & -2.4015\\ -1.0660 & -0.8425 & 1.1057 & 0.7424\end{array}\right].$$

Then one obtains a first-order filter as

$$\begin{cases} \hat{x}(k+1) &= 0.1256\hat{x}(k) + \begin{bmatrix} 1.3812 & -1.7594 & -0.7707 \end{bmatrix} y(k),\ \hat{x}(0) = 0, \\ \hat{z}(k) &= \begin{bmatrix} -0.0637 \\ 0.1001 \\ -1.0660 \end{bmatrix} \hat{x}(k) + \begin{bmatrix} 0.0899 & -0.5044 & -0.8806 \\ 1.8047 & -1.3853 & -2.4015 \\ -0.8425 & 1.1057 & 0.7424 \end{bmatrix} y(k), \end{cases}$$

and the parameters of the filtering error system in (2.32) are given as follows:

$$\begin{aligned} \breve{E} &= \begin{bmatrix} 1 & 1 & 0 & 0 \\ 1 & -1 & 1 & 0 \\ 2 & 0 & 1 & 0 \\ 0 & 0 & 0 & 1 \end{bmatrix}, \\ \breve{A} &= \begin{bmatrix} -1.0000 & 0.5000 & 1.0000 & 0 \\ -1.0000 & -0.3000 & 1.0000 & 0 \\ 0.5000 & 0 & 1.0000 & 0 \\ -0.3782 & -0.3782 & -0.7707 & 0.1256 \end{bmatrix}; \\ \breve{B}_w &= \begin{bmatrix} -0.1000 & 0 & 0.1000 & 0.0610 \end{bmatrix}^T; \\ \breve{C} &= \begin{bmatrix} -2.7855 & 0.4145 & 4.0806 & 0.0637 \\ 2.7806 & -0.4194 & 4.0015 & -0.1001 \\ -0.2632 & -0.2632 & 2.4576 & 1.0660 \end{bmatrix},\ \breve{D}_w = \begin{bmatrix} -0.0209 \\ 0.5597 \\ 0.1100 \end{bmatrix}. \end{aligned} \tag{2.38}$$

To check whether the obtained filtering error system is admissible and strictly (Q, S, R)-dissipative, Theorem 2.17 is utilized again. By solving the LMI in (2.22), a feasible solution is found, which shows the applicability and effectiveness of the method.

By giving the initial condition with

$$\breve{x}(0) = \begin{bmatrix} -1.6756 & -0.2870 & 0.9170 & 0 \end{bmatrix}^T \quad \text{and} \quad w(k) = 0,$$

the state responses of system (2.32) are given in Fig. 2.2, which illustrates the stability of the system. The characteristic polynomial $C(z) = \frac{-11750z^3+136008z^2-45229z+3768}{25000}$ shows the regularity and causality of the system from the first two items of Definition 2.13; the admissibility of the system is obtained. To demonstrate the dissipativity of the system, one chooses

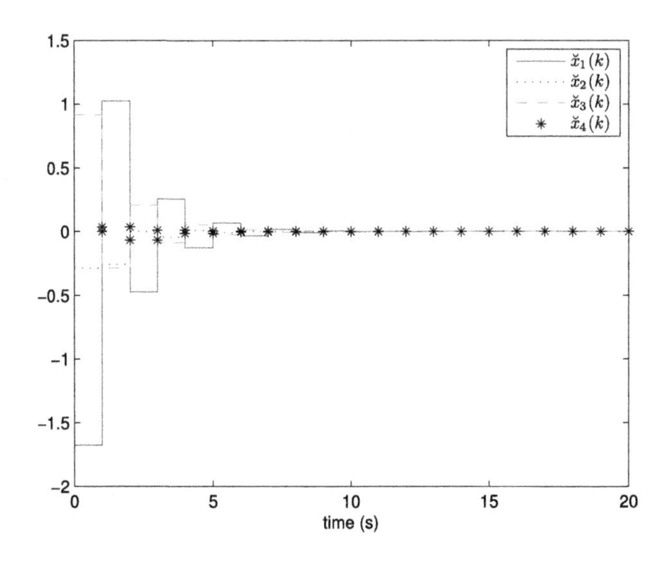

Figure 2.2 State responses.

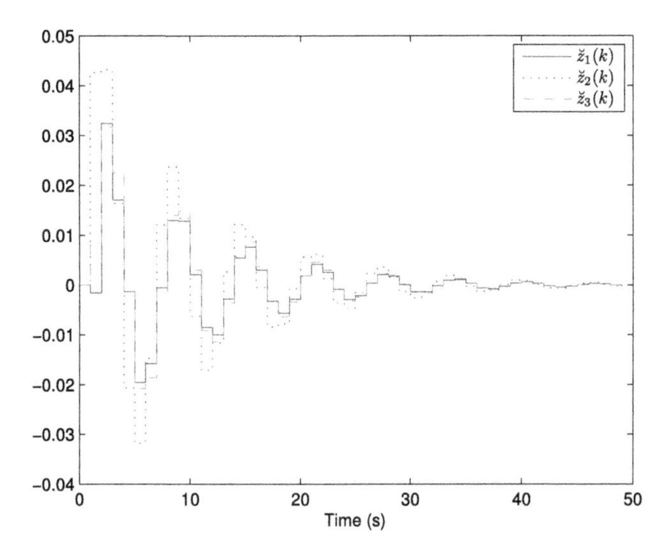

Figure 2.3 Output signal $\breve{z}(k)$.

$w(k) = 0.1e^{-0.1k}\sin(k)$ and zero initial conditions, the output signal $\breve{z}(k)$, and the performance signal $G(\breve{z}, w, \tau) = \langle z, Qz\rangle_\tau + 2\langle z, Sw\rangle_\tau + \langle w, Rw\rangle_\tau$ are proposed in Figs. 2.3 and 2.4, respectively. From Fig. 2.4, one can see that $G(\breve{z}, w, \tau)$ is greater than or equal to zero when $\tau \geq 0$. Then a suf-

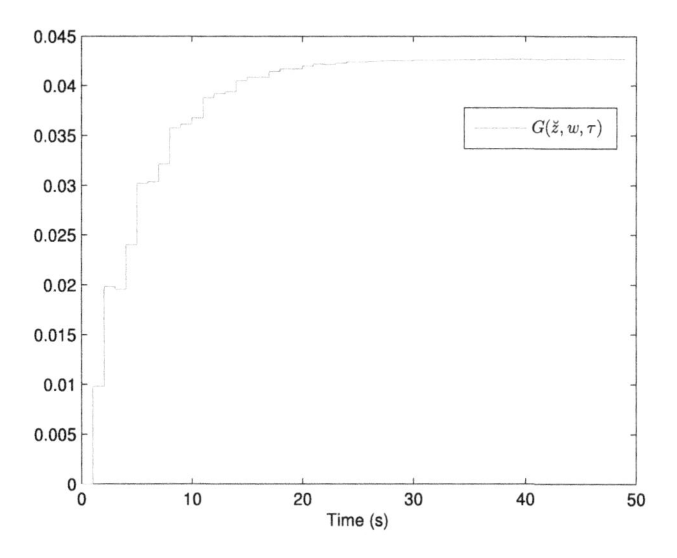

Figure 2.4 Performance signal $G(\check{z}, w, \tau) = \langle \check{z}, Q\check{z} \rangle_{\mathcal{T}} + 2\langle \check{z}, Sw \rangle_{\mathcal{T}} + \langle w, Rw \rangle_{\mathcal{T}}$.

ficiently small scalar $\alpha > 0$ can always be found such that the inequality in (2.18) holds, which shows the dissipativity of the system in (2.32).

2.3.3 Conclusion

In this section, the augmentation system approach is utilized to solve the filtering problem, which aims to guarantee the augmentation singular systems to be admissible and strictly (Q, S, R)-dissipative. The results presented in this section are in terms of strict LMIs, which make the conditions more tractable numerically. A numerical example is given to demonstrate the effectiveness of the proposed method.

CHAPTER 3

H_∞ control with transients for singular systems

In this chapter, the problem of a generalized type of H_∞ control is investigated for continuous-time singular systems, which treats a mixed attenuation of exogenous inputs and initial conditions. First, a performance measure that is essentially the worst-case norm of the regulated outputs over all exogenous inputs and initial conditions is introduced. Necessary and sufficient conditions are obtained to ensure the singular system to be admissible and the performance measure to be less than a prescribed scalar. Based on the criteria, a sufficient condition for the existence of a state-feedback controller is established in terms of LMIs. Moreover, the relationship between the performance measure and the standard H_∞ norm of the system is provided. Two numerical examples are given to demonstrate the properties of the obtained results.

3.1 Performance measure

Consider a class of linear continuous singular systems described by

$$\begin{cases} E\dot{x}(t) = Ax(t) + Bw(t), \quad x(0_-) = x_0, \\ z(t) = Cx(t) + Dw(t), \end{cases} \tag{3.1}$$

where $x(t) \in \mathbb{R}^n$ is the state vector; x_0 is the initial condition; $w(t) \in \mathbb{R}^p$ represents a set of exogenous inputs, which includes disturbances to be rejected, and $z(t) \in \mathbb{R}^s$ is the controlled output; A, B, C, and D denote constant matrices with appropriate dimensions. In contrast with standard linear systems with $E = I$, the matrix $E \in \mathbb{R}^{n\times n}$ has $0 < \text{rank}(E) = r < n$. Since system (3.1) can be rewritten as

$$\begin{cases} \begin{bmatrix} E & 0 \\ 0 & 0 \end{bmatrix}\begin{bmatrix} \dot{x} \\ \dot{\eta} \end{bmatrix} = \begin{bmatrix} A & 0 \\ 0 & -I \end{bmatrix}\begin{bmatrix} x \\ \eta \end{bmatrix} + \begin{bmatrix} B \\ D \end{bmatrix} w, \\ z(t) = \begin{bmatrix} C & I \end{bmatrix}\begin{bmatrix} x \\ \eta \end{bmatrix}, \end{cases}$$

Analysis and Synthesis of Singular Systems
https://doi.org/10.1016/B978-0-12-823739-7.00010-0

we will assume $D = 0$ from now on without loss of generality. Then, the plant under investigation is as follows:

$$\begin{cases} E\dot{x}(t) = Ax(t) + Bw(t), \quad x(0_-) = x_0, \\ z(t) = Cx(t). \end{cases} \tag{3.2}$$

First, we give some definitions and lemmas concerning the unforced system of (3.2):

Lemma 3.1. [216] *Suppose the pair* (E, A) *is regular and impulse-free, then the solution to system (3.2) is impulse-free and unique on* $[0, \infty)$.

Remark 3.2. As there exist nonsingular matrices M and N, such that

$$MEN = \mathrm{diag}(I, 0), \; MAN = \begin{bmatrix} \bar{A}_{11} & \bar{A}_{12} \\ \bar{A}_{21} & \bar{A}_{22} \end{bmatrix}, \; MB = \begin{bmatrix} \bar{B}_1 \\ \bar{B}_2 \end{bmatrix},$$
$$N^{-1}x(t) = \bar{x}(t) = \begin{bmatrix} \bar{x}_1(t) \\ \bar{x}_2(t) \end{bmatrix}, \tag{3.3}$$

system (3.2) is restricted system equivalent to

$$\begin{bmatrix} I & 0 \\ 0 & 0 \end{bmatrix} \dot{\bar{x}}(t) = \begin{bmatrix} \bar{A}_{11} & \bar{A}_{12} \\ \bar{A}_{21} & \bar{A}_{22} \end{bmatrix} \bar{x}(t) + \begin{bmatrix} \bar{B}_1 \\ \bar{B}_2 \end{bmatrix} w(t), \; \bar{x}(0_-) = N^{-1}x_0. \tag{3.4}$$

Notice that singular system (3.2) is impulse-free if and only if $\bar{A}_{22}$ is invertible. Then, under an impulse-free condition, a consistent initial condition of singular system (3.2) is characterized as follows:

$$\begin{aligned} x(0_+) &= \lim_{t\to 0+} x(t) = N \begin{bmatrix} \lim_{t\to 0+} \bar{x}_1(t) \\ \lim_{t\to 0+} \bar{x}_2(t) \end{bmatrix} = N \begin{bmatrix} \bar{x}_1(0_+) \\ \bar{x}_2(0_+) \end{bmatrix} \\ &= N \begin{bmatrix} \bar{x}_1(0_-) \\ -\bar{A}_{22}^{-1}(\bar{A}_{21}\bar{x}_1(0_-) + \bar{B}_2 w(0_+)) \end{bmatrix}. \end{aligned} \tag{3.5}$$

If $x(0_-) = x_0 \neq x(0_+)$, there is a finite jump at $t = 0$. There are some papers dealing with the elimination and minimization of initial jumps, see [107, 108,157].

In view of this, we introduce the following definition for singular system (3.2):

A general excitation performance measure of asymptotically stable system (3.2) is used in this section, which is defined as the worst-case norm of the controlled output over all admissible exogenous signals and initial states:

$$\gamma_g(R) = \sup_{\|w\|^2 + x_0^T E^T R E x_0 \neq 0} \frac{\|z\|}{\left(\|w\|^2 + x_0^T E^T R E x_0\right)^{1/2}}$$

for any $w \in L_2[0, \infty)$ and $x_0 \in \mathbb{R}^n$, where $R > 0$ is a given weighting matrix.

In the following, the problem of H_∞ control with transients (HCT) is stated as follows:

Problem HCT. Establish a necessary and sufficient condition such that the following conditions hold:

1. The singular system in (3.2) with $w(t) = 0$ is admissible.
2. The general excitation performance measure $\gamma_g(R)$ of system (3.2) is less than a prescribed scalar $\gamma > 0$.

Remark 3.3. In this section, both the exogenous input and initial state are taken into account when assessing the performance of the singular system. When $Ex_0 = 0$, the problem considered reduces to the H_∞ control problem of singular system, that is,

$$\gamma_\infty = \sup_{\|w\| \neq 0} \frac{\|z\|}{\|w\|} = \|G\|_\infty = \sup_{\|w\| \neq 0,\ Ex_0 = 0} \frac{\|z\|}{\left(\|w\|^2 + x_0^T E^T R E x_0\right)^{1/2}}.$$

Therefore the definition domain of γ_∞ belongs to that of γ_g, and we get $\gamma_\infty = \|G\|_\infty \leq \gamma_g$. On the other hand, when $Ex_0 \neq 0$ and $w(t) = 0$, γ_g becomes

$$\gamma_0(R) = \sup_{Ex_0 \neq 0} \frac{\|z\|}{\left(x_0^T E^T R E x_0\right)^{1/2}}.$$

Based on the definitions of $\gamma_g(R)$, γ_∞, and $\gamma_0(R)$, some properties about these performance measures are given in following theorem:

Theorem 3.4.

1. *If $E^T R_1 E \geq E^T R_2 E$, then $\gamma_g(R_1) \leq \gamma_g(R_2)$.*
2. *$\gamma_g(R) \geq \max\{\gamma_\infty, \gamma_0(R)\}$.*

Proof. The proving of Items 1 and 2 can be carried out following the similar line as in the proof of Theorem 2 in [3]. □

Before giving our main result, we provide the generalized bounded real lemma for the H_∞ control problem of singular systems.

Lemma 3.5. [3] *The performance measure of system (3.1) with $E = I$ satisfying $\gamma_g(R) < \gamma$ if and only if there exists a matrix $X > 0$ such that the following LMIs hold:*

$$\begin{bmatrix} A^T X + XA & XB & C^T \\ \star & -\gamma^2 I & D^T \\ \star & \star & -I \end{bmatrix} < 0, \; X < \gamma^2 R.$$

Lemma 3.6. *The following sets are equivalent:*

$$\begin{aligned} \Im &= \left\{ X \in \mathbb{R}^{n\times n} : E^T X = X^T E \geq 0, \; \operatorname{rank}(E^T X) = r \right\}, \\ \aleph &= \left\{ X = PE + E_0 \Phi : P = P^T \in \mathbb{R}^{n\times n}, \; E_L^T P E_L > 0, \; \Phi \in \mathbb{R}^{(n-r)\times n} \right\}, \end{aligned}$$

where E_L and E_R are full column rank with $E = E_L E_R^T$, and $E_0 \in \mathbb{R}^{n\times(n-r)}$ with full column rank is the right null matrix of E^T, that is, $E^T E_0 = 0$.

Proof. (Sufficiency) Let $X = PE + E_0\Phi$, we have

$$E^T X = X^T E = E_R E_L^T P E_L E_R^T \geq 0, \quad \text{and} \quad \operatorname{rank}(E^T X) = \operatorname{rank}(E_R E_R^T) = r.$$

(Necessity) Without loss of generality, denote $X = \begin{bmatrix} X_{11} & X_{12} \\ X_{21} & X_{22} \end{bmatrix}$, $E = \begin{bmatrix} I_r & 0 \\ 0 & 0 \end{bmatrix}$, where $X_{11} \in \mathbb{R}^{r\times r}$ and $X_{22} \in \mathbb{R}^{(n-r)\times(n-r)}$, then we have $E_0 = \begin{bmatrix} 0 \\ I_{n-r} \end{bmatrix}$. By using $E^T X = X^T E$, we have $X_{12} = 0$, $X_{11} \geq 0$. Due to $\operatorname{rank}(E^T X) = \operatorname{rank}(X_{11}) = r$, we arrive at $X_{11} > 0$. Based on above discussion, we can find a matrix $P = \begin{bmatrix} X_{11} & 0 \\ 0 & I_{n-r} \end{bmatrix} > 0$ and $\Phi = \begin{bmatrix} X_{21} & X_{22} \end{bmatrix}$ such that $X = PE + E_0\Phi$. Moreover, $E_L^T P E_L > 0$ holds. □

Then Problem HCT can be solved in terms of LMIs as follows:

Theorem 3.7. *Given a scalar $\gamma > 0$ and $E_0 \in \mathbb{R}^{n\times(n-r)}$ with full column rank is the right null matrix of E^T, that is, $E^T E_0 = 0$. Then the following statements are equivalent:*

1. *The system in (3.2) is admissible and its general excitation performance measure $\gamma_g(R) < \gamma$.*
2. *There exists a matrix X such that the following LMIs hold:*

$$E^T X = X^T E \geq 0, \tag{3.6}$$

$$\begin{bmatrix} A^TX+X^TA & X^TB & C^T \\ \star & -\gamma^2 I & 0 \\ \star & \star & -I \end{bmatrix} < 0, \tag{3.7}$$

$$E^TX \leq \gamma^2 E^T RE. \tag{3.8}$$

3. *There exist matrices $P = P^T$ and Φ such that the following LMIs hold:*

$$\begin{bmatrix} A^T(PE+E_0\Phi)+(PE+E_0\Phi)^TA & (PE+E_0\Phi)^TB & C^T \\ \star & -\gamma^2 I & 0 \\ \star & \star & -I \end{bmatrix} < 0, \tag{3.9}$$

$$E^TPE \leq \gamma^2 E^T RE, \tag{3.10}$$

$$E_L^TPE_L > 0, \tag{3.11}$$

where E_L and E_R are full column rank with $E = E_L E_R^T$.

Proof. 3⇒1 Suppose that there exist matrices P and Φ such that the LMIs in (3.9) and (3.11) holds, then

$$\begin{aligned} &A^T(PE+E_0\Phi)+(PE+E_0\Phi)^TA < 0, \\ &E^T(PE+E_0\Phi) = (PE+E_0\Phi)^TE = E^TPE \geq 0 \end{aligned}$$

are obtained which gives the necessary and sufficient condition for the admissibility of the singular system in (3.2) [209]. On the other hand, from (3.9), the following inequality holds for $\begin{bmatrix} x(t) \\ w(t) \end{bmatrix} \not\equiv 0$ with $t \in [0, \infty)$:

$$\begin{aligned} &x^T(t)\left(A^T(PE+E_0\Phi)+(PE+E_0\Phi)^TA+C^TC\right)x(t) \\ &+2x^T(t)(PE+E_0\Phi)^TBw(t) - \gamma^2 w^T(t)w(t) \leq 0. \end{aligned} \tag{3.12}$$

We choose a Lyapunov candidate as

$$V(t) = x^T(t)E^TPEx(t) = x^T(t)E^T(PE+E_0\Phi)x(t).$$

Hence, the derivative of the Lyapunov function along the trajectory of system (3.2) is given by

$$\begin{aligned} \dot{V}(t) = {} & x^T(t)\left(A^T(PE+E_0\Phi)+(PE+E_0\Phi)^TA\right)x(t) \\ & + 2x^T(t)(PE+E_0\Phi)^TBw(t), \end{aligned}$$

and (3.12) implies that

$$\dot{V}(t) + |z(t)|^2 - \gamma^2 |w(t)|^2 \leq 0.$$

Integrating the above inequality from zero to infinity and noting that $\dot{V}(t) + |z(t)|^2 - \gamma^2 |w(t)|^2 \not\equiv 0$ with $t \in [0, \infty)$, we get

$$\|z\|^2 \quad < \quad \gamma^2 \|w\|^2 + x(0_+)^T E^T PEx(0_+).$$

By using $E^T PE \leq \gamma^2 E^T RE$ in (3.10) and $\bar{x}_1(0_+) = \bar{x}_1(0_-)$ in (3.4), it yields

$$\begin{aligned}
\|z\|^2 &< \gamma^2 \|w\|^2 + \gamma^2 x(0_+)^T E^T REx(0_+) \\
&= \gamma^2 \|w\|^2 + \gamma^2 \bar{x}(0_+)^T N^T E^T REN\bar{x}(0_+) \\
&= \gamma^2 \|w\|^2 + \gamma^2 \bar{x}(0_+)^T \begin{bmatrix} I & 0 \\ 0 & 0 \end{bmatrix} M^{-T} RM^{-1} \begin{bmatrix} I & 0 \\ 0 & 0 \end{bmatrix} \bar{x}(0_+) \\
&= \gamma^2 \|w\|^2 + \gamma^2 \bar{x}(0_-)^T \begin{bmatrix} I & 0 \\ 0 & 0 \end{bmatrix} M^{-T} RM^{-1} \begin{bmatrix} I & 0 \\ 0 & 0 \end{bmatrix} \bar{x}(0_-) \\
&= \gamma^2 \|w\|^2 + \gamma^2 x_0^T E^T REx_0,
\end{aligned} \tag{3.13}$$

where M and N are defined in (3.3). Hence, $\gamma_g(R) < \gamma$ holds true and the result is proved.

1⇒2 Suppose that the singular system in (3.2) is admissible and $\gamma_g(R) < \gamma$, then there exist two nonsingular matrices S and T such that

$$SET = \begin{bmatrix} I & 0 \\ 0 & 0 \end{bmatrix}, \ SAT = \begin{bmatrix} A_1 & 0 \\ 0 & I \end{bmatrix}, \ SB = \begin{bmatrix} B_1 \\ B_2 \end{bmatrix}, \ CT = \begin{bmatrix} C_1 & C_2 \end{bmatrix},$$

where A_1 is stable. We set $x(t) = T\tilde{x}(t) = T \begin{bmatrix} \tilde{x}_1(t) \\ \tilde{x}_2(t) \end{bmatrix}$, and the system in (3.2) is restricted system equivalent to

$$\begin{cases} \begin{bmatrix} I & 0 \\ 0 & 0 \end{bmatrix} \dot{\tilde{x}}(t) = \begin{bmatrix} A_1 & 0 \\ 0 & I \end{bmatrix} \tilde{x}(t) + \begin{bmatrix} B_1 \\ B_2 \end{bmatrix} w(t), \ \tilde{x}_0 = T^{-1} x_0 \equiv \begin{bmatrix} \tilde{x}_{10} \\ \tilde{x}_{20} \end{bmatrix}, \\ z(t) = \begin{bmatrix} C_1 & C_2 \end{bmatrix} \tilde{x}(t), \end{cases}$$

that is,

$$\begin{cases} \dot{\tilde{x}}_1(t) = A_1 \tilde{x}_1(t) + B_1 w(t), \ \ \tilde{x}_{10} = \tilde{x}_1(0), \\ z(t) = C_1 \tilde{x}_1(t) - C_2 B_2 w(t). \end{cases} \tag{3.14}$$

$$
\begin{aligned}
\gamma_g(R) &= \sup_{\|w\|^2+x_0^T E^T R E x_0 \neq 0} \frac{\|z\|}{\left(\|w\|^2 + x_0^T E^T R E x_0\right)^{1/2}} \\
&= \sup_{\|w\|^2+\tilde{x}_0^T T^T E^T R E T \tilde{x}_0 \neq 0} \frac{\|z\|}{\left(\|w\|^2 + \tilde{x}_0^T T^T E^T R E T \tilde{x}_0\right)^{1/2}} \\
&= \sup_{\|w\|^2+\tilde{x}_0^T T^T E^T S^T S^{-T} R S^{-1} S E T \tilde{x}_0 \neq 0} \frac{\|z\|}{\left(\|w\|^2 + \tilde{x}_0^T T^T E^T S^T S^{-T} R S^{-1} S E T \tilde{x}_0\right)^{1/2}} \\
&= \sup_{\|w\|^2+\tilde{x}_{10}^T R_1 \tilde{x}_{10} \neq 0} \frac{\|z\|}{\left(\|w\|^2 + \tilde{x}_{10}^T R_1 \tilde{x}_{10}\right)^{1/2}} \triangleq \gamma_N,
\end{aligned}
$$

where $R_1 \in \mathbb{R}^{r\times r}$ satisfying $R_1 > 0$ is the (1, 1) block of matrix $S^{-T}RS^{-1}$. It can be seen that $\gamma_g(R)$ can be transformed to the performance measure γ_N of the system in (3.14) with a given weighting matrix $R_1 > 0$. Based on Lemma 3.5, for $\gamma_N < \gamma$, we obtain that there exists a matrix $X_1 > 0$ such that the following LMIs hold:

$$
\begin{bmatrix} A_1^T X_1 + X_1 A_1 & X_1 B_1 & C_1^T \\ \star & -\gamma^2 I & -(C_2 B_2)^T \\ \star & \star & -I \end{bmatrix} < 0, \tag{3.15}
$$

$$
X_1 < \gamma^2 R_1. \tag{3.16}
$$

Denoting $W = C_2^T C_2 + \alpha I$ with a sufficiently small scalar $\alpha > 0$, it yields the following inequality from (3.15):

$$
\begin{bmatrix} A_1^T X_1 + X_1 A_1 + C_1^T C_1 & X_1 B_1 - C_1^T C_2 B_2 \\ \star & -\gamma^2 I + B_2^T W B_2 \end{bmatrix} < 0,
$$

which is equivalent to

$$
\begin{bmatrix} A_1^T X_1 + X_1 A_1 + C_1^T C_1 & X_1 B_1 - C_1^T C_2 B_2 & 0 \\ \star & -\gamma^2 I & B_2^T W \\ \star & \star & -W \end{bmatrix} < 0.
$$

Then there exists

$$
\begin{bmatrix} A_1^T X_1 + X_1 A_1 + C_1^T C_1 & X_1 B_1 - C_1^T C_2 B_2 & 0 \\ \star & -\gamma^2 I & B_2^T W \\ \star & \star & -W - \alpha I \end{bmatrix} < 0,
$$

which implies

$$\Pi = \begin{bmatrix} A_1^T X_1 + X_1 A_1 + C_1^T C_1 & 0 & X_1 B_1 - C_1^T C_2 B_2 \\ \star & -W - \alpha I & -WB_2 \\ \star & \star & -\gamma^2 I \end{bmatrix} < 0.$$

Then following the same lines as in the proof of Theorem 5.1 in [209], we can construct a matrix X given by

$$X = S^T \begin{bmatrix} X_1 & 0 \\ -C_2^T C_1 & -C_2^T C_2 - \alpha I \end{bmatrix} T^{-1}, \tag{3.17}$$

such that the following inequality holds:

$$\begin{bmatrix} T^T(A^T X + X^T A + C^T C)T & T^T X^T B \\ \star & -\gamma^2 I \end{bmatrix} = \Pi < 0,$$

which is equivalent to (3.7). Moreover, X in (3.17) also satisfies the equality in (3.6) and the inequality in (3.8), because of that

$$0 \leq E^T X = X^T E = T^{-T} \begin{bmatrix} X_1 & 0 \\ 0 & 0 \end{bmatrix} T^{-1} \leq T^{-T} \begin{bmatrix} \gamma^2 R_1 & 0 \\ 0 & 0 \end{bmatrix} T^{-1} = \gamma^2 E^T R E.$$

2⇒3 It is easy to establish the proof by replacing $X = PE + E_0 \Phi$ based on Lemma 3.6. □

Remark 3.8. The condition $\operatorname{rank}\left(E^T(P - \gamma^2 R)E\right) = \operatorname{rank}(E)$ in (3.9) and (3.10) also can be satisfied without loss of generality. If it is not the case, then we can choose a sufficient small scalar $\alpha > 0$ and a matrix $\tilde{P} > 0$ such that $\bar{P} = P - \alpha \tilde{P} > 0$ satisfying $\operatorname{rank}\left(E^T(\bar{P} - \gamma^2 R)E\right) = \operatorname{rank}(E)$, (3.9) and (3.10). The condition $\operatorname{rank}\left(E^T(X - \gamma^2 RE)\right) = \operatorname{rank}(E)$ in (3.6), (3.7), and (3.8) can be satisfied following a similar argument.

When $w(t) \equiv 0$, the performance measure $\gamma_0(R)$ can be characterized by the following LMIs based on Theorem 3.7:

Corollary 3.9. *Given a scalar $\gamma > 0$, the system in (3.2) is admissible and its performance measure $\gamma_0(R) < \gamma$ if and only if any of the following conditions holds:*
1) *there exists a matrix X such that the following LMIs hold:*

$$E^T X = X^T E \geq 0, \quad \begin{bmatrix} A^T X + X^T A & C^T \\ \star & -I \end{bmatrix} < 0,$$

$$E^T X \leq \gamma^2 E^T RE;$$

2) *there exist matrices $P = P^T$ and Φ such that the following LMIs hold:*

$$\begin{bmatrix} A^T(PE + E_0\Phi) + (PE + E_0\Phi)^T A & C^T \\ \star & -I \end{bmatrix} < 0, \tag{3.18}$$

$$E^T PE \leq \gamma^2 E^T RE, \tag{3.19}$$

$$E_L^T PE_L > 0, \tag{3.20}$$

where E_0 and E_L are defined in Lemma 3.6.

Note that $\gamma_g(R)$ is the infimum of γ, for which the LMIs in (3.9)–(3.11) are feasible; $\gamma_0(R)$ is the infimum of γ, for which the LMIs in (3.18)–(3.20) are feasible, whereas γ_∞ is the infimum of γ, for which the inequality in (3.9) is feasible.

3.2 Controller design

Consider the following singular system:

$$\begin{cases} E\dot{x}(t) = Ax(t) + Bw(t) + Fu(t), \quad x(0_-) = x_0, \\ z(t) = Cx(t) + Gu(t), \end{cases} \tag{3.21}$$

where $x(t)$, $w(t)$, A, B, C are defined in (3.1); $u(t) \in \mathbb{R}^m$ is the control input, F and G are constant matrices. A state-feedback controller in the form of

$$u(t) = Kx(t), \;\; K \in \mathbb{R}^{m \times n}, \tag{3.22}$$

will be designed for the singular system in (3.21) such that the general excitation performance measure of following closed-loop system is less than a prescribed positive scalar:

$$\begin{cases} E\dot{x}(t) = (A + FK)x(t) + Bw(t), \quad x(0_-) = x_0, \\ z(t) = (C + GK)x(t). \end{cases} \tag{3.23}$$

If we use item 3 in Theorem 3.7 to design the state feedback controller, matrices A and C should be replaced by $A + FK$ and $C + GK$, respectively. The inequality in (3.9) becomes

$$\begin{bmatrix} (A + FK)^T \Xi + \Xi^T (A + FK) & \Xi^T B & (C + GK)^T \\ \star & -\gamma^2 I & 0 \\ \star & \star & -I \end{bmatrix} < 0, \tag{3.24}$$

where $\Xi = PE + E_0\Phi$. Then performing congruence transformation to inequality (3.24) with diag$\{\Xi^{-T}, I, I\}$ and its transpose, we have

$$\begin{bmatrix} \Xi^{-T}(A+FK)^T + (A+FK)\Xi^{-1} & B & \Xi^{-T}(C+GK)^T \\ \star & -\gamma^2 I & 0 \\ \star & \star & -I \end{bmatrix} < 0. \tag{3.25}$$

If the inequality in (3.10) is not present, the controller can be obtained as $K = Y\Xi$ by setting $Y = K\Xi^{-1}$ in inequality (3.25). However, the inequality in (3.10) should also be solved. The matrix P appears in (3.10), and $(PE + E_0\Phi)^{-1}$ appears in (3.25). On the other hand, there is no expansion formula for $(PE + E_0\Phi)^{-1}$ as that given in Lemma 2.4 for $(PE + U^T Q \Lambda^T)^{-1}$, which makes the matrix inequalities (3.10) and (3.25) nonlinear, and hence difficult to solve.

Before giving the controller design method, a sufficient condition guaranteeing the general excitation performance of closed-loop system (3.23) to be less than a prescribed positive scalar can be obtained by applying the LMIs in (3.9) and (3.10) of Theorem 3.7, in which E_0 and Φ are replaced by U^T and $Q\Lambda^T$ with $P = P^T$, Q unknown, that is, the following two LMIs hold:

$$\begin{bmatrix} \Theta + \Theta^T & (PE + U^T Q \Lambda^T)^T B & (C+GK)^T \\ \star & -\gamma^2 I & 0 \\ \star & \star & -I \end{bmatrix} < 0, \tag{3.26}$$

$$E^T PE \le \gamma^2 E^T RE, \tag{3.27}$$

$$E_L^T PE_L > 0, \tag{3.28}$$

where $\Theta = (A + FK)^T(PE + U^T Q\Lambda^T)$, $U \in \mathbb{R}^{(n-r)\times n}$ with full row rank, and $\Lambda \in \mathbb{R}^{n\times(n-r)}$ with full column rank are the left and right null matrices of matrix E, respectively, that is, $UE = 0$ and $E\Lambda = 0$.

Based on discussion below Eq. (3.23) and Lemma 2.4, we are now in a position to present the state-feedback controller design method for system (3.21).

Theorem 3.10. *There exists a state-feedback controller such that closed-loop system (3.23) is admissible and its general excitation performance measure $\gamma_g(R) < \gamma$ if there exist matrices $\bar{P} = \bar{P}^T$ satisfying $E_R^T \bar{P} E_R > 0$, $\bar{Q}$ and W such that the* *following LMIs hold:*

$$\begin{bmatrix} sym(A\bar{P}E^T + A\Lambda\bar{Q}U + FW) & B & E\bar{P}C^T + U^T\bar{Q}^T\Lambda^T C^T + W^T G^T \\ \star & -\gamma^2 I & 0 \\ \star & \star & -I \end{bmatrix} < 0, \tag{3.29}$$

$$\begin{bmatrix} -\gamma^2 E^T RE & E_R \\ E_R^T & -E_R^T \bar{P} E_R \end{bmatrix} \le 0, \tag{3.30}$$

where U, Λ, E_L, E_R are defined in Lemma 2.4. Under the conditions, a desired controller can be obtained by

$$K = W(\bar{P}E^T + \Lambda\bar{Q}U)^{-1}. \tag{3.31}$$

Proof. Performing the congruence transformation to the inequality in (3.26) by $\mathrm{diag}(H, I, I)$, with $H = (PE + U^T Q\Lambda^T)^{-1}$, we have

$$\begin{bmatrix} H^T(A + FK)^T + (A + FK)H & B & H^T(C + GK)^T \\ \star & -\gamma^2 I & 0 \\ \star & \star & -I \end{bmatrix} < 0 \ . \tag{3.32}$$

Let $W = KH$, then the inequality in (3.29) is obtained. For the inequality in (3.27), $E^T PE \le \gamma^2 E^T RE$ is equivalent to $E_R E_L^T PE_L E_R^T - \gamma^2 E^T RE \le 0$. By Schur complement, it is equal to

$$\begin{bmatrix} -\gamma^2 E^T RE & E_R \\ E_R^T & -(E_L^T PE_L)^{-1} \end{bmatrix} = \begin{bmatrix} -\gamma^2 E^T RE & E_R \\ E_R^T & -E_R^T \bar{P} E_R \end{bmatrix} \le 0. \tag{3.33}$$

Considering $E_L^T PE_L = (E_R^T \bar{P} E_R)^{-1}$, we have $E_L^T PE_L = (E_R^T \bar{P} E_R)^{-1} > 0$, which is the inequality in (3.28). In addition, based on the definition of W, $K = W(\bar{P}E^T + \Lambda\bar{Q}U)^{-1}$ is obtained. Then the desired result follows immediately. □

Remark 3.11. Note that an H_∞ state-feedback controller can be obtained by $K_\infty = W_\infty(\bar{P}_\infty E^T + \Lambda\bar{Q}_\infty U)^{-1}$, where W_∞, $\bar{P}_\infty$, and $\bar{Q}_\infty$ are the values of W, $\bar{P}$, and $\bar{Q}$ obtained by solving only inequality (3.29) with the minimal value of γ. A $\gamma_g(R)$ state-feedback controller can be obtained by $K_w(R) = W_w(\bar{P}_w E^T + \Lambda\bar{Q}_w U)^{-1}$, where W_w, $\bar{P}_w$, and $\bar{Q}_w$ are the values of W, $\bar{P}$, and $\bar{Q}$ obtained by solving inequality (3.29) and inequality (3.30), with the minimal value of γ. Qualitatively, the inequality in (3.30) is relaxed to be always satisfied for sufficiently large positive definite R. Hence, the performance measure $\gamma_g(R)$ will be close to γ_∞ for large R, and a state-feedback controller for $\gamma_g(R)$ obtained by Theorem (3.10) will approach an H_∞ state-feedback controller. Moreover, a $\gamma_0(R)$ state-feedback controller can be obtained by $K_0(R) = W_0(\bar{P}_0 E^T + \Lambda \bar{Q}_0 U)^{-1}$, where W_0, $\bar{P}_0$, and $\bar{Q}_0$ are the values of W, $\bar{P}$, and $\bar{Q}$ obtained by solving inequality (3.29) with the second row and column omitted and inequality (3.30), with the minimal value of γ.

Remark 3.12. For the static output feedback (SOF) control problem, let $y(t) = C_y x(t)$ and $u(t) = Ky(t)$. Then, the terms FKC_yH and GKC_yH will appear by using the method in Theorem 3.10. These nonlinear terms cannot be solved by using LMI toolbox directly. Recently, an augmented system approach is proposed to address the SOF problem in [158]. The advantages of the approach can separate the input matrix F or G, and gain-out matrix KC_y and decouple the Lyapunov matrix P and the controller matrix K, which may help to solve the SOF control problem easily.

3.3 Illustrative examples

In this section, we use numerical examples to illustrate the properties of obtained results. The following example is given to illustrate the relation obtained in Theorem 3.4.

Example 3.1. Given a singular system with following parameters:

$$E = \begin{bmatrix} 1 & 1 \\ 1 & 1 \end{bmatrix}, \quad A = \begin{bmatrix} -1 & 0 \\ 0 & 2 \end{bmatrix}, \quad B = \begin{bmatrix} \frac{1}{\sqrt{2}} \\ 0 \end{bmatrix}, \quad C = \begin{bmatrix} 1 & 0 \\ 0 & 1 \end{bmatrix}, \quad E_0 = \begin{bmatrix} 1 \\ -1 \end{bmatrix},$$

we have $\gamma_\infty = 1$. Let $R = \rho^2 I$, then $\gamma_g(\rho^2 I)$ and $\gamma_0(\rho^2 I)$ can be computed according to description given in the paragraph above Theorem 3.4. In Fig. 3.1, Curve 1 presents the plot of $\gamma_g(\rho^2 I)$ versus ρ; Curve 2 gives the plot of γ_∞; Curve 3 denotes the plot of $\gamma_0(\rho^2 I)$. From Fig. 3.1, we can see $\gamma_g(\rho^2 I) = \gamma_\infty$ when ρ becomes larger, while the curve $\gamma_g(\rho^2 I)$ asymptotically approaches $\gamma_0(\rho^2 I)$ for small ρ.

The following example is given to illustrate the property mentioned in Remark 3.11.

Example 3.2. Given a singular system with following parameters:

$$E = \begin{bmatrix} 1 & 1 \\ 1 & 1 \end{bmatrix}, \quad A = \begin{bmatrix} 0 & 2 \\ -1 & 1 \end{bmatrix}, \quad B = \begin{bmatrix} 4 \\ 4 \end{bmatrix}, \quad C = \begin{bmatrix} 1 & 0 \\ 0 & 1 \end{bmatrix},$$

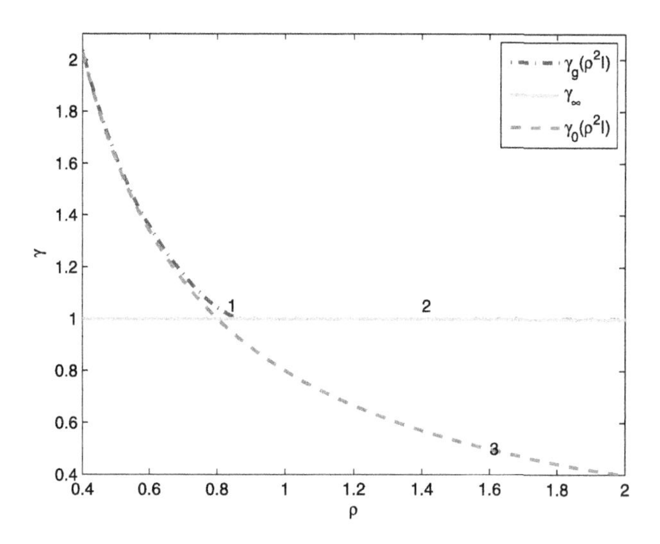

Figure 3.1 Performance measure versus parameter ρ.

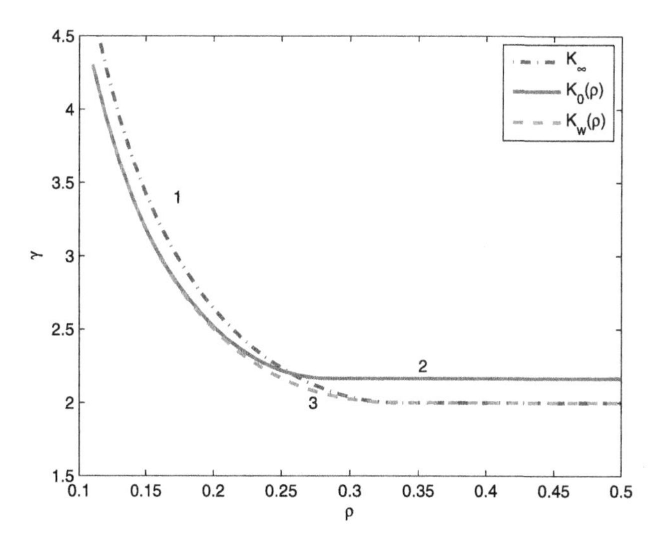

Figure 3.2 Performance measure $\gamma = \gamma_g(\rho^2 I)$ of the closed-loop system versus parameter ρ under different state-feedback controllers.

$$F = \begin{bmatrix} 0 \\ 0.3 \end{bmatrix}, \; G = \begin{bmatrix} -0.3 \\ 0.5 \end{bmatrix}, \; \Lambda = \begin{bmatrix} 1 \\ -1 \end{bmatrix}, \; U = \begin{bmatrix} -1 & 1 \end{bmatrix},$$

since $\deg\{\det(sE - A)\} \neq \text{rank}(E)$, the singular system is not impulse-free. Based on the LMI in (3.29), we obtain a standard H_∞ state-feedback controller $K_\infty = \begin{bmatrix} -0.8332 & -3.3333 \end{bmatrix}$ and $\gamma_\infty = 2.0000$. Similarly, let $R = \rho^2 I$, the state-feedback controller $K_w(\rho) = \begin{bmatrix} -0.0002 & -2.5020 \end{bmatrix}$ and $\gamma_g(\rho^2 I) = 2.2714$ can be obtained by solving the LMIs in (3.29) and (3.30) with $\rho = 0.22$. Fig. 3.2 shows three curves: Curve 1 denotes the general excitation performance measure $\gamma_g(\rho^2 I)$ for closed-loop system with K_∞; Curve 2 is the general excitation performance measure $\gamma_g(\rho^2 I)$ for the closed-loop system with $K_0(\rho)$; Curve 3 refers to the general excitation performance measure $\gamma_g(\rho^2 I)$ for the closed-loop system with $K_w(\rho)$. From this figure, we can see that the general excitation performance measure $\gamma_g(\rho^2 I)$ is close to γ_∞ for large ρ, and is close to $\gamma_0(\rho^2 I)$ for small ρ.

3.4 Conclusion

In this chapter, the problem of H_∞ control with transients for continuous-time singular systems has been studied. The necessary and sufficient conditions in terms of LMIs were proposed for performance analysis at first. Based on this, a state-feedback controller has been designed to guarantee the closed-loop system to be admissible and minimize the general excitation performance measure. Moreover, the relationship between the H_∞ performance measure with transients and the standard H_∞ performance has been revealed, which has been illustrated by numerical examples.

CHAPTER 4

Delay-dependent admissibility and H_∞ control of discrete singular delay systems

In this chapter, the issues of admissibility and H_∞ control for discrete singular systems with time-varying delay are addressed, respectively. Firstly, to reduce the conservatism of existing admissibility conditions, we adopt an improved reciprocally convex combination approach to bound the forward difference of the double summation term, and utilize the reciprocally convex combination approach to bound the forward difference of the triple summation term in the Lyapunov function. Without employing decomposition and equivalent transformation of the considered system, a strict delay-dependent LMI criterion is built to guarantee the considered system to be regular, causal, and stable. Secondly, with the introduction of the delay partitioning technique, strict LMI sufficient criteria are obtained for discrete-time singular systems to be regular, causal, and stable. Based on these criteria, the robust stabilization and robust H_∞ control problems are addressed and the desired state-feedback controllers are given. Numerical examples are given to illustrate the reduced conservatism of the developed results.

4.1 New admissibility analysis for discrete singular systems with time-varying delay

In this section, the problem of admissibility analysis for discrete-time singular system with time-varying delay is investigated. By employing the reciprocally convex combination approach to bound the forward difference of a triple-summation term, a sufficient criterion is presented in terms of LMIs to guarantee the considered system to be regular, causal, and stable. Finally, a numerical example is exhibited to illustrate the effectiveness and the reduced conservatism of the proposed result.

Analysis and Synthesis of Singular Systems
https://doi.org/10.1016/B978-0-12-823739-7.00011-2

4.1.1 Problem formulation

Consider discrete-time singular systems with time-varying delay described by

$$Ex(k+1) = Ax(k) + A_d x(k-d(k)); \tag{4.1}$$
$$x(k) = \phi(k), \ \ k \in [-d_2, 0],$$

where $x(k) \in \mathbb{R}^n$ is the state vector and $d(k)$ is a time-varying delay satisfying $1 \le d_1 \le d(k) \le d_2$, where d_1 and d_2 are prescribed positive integers representing the lower and upper bounds of the time delay, respectively. $\phi(k)$ is the compatible initial condition. The matrix $E \in \mathbb{R}^{n\times n}$ may be singular, and it is assumed that $\text{rank}(E) = r \le n$. A and A_d are known real constant matrices with appropriate dimensions.

The following definitions and lemmas will be used in the proof of the main results:

Definition 4.1. [25]
1. The pair (E, A) is said to be regular if $\det(zE - A)$ is not identically zero.
2. The pair (E, A) is said to be causal if $\deg(\det(zE - A)) = \text{rank}(E)$.

Definition 4.2. [191] The singular system in (4.1) is said to be stable if for any scalar $\varepsilon > 0$, there exists a scalar $\delta(\varepsilon) > 0$ such that, for any compatible initial conditions $\phi(k)$ satisfying $\sup_{-d_2\le k\le 0} \|\phi(k)\| \le \delta(\varepsilon)$, the solution $x(k)$ of system (4.1) satisfies $\|x(k)\| \le \varepsilon$ for any $k \ge 0$, moreover $\lim_{k\to\infty} x(k) = 0$.

Definition 4.3. [121]
1. The singular system in (4.1) is said to be regular and causal if the pairs (E, A) is regular and causal.
2. The singular system in (4.1) is said to be admissible if it is regular, causal, and stable.

Lemma 4.4. [140] *Let $f_1, f_2, ..., f_N : \mathbb{R}^m \to \mathbb{R}$ have positive values in an open subset D of $\mathbb{R}^m$. Then the reciprocally convex combination of f_i over D satisfies*

$$\min_{\{\alpha_i | \alpha_i > 0, \sum_i \alpha_i = 1\}} \sum_i \frac{1}{\alpha_i} f_i(k) = \sum_i f_i(k) + \max_{g_{i,j}(k)} \sum_{i\neq j} g_{i,j}(k), \tag{4.2}$$

$$\textit{subject to } \left\{ g_{i,j} : \mathbb{R}^m \to \mathbb{R}, g_{j,i}(k) = g_{i,j}(k), \begin{bmatrix} f_i(k) & g_{j,i}(k) \\ g_{i,j}(k) & f_j(k) \end{bmatrix} \ge 0 \right\}.$$

Lemma 4.5. *For any matrix $M > 0$, integers $a < b \leq c$, vector function $x(i)$: $[k+a, k+c-1] \to \mathbb{R}^n$, there hold*

$$-(b-a)\sum_{i=a}^{b-1} x^T(i)Mx(i) \leq -\left(\sum_{i=a}^{b-1} x(i)\right)^T M\left(\sum_{i=a}^{b-1} x(i)\right), \tag{4.3}$$

$$-\beta\sum_{j=a}^{b-1}\sum_{i=k+j}^{k+c-1} x^T(i)Mx(i) \leq -\vartheta^T(k)M\vartheta(k), \tag{4.4}$$

where $\beta = \frac{(2c+1)(b-a)-(b^2-a^2)}{2}$, $\vartheta(k) = \sum_{j=a}^{b-1}\sum_{i=k+j}^{k+c-1} x(i)$.

Remark 4.6. Based on similar arguments as those of Lemma 1 in [84], Lemma 4.5 can be established. Moreover, Lemma 4.5 is more general than Lemma 1 in [84], since the latter can be derived by setting $a = -d_2$, $b = c = -d_1$ in Lemma 4.5.

Lemma 4.7. *For any constant matrix* $\begin{bmatrix} M & S \\ \star & M \end{bmatrix} > 0$, *integers* $a \leq d(k) \leq b \leq c$, *vector function* $w(i) = x(i+1) - x(i)$, $x(i) : [k+a, k+c] \to \mathbb{R}^n$, *then*

$$-\beta\sum_{j=a}^{b-1}\sum_{i=k+j}^{k+c-1} w^T(i)Mw(i) \leq -\eta^T(k)\begin{bmatrix} M & S \\ \star & M \end{bmatrix}\eta(k),$$

where β is defined in Lemma 4.5, and

$$\eta(k) = \begin{bmatrix} \eta_1(k) \\ \eta_2(k) \end{bmatrix} = \begin{bmatrix} (d(k)-a)x(k+c) - \sum_{j=a}^{d(k)-1} x(k+j) \\ (b-d(k))x(k+c) - \sum_{j=d(k)}^{b-1} x(k+j) \end{bmatrix}.$$

Proof. When $a < d(k) < b$, by utilizing (4.4), we have

$$\begin{aligned} &-\beta\sum_{j=a}^{b-1}\sum_{i=k+j}^{k+c-1} w^T(i)Mw(i) \\ &= -\beta\sum_{j=a}^{d(k)-1}\sum_{i=k+j}^{k+c-1} w^T(i)Mw(i) - \beta\sum_{j=d(k)}^{b-1}\sum_{i=k+j}^{k+c-1} w^T(i)Mw(i) \\ &\leq -\frac{\beta}{\beta_1}\eta_1^T(k)M\eta_1(k) - \frac{\beta}{\beta_2}\eta_2^T(k)M\eta_2(k)w, \end{aligned} \tag{4.5}$$

where $\beta_1 = \frac{(2c+1)(d(k)-a)-(d^2(k)-a^2)}{2}$, $\beta_2 = \frac{(2c+1)(b-d(k))-(b^2-d^2(k))}{2}$.

Noting $\frac{\beta_1}{\beta}+\frac{\beta_2}{\beta}=1$ and that according to Lemma 4.4, it follows from the inequality in (4.5) that

$$\eta(k)=\begin{bmatrix}\eta_1(k)\\ \eta_2(k)\end{bmatrix}=\begin{bmatrix}(d(k)-a)x(k+c)-\sum_{j=a}^{d(k)-1}x(k+j)\\ (b-d(k))x(k+c)-\sum_{j=d(k)}^{b-1}x(k+j)\end{bmatrix}. \tag{4.6}$$

When $d(k)=a$ or $d(k)=b$, we have $(d(k)-a)x(k+c)-\sum_{j=a}^{d(k)-1}x(k+j)$ $=0$ or $(b-d(k))x(k+c)-\sum_{j=d(k)}^{b-1}x(k+j)=0$, and the inequality in (4.6) still holds by using Lemma 4.5. Hence, the lemma is proved. □

Remark 4.8. In [190], the reciprocally convex combination approach is extended to discrete-time systems, where it is used to bound the single summation term. Here the approach is applied in Lemma 4.7 to bound the double-summation term.

Lemma 4.9. [104] *For a matrix $R>0$, a matrix Γ, and a matrix Ξ, the following statements are equivalent:*

- $\Xi-\Gamma^TR\Gamma<0$.
- *There exists a matrix Ψ such that*

$$\begin{bmatrix}\Xi+\mathrm{sym}(\Gamma^T\Psi) & \Psi^T\\ \star & -R\end{bmatrix}<0.$$

The purpose of this section is to analyze the admissibility of system (4.1), and a new stability criterion with less conservatism than some existing ones will be proposed.

4.1.2 Main results

Theorem 4.10. *For given positive integers d_1, d_2, the discrete-time singular time-delay system in (4.1) is admissible for any time-varying delay $d(k)$ satisfying $1\le d_1\le d(k)\le d_2$, if there exist matrices $P>0$, $Q_i>0$, $P_i=P_i^T$, $i=1,2,3$, $R_j>0$, $S_j>0$, $j=1,2$, P_4, X, W, and Ψ such that the following LMIs hold:*

$$\mathfrak{R}_1>0,\ \mathfrak{R}_2>0,\ \begin{bmatrix}S_2 & P_4\\ \star & S_2\end{bmatrix}>0; \tag{4.7}$$

$$\begin{bmatrix}\Xi+\mathrm{sym}(\Lambda_1^T\Psi) & \Psi^T\\ \star & -\begin{bmatrix}S_2 & P_4\\ \star & S_2\end{bmatrix}\end{bmatrix}<0; \tag{4.8}$$

$$\begin{bmatrix} \Xi + \mathrm{sym}(\Lambda_2^T \Psi) & \Psi^T \\ \star & -\begin{bmatrix} S_2 & P_4 \\ \star & S_2 \end{bmatrix} \end{bmatrix} < 0, \tag{4.9}$$

where $R \in \mathbb{R}^{n\times(n-r)}$ is any matrix with full column rank and satisfies $E^T R = 0$, and

$$\begin{aligned}
\Xi =& W_{P1}^T P W_{P1} - W_{P2}^T P W_{P2} + W_1^T Q_1 W_1 - W_2^T Q_1 W_2 \\
&+ W_2^T Q_2 W_2 - d_1 W_2^T E^T P_1 E W_2 - W_3^T Q_3 W_3 + d_1^2 W_{R1}^T R_1 W_{R1} \\
&+ d_{12}^2 W_{R1}^T R_2 W_{R1} + d_1 W_1^T E^T P_1 E W_1 - W_4^T Q_2 W_4 + d_{12} W_2^T E^T P_2 E W_2 \\
&- d_{12} W_3^T E^T P_2 E W_3 + d_{12} W_3^T E^T P_3 E W_3 - d_{12} W_4^T E^T P_3 E W_4 \\
&- W_{\mathfrak{R}1}^T \mathfrak{R}_1 W_{\mathfrak{R}1} - W_{\mathfrak{R}2}^T \mathfrak{R}_2 W_{\mathfrak{R}2} + \alpha_1^2((A-E)W_1 + A_d W_3)^T \\
&\times S_1((A-E)W_1 + A_d W_3) + \alpha_2^2((A-E)W_1 + A_d W_3)^T \\
&\times S_2((A-E)W_1 + A_d W_3) - (d_1 E W_1 + (d_{12}+1) W_1^T Q_3 W_1 \\
&+ \mathrm{sym}(W_1^T W R^T (A W_1 + A_d W_3)) - W_5)^T S_1 (d_1 E W_1 - W_5),
\end{aligned}$$

$$P = \begin{bmatrix} P_{11} P_{12} P_{13} \\ \star \ P_{22} P_{23} \\ \star \ \star \ P_{33} \end{bmatrix}, \quad W_{P1} = \begin{bmatrix} A W_1 + A_d W_3 \\ E W_1 - E W_2 + W_5 \\ E W_2 - E W_4 + W_6 + W_7 \end{bmatrix},$$

$$W_{P2} = \begin{bmatrix} E W_1 \\ W_5 \\ W_6 + W_7 \end{bmatrix}, \quad W_{R1} = \begin{bmatrix} E W_1 \\ (A-E) W_1 + A_d W_3 \end{bmatrix},$$

$$R_1 = \begin{bmatrix} R_{11a} R_{12a} \\ \star \ R_{22a} \end{bmatrix}, \quad R_2 = \begin{bmatrix} R_{11b} R_{12b} \\ \star \ R_{22b} \end{bmatrix}, \quad W_{\mathfrak{R}1} = \begin{bmatrix} W_5 \\ E W_1 - E W_2 \end{bmatrix},$$

$$W_{\mathfrak{R}2} = \begin{bmatrix} W_6 \\ E W_2 - E W_3 \\ W_7 \\ E W_3 - E W_4 \end{bmatrix}, \quad \Lambda_1 = \begin{bmatrix} -W_6 \\ d_{12} E W_1 - W_7 \end{bmatrix},$$

$$\Lambda_2 = \begin{bmatrix} d_{12} E W_1 - W_6 \\ -W_7 \end{bmatrix}, \quad X = \begin{bmatrix} X_{11} X_{12} \\ X_{21} X_{22} \end{bmatrix}, \quad \mathfrak{R}_1 = \begin{bmatrix} R_{11a} R_{12a} + P_1 \\ \star \ R_{22a} + P_1 \end{bmatrix},$$

$$\mathfrak{R}_2 = \begin{bmatrix} R_{11b} R_{12b} + P_2 \, X_{11} & X_{12} \\ \star \ R_{22b} + P_2 \, X_{21} & X_{22} \\ \star \quad \star & R_{11b} R_{12b} + P_3 \\ \star \quad \star & \star \ R_{22b} + P_3 \end{bmatrix},$$

$$W_i = [0_{n,(i-1)n} I_n 0_{n,(7-i)n}], \ i = 1,2,\ldots,7, \ d_{12} = d_2 - d_1.$$

Proof. According to the given condition, we first prove that system (4.1) is regular and causal. Since rank $(E) = r$, we choose two nonsingular matrices M and N such that

$$MEN = \begin{bmatrix} I_r & 0 \\ 0 & 0 \end{bmatrix}. \tag{4.10}$$

Set

$$MAN = \begin{bmatrix} A_1 & A_2 \\ A_3 & A_4 \end{bmatrix}, N^T W = \begin{bmatrix} W_1 \\ W_2 \end{bmatrix}, M^{-T} R = \begin{bmatrix} 0 \\ I \end{bmatrix} F, \tag{4.11}$$

where $F \in \mathbb{R}^{(n-r)\times(n-r)}$ is nonsingular. According to Lemma 4.9, the equivalent representation of condition (4.8) is

$$\Theta_1 = \Xi - \Lambda_1^T \begin{bmatrix} S_2 & P_4 \\ \star & S_2 \end{bmatrix} \Lambda_1 < 0.$$

Expand Ξ as

$$\Xi = \begin{bmatrix} \Xi_{11} & \bullet \\ \bullet & \bullet \end{bmatrix},$$

where $\bullet$ represents the elements of the matrix that are not relevant in our discussion, $\Xi_{11} \in \mathbb{R}^{n\times n}$ and

$$\begin{aligned}
\Xi_{11} = {} & Q_1 + (d_{12}+1)Q_3 + d_1 E^T P_1 E + \alpha_1^2 (A-E)^T S_1 (A-E) \\
& + \alpha_2^2 (A-E)^T S_2 (A-E) + WR^T A + A^T R W^T - E^T P_{11} E \\
& + A^T P_{11} A + A^T P_{12} E + E^T P_{12}^T A + E^T P_{22} E + d_1^2 E^T R_{12a}(A-E) \\
& + d_1^2 (A-E)^T R_{12a}^T E + d_1^2 (A-E)^T R_{22a}(A-E) + d_{12}^2 E^T R_{12b}(A-E) \\
& + d_{12}^2 (A-E)^T R_{12b}^T E + d_{12}^2 (A-E)^T R_{22b}(A-E) - d_1^2 E^T S_1 E \\
& + d_1^2 E^T R_{11a} E + d_{12}^2 E^T R_{11b} E - E^T (R_{22a} + P_1) E.
\end{aligned}$$

Due to $\Theta_1 < 0$, we have $\Xi_{11} - d_{12}^2 E^T S_2 E < 0$, which implies

$$\begin{aligned}
\Omega = {} & d_1 E^T P_1 E - d_1^2 E^T S_1 E + WR^T A + A^T R W^T - E^T P_{11} E + A^T P_{12} E \\
& + E^T P_{12}^T A + E^T P_{22} E + d_1^2 E^T R_{12a}(A-E) + d_1^2 (A-E)^T R_{12a}^T E \\
& + d_{12}^2 E^T R_{12b}(A-E) + d_{12}^2 (A-E)^T R_{12b}^T E - E^T (R_{22a} + P_1) E \\
& - d_{12}^2 E^T S_2 E < 0.
\end{aligned}$$

Premultiplying and postmultiplying $\Omega < 0$ by N^T and N, respectively, substituting (4.10) and (4.11) into the above inequality give

$$\begin{bmatrix} \bullet & \bullet \\ \bullet & W_2F^TA_4 + A_4^TFW_2^T \end{bmatrix} < 0.$$

From the above inequality, it is easy to see that $W_2F^TA_4 + A_4^TFW_2^T < 0$, which implies A_4 is nonsingular. Thus the pair (E, A) is regular and causal. According to Definition 4.1 and Definition 4.3, system (4.1) is regular and causal.

To prove the stability of the system in (4.1), design a Lyapunov function as

$$V(k) = V_1(k) + V_2(k) + V_3(k) + V_4(k),$$

where

$$\begin{aligned} V_1(k) &= \varepsilon^T(k)P\varepsilon(k), \\ V_2(k) &= \sum_{i=k-d_1}^{k-1} x^T(i)Q_1x(i) + \sum_{i=k-d_2}^{k-d_1-1} x^T(i)Q_2x(i) + \sum_{i=k-d(k)}^{k-1} x^T(i)Q_3x(i) \\ &\quad + \sum_{i=-d_2+1}^{-d_1}\sum_{j=k+i}^{k-1} x^T(j)Q_3x(j), \\ V_3(k) &= d_1\sum_{i=-d_1}^{-1}\sum_{j=k+i}^{k-1} \eta^T(j)R_1\eta(j) + d_{12}\sum_{i=-d_2}^{-d_1-1}\sum_{j=k+i}^{k-1} \eta^T(j)R_2\eta(j), \\ V_4(k) &= \alpha_1\sum_{i=-d_1}^{-1}\sum_{j=i}^{-1}\sum_{l=k+j}^{k-1} \theta^T(l)E^TS_1E\theta(l) + \alpha_2\sum_{i=-d_2}^{-d_1-1}\sum_{j=i}^{-1}\sum_{l=k+j}^{k-1} \theta^T(l)E^TS_2E\theta(l), \end{aligned}$$

with $\theta(i) = x(i+1) - x(i)$ and

$$\varepsilon(k) = \begin{bmatrix} Ex(k) \\ \sum_{i=k-d_1}^{k-1} Ex(i) \\ \sum_{i=k-d_2}^{k-d_1-1} Ex(i) \end{bmatrix}, \quad \eta(k) = \begin{bmatrix} Ex(k) \\ E\theta(k) \end{bmatrix},$$

$$\alpha_1 = \frac{d_1(d_1+1)}{2}, \quad \alpha_2 = \frac{d_{12}(d_1+d_2+1)}{2}.$$

By denoting the forward difference of $V(k)$ as $\Delta V(k) = V(k+1) - V(k)$ and calculating it along the solution of system (4.1), we have

$$\Delta V_1(k) = \varepsilon^T(k+1)P\varepsilon(k+1) - \varepsilon^T(k)P\varepsilon(k),$$

where

$$\varepsilon(k+1) = \begin{bmatrix} Ex(k+1) \\ Ex(k) - Ex(k-d_1) + \sum_{i=k-d_1}^{k-1} Ex(i) \\ Ex(k-d_1) - Ex(k-d_2) + \sum_{i=k-d_2}^{k-d_1-1} Ex(i) \end{bmatrix},$$

so

$$\Delta V_1(k) = \xi^T(k)(W_{P1}^T P W_{P1} - W_{P2}^T P W_{P2})\xi(k), \tag{4.12}$$

where

$$\begin{aligned}\xi^T(k) =[\ & x^T(k) \quad x^T(k-d_1) \quad x^T(k-d(k)) \quad x^T(k-d_2) \\ & \textstyle\sum_{i=k-d_1}^{k-1}(Ex(i))^T \quad \sum_{i=k-d(k)}^{k-d_1-1}(Ex(i))^T \quad \sum_{i=k-d_2}^{k-d(k)-1}(Ex(i))^T\ ].\end{aligned}$$

The estimation of the forward difference of $V_2(k)$ is

$$\begin{aligned}\Delta V_2(k) =& x^T(k)Q_1x(k) - x^T(k-d_1)Q_1x(k-d_1) + x^T(k-d_1)Q_2x(k-d_1) \\ &+ x^T(k)Q_3x(k) - x^T(k-d(k))Q_3x(k-d(k)) + d_{12}x^T(k)Q_3x(k) \\ &- x^T(k-d_2)Q_2x(k-d_2) - \sum_{i=k-d_2+1}^{k-d_1} x^T(i)Q_3x(i) \\ \le& \xi^T(k)(W_1^TQ_1W_1 - W_2^TQ_1W_2 + W_2^TQ_2W_2 - W_4^TQ_2W_4 \\ &+ (d_{12}+1)W_1^TQ_3W_1 - W_3^TQ_3W_3)\xi(k).\end{aligned} \tag{4.13}$$

Introducing three equalities, just as those in [84] with any real symmetric matrices P_1, P_2, P_3, according to (4.3) in Lemma 4.5, and considering $\mathfrak{R}_1 > 0$ in (4.7), we have

$$\begin{aligned}\Delta V_3(k) =& d_1^2\eta^T(k)R_1\eta(k) - d_1\sum_{i=k-d_1}^{k-1}\eta^T(i)R_1\eta(i) + d_{12}^2\eta^T(k)R_2\eta(k) \\ &+ d_1x^T(k)E^TP_1Ex(k) - d_1x^T(k-d_1)E^TP_1Ex(k-d_1) \\ &- d_{12}\sum_{i=k-d_2}^{k-d_1-1}\eta^T(i)R_2\eta(i) - d_1\sum_{i=k-d_1}^{k-1}\eta^T(i)\begin{bmatrix} 0_n & P_1 \\ P_1 & P_1 \end{bmatrix}\eta(i) \\ &+ d_{12}x^T(k-d_1)E^TP_2Ex(k-d_1) \\ &- d_{12}x^T(k-d(k))E^TP_2Ex(k-d(k)) \\ &- d_{12}\sum_{i=k-d(k)}^{k-d_1-1}\eta^T(i)\begin{bmatrix} 0_n & P_2 \\ P_2 & P_2 \end{bmatrix}\eta(i)\end{aligned}$$

$$
\begin{aligned}
&+ d_{12}x^T(k-d(k))E^TP_3Ex(k-d(k)) \\
&- d_{12}x^T(k-d_2)E^TP_3Ex(k-d_2) - d_{12}\sum_{i=k-d_2}^{k-d(k)-1}\eta^T(i)\begin{bmatrix} 0_n & P_3 \\ P_3 & P_3 \end{bmatrix}\eta(i) \\
&\le d_1^2\eta^T(k)R_1\eta(k) + d_{12}^2\eta^T(k)R_2\eta(k) + d_1x^T(k)E^TP_1Ex(k) \\
&- d_1x^T(k-d_1)E^TP_1Ex(k-d_1) \\
&+ d_{12}x^T(k-d_1)E^TP_2Ex(k-d_1) - d_{12}x^T(k \\
&- d(k))E^TP_2Ex(k-d(k)) \\
&+ d_{12}x^T(k-d(k))E^TP_3Ex(k-d(k)) \\
&- d_{12}x^T(k-d_2)E^TP_3Ex(k-d_2) \\
&- \begin{bmatrix} \sum_{i=k-d_1}^{k-1} Ex(i) \\ Ex(k) - Ex(k-d_1) \end{bmatrix}^T \mathfrak{R}_1 \begin{bmatrix} \sum_{i=k-d_1}^{k-1} Ex(i) \\ Ex(k) - Ex(k-d_1) \end{bmatrix} \\
&- d_{12}\sum_{i=k-d(k)}^{k-d_1-1}\eta^T(i)(R_2+\mathfrak{P}_2)\eta(i) - d_{12}\sum_{i=k-d_2}^{k-d(k)-1}\eta^T(i)(R_2+\mathfrak{P}_3)\eta(i),
\end{aligned}
\tag{4.14}
$$

where

$$
\mathfrak{P}_2 = \begin{bmatrix} 0_n & P_2 \\ P_2 & P_2 \end{bmatrix}, \mathfrak{P}_3 = \begin{bmatrix} 0_n & P_3 \\ P_3 & P_3 \end{bmatrix}.
$$

Then by employing Lemmas 4.5 and 4.7, the following inequality holds for any matrix $\mathfrak{R}_2 > 0$ when $d_1 < d(k) < d_2$:

$$
\begin{aligned}
&- d_{12}\sum_{i=k-d(k)}^{k-d_1-1}\eta^T(i)(R_2+\mathfrak{P}_2)\eta(i) - d_{12}\sum_{i=k-d_2}^{k-d(k)-1}\eta^T(i)(R_3+\mathfrak{P}_3)\eta(i) \\
&\le -\frac{d_{12}^2}{d(k)-d_1}\sum_{i=k-d(k)}^{k-d_1-1}\eta^T(i)(R_2+\mathfrak{P}_2)\eta(i) \\
&- \frac{d_{12}^2}{d_2-d(k)}\sum_{i=k-d_2}^{k-d(k)-1}\eta^T(i)(R_3+\mathfrak{P}_3)\eta(i) \\
&\le -\frac{d_{12}}{d(k)-d_1}\psi_1^T(R_2+\mathfrak{P}_2)\psi_1 - \frac{d_{12}}{d_2-d(k)}\psi_2^T(R_2+\mathfrak{P}_3)\psi_2 \\
&\le -\begin{bmatrix} \psi_1 \\ \psi_2 \end{bmatrix}^T \mathfrak{R}_2 \begin{bmatrix} \psi_1 \\ \psi_2 \end{bmatrix},
\end{aligned}
\tag{4.15}
$$

where

$$\psi_1 = \begin{bmatrix} \sum_{i=k-d(k)}^{k-d_1-1} Ex(i) \\ Ex(k-d_1) - Ex(k-d(k)) \end{bmatrix}, \quad \psi_2 = \begin{bmatrix} \sum_{i=k-d_2}^{k-d(k)-1} Ex(i) \\ Ex(k-d(k)) - Ex(k-d_2) \end{bmatrix}.$$

When $d(k) = d_1$ or $d(k) = d_2$, the inequality (4.15) still holds based on the notation. From (4.14) and (4.15), ΔV_3 can be bounded as

$$\begin{aligned} \Delta V_3(k) \leq \xi^T(k)(& d_{12} W_2^T E^T P_2 E W_2 - d_{12} W_3^T E^T P_2 E W_3 \\ &+ d_{12} W_3^T E^T P_3 E W_3 + d_1^2 W_{R1}^T R_1 W_{R1} + d_{12}^2 W_{R1}^T R_2 W_{R1} \\ &+ d_1 W_1^T E^T P_1 E W_1 - d_1 W_2^T E^T P_1 E W_2 \\ &- d_{12} W_4^T E^T P_3 E W_4 - W_{\Re 1}^T \Re_1 W_{\Re 1} \\ &- W_{\Re 2}^T \Re_2 W_{\Re 2})\xi(k). \end{aligned} \tag{4.16}$$

The forward difference of $V_4(k)$ is calculated as

$$\begin{aligned} \Delta V_4(k) =& \alpha_1 \sum_{i=-d_1}^{-1} \sum_{j=i}^{-1} (\theta^T(k) E^T S_1 E \theta(k) - \theta^T(k+j) E^T S_1 E \theta(k+j)) \\ &+ \alpha_2 \sum_{i=-d_2}^{-d_1-1} \sum_{j=i}^{-1} (\theta^T(k) E^T S_2 E \theta(k) - \theta^T(k+j) E^T S_2 E \theta(k+j)) \\ =& \alpha_1^2 \theta^T(k) E^T S_1 E \theta(k) - \alpha_1 \sum_{i=-d_1}^{-1} \sum_{j=k+i}^{k-1} \theta^T(j) E^T S_1 E \theta(j) \\ &+ \alpha_2^2 \theta^T(k) E^T S_2 E \theta(k) - \alpha_2 \sum_{i=-d_2}^{-d_1-1} \sum_{j=k+i}^{k-1} \theta^T(j) E^T S_2 E \theta(j). \end{aligned}$$

By using (4.4) in Lemma 4.5, yields

$$\begin{aligned} &-\alpha_1 \sum_{i=-d_1}^{-1} \sum_{j=k+i}^{k-1} \theta^T(j) E^T S_1 E \theta(j) \\ &\leq -(d_1 Ex(k) - \sum_{i=k-d_1}^{k-1} Ex(i))^T S_1 (d_1 Ex(k) - \sum_{i=k-d_1}^{k-1} Ex(i)) \end{aligned}$$

Considering Lemma 4.9, we have the following inequality for $d_1 < d(k) < d_2$ and a matrix P_4 satisfying $\begin{bmatrix} S_2 & P_4 \\ \star & S_2 \end{bmatrix} > 0$:

$$-\alpha_2 \sum_{i=-d_2}^{-d_1-1} \sum_{j=k+i}^{k-1} \theta^T(j)E^T S_2 E\theta(j) \leq -\gamma^T(k) \begin{bmatrix} S_2 & P_4 \\ \star & S_2 \end{bmatrix} \gamma(k), \tag{4.17}$$

where

$$\gamma(k) = \begin{bmatrix} (d(k) - d_1)Ex(k) - \sum_{i=k-d(k)}^{k-d_1-1} Ex(i) \\ (d_2 - d(k))Ex(k) - \sum_{i=k-d_2}^{k-d(k)-1} Ex(i) \end{bmatrix}.$$

When $d(k) = d_1$ or $d(k) = d_2$, according to the notation, we have $(d(k) - d_1)Ex(k) - \sum_{i=k-d(k)}^{k-d_1-1} Ex(i) = 0$ or $(d_2 - d(k))Ex(k) - \sum_{i=k-d_2}^{k-d(k)-1} Ex(i) = 0$, and the inequality (4.17) still holds by using inequality (4.4) in Lemma 4.5. One can obtain

$$\begin{aligned} \Delta V_4(k) \leq & \xi^T(k)[\alpha_1^2((A - E)W_1 + A_d W_3)^T S_1((A - E)W_1 + A_d W_3) \\ & + \alpha_2^2((A - E)W_1 + A_d W_3)^T S_2((A - E)W_1 + A_d W_3) \\ & - (d_1 EW_1 - W_5)^T S_1(d_1 EW_1 - W_5) \\ & - \Lambda^T(k) \begin{bmatrix} S_2 & P_4 \\ \star & S_2 \end{bmatrix} \Lambda(k)]\xi(k), \end{aligned} \tag{4.18}$$

where

$$\Lambda(k) = \begin{bmatrix} (d(k) - d_1)EW_1 - W_6 \\ (d_2 - d(k))EW_1 - W_7 \end{bmatrix}.$$

On the other hand, it is clear that

$$f(k) = 2x^T(k)WR^T Ex(k+1) \equiv 0. \tag{4.19}$$

Then we can obtain

$$\Delta V(k) = \Delta V_1(k) + \Delta V_2(k) + \Delta V_3(k) + \Delta V_4(k) + f(k) \leq \xi^T(k)\Theta\xi(k),$$

where

$$\Theta = \Xi - \Lambda^T(k) \begin{bmatrix} S_2 & P_4 \\ \star & S_2 \end{bmatrix} \Lambda(k). \tag{4.20}$$

Based on Lemma 4.9, the equivalent condition $\Theta < 0$ is that there exists a matrix Ψ such that

$$\Theta(k) = \begin{bmatrix} \Xi + \mathrm{sym}(\Lambda^T(k)\Psi) & \Psi^T \\ \star & -\begin{bmatrix} S_2 & P_4 \\ \star & S_2 \end{bmatrix} \end{bmatrix} < 0. \tag{4.21}$$

Due to the convexity of $\Theta(k)$ with respect to $d(k)$, the conditions (4.7) and (4.8) can guarantee the condition (4.21) holds. Hence, there exists a scalar $\alpha > 0$ such that $\Delta V(k) \le -\alpha||x(k)||^2$. Therefore (4.8) and (4.9) guarantee that the system (4.1) is asymptotically stable for any time-varying delay $d(k)$ that satisfies $0 < d_1 \le d(k) \le d_2$. The proof of Theorem 4.10 has been completed. □

Remark 4.11. The main difference from other papers, such as [188], is that the improved reciprocally convex combination approach is applied to bound the forward difference of the double summation term in $V_3(k)$, and the reciprocally convex combination approach to bound the forward difference of the triple summation term in $V_4(k)$. This technique leads to a less conservative result. On the other hand, without decomposition and equivalent transformation of the considered system, a strict delay-dependent LMI criterion is established, which reduces the computational complexity compared with nonstrict LMI conditions [33]. The number of decision variables needed in Theorem 4.10 is given by $31.5n^2 + (7.5 - r)n$.

4.1.3 Numerical example

The following example illustrates the effectiveness of our method and the advantage over some previous ones:

Example 4.1. Consider discrete singular time-delay system (4.1) with

$$E = \begin{bmatrix} 1 & 0 \\ 0 & 0 \end{bmatrix}, A = \begin{bmatrix} 0.8 & 0 \\ 0.05 & 0.9 \end{bmatrix}, A_d = \begin{bmatrix} -0.1 & 0 \\ -0.2 & -0.1 \end{bmatrix}.$$

Applying the approaches of [33] [36] [188] and the result presented in this section, the admissible maximum values of d_2 that guarantees the system (4.1) to be admissible, with various d_1 are shown in Table 4.1. From the results, it is clear to see that the criterion in this section has reduced conservatism for admissibility condition of discrete-time singular systems with time-varying delay.

Table 4.1 Comparison of the admissible maximum values of d_2 for various d_1.

d_1	0	3	6	9	12
Theorem 2 [121]	7	8	10	13	15
Theorem 2 [33]	12	13	14	15	17
Theorem 2 [36]	15	16	19	22	25
Theorem 1 [73]	18	18	21	24	27
Corollary 2 [188]	18	19	21	24	27
Theorem 4.10	35	37	39	42	44

4.1.4 Conclusion

In this section, the delay-dependent admissibility problem of discrete-time singular system with time-varying delay is investigated. A triple-summation term is introduced into the Lyapunov function, and the improved reciprocally convex combination approach is utilized to bound the forward difference of the double summation term. The new admissibility criterion, given in terms of LMIs, ensures the regularity, causality, and stability of the considered system. The result reduces the conservatism of existing results significantly. A numerical example is given to demonstrate the effectiveness and advantage of the proposed result. Further research is to find a trade-off between conservatism and computational complexity, and design an H_∞ or dissipativity controller with reliability condition.

4.2 Delay-dependent robust H_∞ controller synthesis for discrete singular delay systems

In this section, the problems of delay-dependent robust stability analysis, robust stabilization, and robust H_∞ control are investigated for uncertain discrete-time singular systems with state delay. First, by making use of the delay-partitioning technique, a new delay-dependent criterion is given to ensure the nominal system to be regular, causal, and stable. This new criterion is further extended to singular systems with both delay and parameter uncertainties. Then, without the assumption that the considered systems being regular and causal, robust controllers are designed for discrete-time singular time-delay systems such that the closed-loop systems have the characteristics of regularity, causality, and asymptotic stability. Moreover, the problem of robust H_∞ control is solved following a similar line. The obtained results are dependent not only on the delay, but also the partitioning size, and the conservatism are nonincreasing with reducing partitioning size.

These results are shown, via extensive numerical examples, to be much less conservative than existing results in the literature.

4.2.1 Problem formulation

Consider a class of linear discrete-time uncertain singular systems with state delay described by

$$\left\{\begin{array}{rcl} Ex(k+1) & = & (A+\Delta A(k))x(k)+(A_d+\Delta A_d(k))x(k-d) \\ & + & B_w w(k)+(B+\Delta B(k))u(k), \\ z(k) & = & Cx(k)+Du(k), \\ x(k) & = & \phi(k),\ k\in[-\bar{d},\ 0], \end{array}\right. \tag{4.22}$$

where $x(k)\in\mathbb{R}^n$ is the state vector; $u(k)\in\mathbb{R}^q$ is the control input; $w(k)\in\mathbb{R}^p$ is the disturbance input, and $z(k)\in\mathbb{R}^s$ is the controlled output; A, A_d, B, B_w, C, and D are constant matrices with appropriate dimensions; d is a constant positive integer satisfying $0<d\le\bar{d}$ (d can always be described by $d=m\tau$, where m and τ are integers), where $\bar{d}$ is a positive integer representing the upper bound of the delay; matrix E may be singular and $\operatorname{rank}E=r\le n$; $\phi(k)$ is a compatible vector valued initial function; ΔA, ΔA_d, and ΔB are time-varying uncertain matrices of the form

$$[\Delta A\ \ \Delta A_d\ \ \Delta B]=MF(k)[N_1\ \ N_2\ \ N_3],$$

where M, N_1, N_2, and N_3 are constant matrices, and $F(k)\in\mathbb{R}^{l\times b}$ is an unknown real matrix satisfying $F(k)F(k)^T\le I$.

Before moving on, we give some definitions and lemmas concerning the following nominal unforced counterpart of the system in (4.22):

$$\left\{\begin{array}{r} Ex(k+1)=Ax(k)+A_dx(k-d), \\ x(k)=\phi(k),\ k\in[-\bar{d},\ 0] \end{array}\right. \tag{4.23}$$

Lemma 4.12. [72] *Suppose the pair (E,A) is regular and causal, then the solution to system (4.23) is causal and unique on $[0,\infty)$ for any constant time-delay d satisfying $0<d\le\bar{d}$.*

In view of this, we introduce the following definition for singular delay system (4.23):

Definition 4.13.

1. The singular delay system in (4.23) is said to be regular and causal if the pair (E,A) is regular and causal.

2. The singular system in (4.23) is said to be asymptotically stable if, for any $\varepsilon > 0$, there exists a scalar $\delta(\varepsilon) > 0$, such that for any compatible initial conditions $\phi(k)$ satisfying $\sup_{-\bar{d}\leq k\leq -1} \|\phi(k)\| \leq \delta(\varepsilon)$, the solution x(k) of (4.23) satisfies $\|x(k)\| \leq \varepsilon$ for $k \geq 0$; furthermore, $x(k) \rightarrow 0$, when $k \rightarrow \infty$.
3. The singular time-delay system in (4.23) is said to be admissible if it is regular, causal, and asymptotically stable.

Before giving the next definition, we present the formulation of the closed-loop system with a state feedback controller,

$$u(k) = Kx(k), \ \ K \in \mathbb{R}^{q\times n}, \tag{4.24}$$

$$\left\{ \begin{array}{rcl} Ex(k+1) & = & (A + BK + \Delta A(k) + \Delta B(k)K)x(k) \\ & + & (A_d + \Delta A_d(k))x(k-d) + B_w w(k), \\ z(k) & = & (C + DK)x(k), \\ x(k) & = & \phi(k), \ k \in [-\bar{d}, \ 0]. \end{array} \right. \tag{4.25}$$

Definition 4.14. System (4.25) is said to be robust asymptotically stable with γ disturbance attenuation if the following requirements are satisfied:
1. With $w(k) = 0$, the closed-loop system in (4.25) is asymptotically stable for all uncertainties.
2. Under zero initial condition, the closed-loop system in (4.25) satisfies $||z||_2 < \gamma||w||_2$ for any nonzero $w \in l_2[0, \infty)$, where $\gamma > 0$ is a prescribed scalar.

Lemma 4.15. [175] *The system in (4.23) is asymptotically stable if and only if* $\det(zE - A - z^{-d}A_d) \neq 0$ *for* $|z| \geq 1$.

Lemma 4.16. [209] *Given matrices Ω, Γ, and Φ with appropriate dimensions, and with Ω symmetric, then*

$$\Omega + \Gamma F \Phi + \Phi^T F^T \Gamma^T < 0$$

for any F satisfying $F^T F \leq I$, if and only if there exists a scalar $\varepsilon > 0$ such that

$$\Omega + \varepsilon \Gamma \Gamma^T + \varepsilon^{-1} \Phi^T \Phi < 0.$$

In this subchapter, three problems for uncertain discrete singular delay system (4.22) are investigated.
1. The robust stability problem: the objective is to establish new robust stability criterion such that the discrete-time singular system in (4.22) with $u(k) = 0$ and $w(k) = 0$ is admissible.

2. The robust stabilization problem: the purpose is to design a state feedback controller such that the resulting closed-loop system is admissible.
3. The robust H_∞ control problem: the aim is the design of a state feedback controller such that the resulting closed-loop system is admissible with an H_∞ disturbance attenuation.

4.2.2 Robust stability

In this subsection, we obtain a solution to the robust stability analysis problem formulated previously by using a strict LMI approach. First, we present the following result for the nominal singular delay systems, which will play a key role in solving the aforementioned problems:

4.2.2.1 Stability analysis: nominal case

In this subsection, we will present a new delay-dependent sufficient condition guaranteeing the nominal system in (4.23) is admissible. The main idea is based on the delay-partitioning technique, which constitutes the major difference from most of the existing results. Our new result is given as follows:

Theorem 4.17. *Given positive integers m, τ, the system in (4.23) is admissible if there exist matrices $P_1 > 0$, $Q > 0$, $Z > 0$, S_1, S_2, S_3, P_2, P_3, and P_4, such that*

$$\Theta < 0, \tag{4.26}$$

where $R \in \mathbb{R}^{n\times(n-r)}$ is any full-column rank matrix satisfying $E^T R = 0$ and

$$\begin{aligned}
\Theta &= W_P^T(P_1+\tau^2 Z)W_P + W_Q^T\tilde{Q}W_Q - W_Z^T E^T Z E W_Z \\
&\quad + sym(W_{P1}^T E^T P_1 W_P + P^T W_{P2} + SR^T W_P), \\
\tilde{Q} &= \begin{bmatrix} Q & 0_{mn,mn} \\ 0_{mn,mn} & -Q \end{bmatrix}, \quad Q = \begin{bmatrix} Q_{11} & \cdots & Q_{1m} \\ \star & \ddots & \vdots \\ \star & \star & Q_{mm} \end{bmatrix}, \\
S &= \begin{bmatrix} S_1^T & S_2^T & 0_{n-r,(m-1)n} & S_3^T \end{bmatrix}^T, \quad W_P = \begin{bmatrix} 0_{n,(m+1)n} & I_n \end{bmatrix}, \\
W_{P1} &= \begin{bmatrix} I_n & 0_{n,(m+1)n} \end{bmatrix}, \quad W_{P2} = \begin{bmatrix} A-E & 0_{n,(m-1)n} & A_d & -I_n \end{bmatrix}, \\
W_Q &= \left[\begin{array}{ccc} I_{mn} & 0_{mn,2n} & \\ \hline 0_{mn,n} & I_{mn} & 0_{mn,n} \end{array}\right], \quad P = \begin{bmatrix} P_2 & P_4 & 0_{n,(m-1)n} & P_3 \end{bmatrix}, \\
W_Z &= \begin{bmatrix} I_n & -I_n & 0_{n,mn} \end{bmatrix}.
\end{aligned}$$

Proof. First, we prove the regularity and causality of the system. Let

$$\begin{aligned}
\bar{E} &= \begin{bmatrix} E & 0 \\ 0 & 0 \end{bmatrix}, \ \bar{A} = \begin{bmatrix} E & I \\ A-E & -I \end{bmatrix}, \ \bar{Z} = \begin{bmatrix} Z & 0 \\ 0 & 0 \end{bmatrix}, \\
\bar{P} &= \begin{bmatrix} P_1 & 0 \\ 0 & 0 \end{bmatrix}, \ \bar{Q} = \begin{bmatrix} Q_{11} & 0 \\ 0 & \tau^2 Z \end{bmatrix}, \\
\bar{S} &= \begin{bmatrix} S_1 & P_2^T \\ S_3 & P_3^T \end{bmatrix}, \ \bar{R} = \begin{bmatrix} R & 0 \\ 0 & I \end{bmatrix}.
\end{aligned}$$

Since $\text{rank}\bar{E} = \text{rank}E = r \leq n$, there exist nonsingular matrices U and V, such that

$$U\bar{E}V = \begin{bmatrix} I_r & 0 \\ 0 & 0 \end{bmatrix}.$$

Denote

$$\begin{aligned}
U\bar{A}V &= \begin{bmatrix} A_{11} & A_{12} \\ A_{21} & A_{22} \end{bmatrix}, \\
V^T\bar{S} &= \begin{bmatrix} S_{11} \\ S_{21} \end{bmatrix}, \ U^{-T}\bar{R} = \begin{bmatrix} 0 \\ I \end{bmatrix} H,
\end{aligned}$$

where $H \in \mathbb{R}^{(2n-r)\times(2n-r)}$ is a nonsingular matrix determined by $U^{-T}\bar{R}$. Define

$$L = \begin{bmatrix} I_n & 0 & 0_{n,(m-1)n} & 0 \\ 0 & 0 & 0_{n,(m-1)n} & I_n \\ 0_{(m-1)n,n} & 0_{(m-1)n,n} & I_{(m-1)n} & 0_{(m-1)n,n} \\ 0 & I_n & 0 & 0 \end{bmatrix}.$$

Then performing a congruence transformation to (4.26) by L, we obtain the following inequality:

$$\begin{bmatrix} \hat{\Theta}_{11} & \hat{\Theta}_{13} & \bullet & \bullet \\ \star & \hat{\Theta}_{33} & \bullet & \bullet \\ \bullet & \bullet & \bullet & \bullet \\ \bullet & \bullet & \bullet & \bullet \end{bmatrix} < 0, \tag{4.27}$$

where

$$\hat{\Theta}_{11} = P_2^T(A-E) + (A-E)^T P_2 - E^T Z E + Q_{11},$$

$$\begin{aligned}\hat{\Theta}_{13} &= -P_2^T + (A-E)^T P_3 + S_1 R^T + E^T P_1, \\ \hat{\Theta}_{33} &= \tau^2 Z + P_1 - P_3 - P_3^T + S_3 R^T + R S_3^T.\end{aligned}$$

From (4.27), we have

$$\bar{A}^T \bar{P} \bar{A} - \bar{E}^T \bar{P} \bar{E} + \bar{S} \bar{R}^T \bar{A} + \bar{A}^T \bar{R} \bar{S}^T - \bar{E}^T \bar{Z} \bar{E} + \bar{Q} < 0,$$

which implies that

$$-\bar{E}^T \bar{P} \bar{E} + \bar{S} \bar{R}^T \bar{A} + \bar{A}^T \bar{R} \bar{S}^T - \bar{E}^T \bar{Z} \bar{E} < 0. \tag{4.28}$$

Performing a congruence transformation to (4.28) by V^T and V, we obtain

$$\begin{bmatrix} \bullet & \bullet \\ \bullet & S_{21} H^T A_{22} + A_{22}^T H S_{21}^T \end{bmatrix} < 0,$$

which implies that A_{22} is nonsingular. Hence,

$$\det(z\bar{E} - \bar{A}) = \det(U^{-1}) \det(zI_r - A_{11} + A_{12} A_{22}^{-1} A_{21}) \det(-A_{22}) \det(V^{-1})$$

is not identically zero, and $\deg \det(z\bar{E} - \bar{A}) = r$. This, together with Definition 4.13, leads to the pair $(\bar{E}, \bar{A})$ being regular and causal. Noticing the fact that

$$\begin{aligned}\det(zE - A) &= \det(z\bar{E} - \bar{A}), \\ \deg(\det(zE - A)) &= \deg(\det(z\bar{E} - \bar{A})),\end{aligned}$$

we can easily see that the pair (E, A) is regular and causal. Then according to Lemma 4.12 and Definition 4.13, the system in (4.23) is regular and causal.

Then we are in the position to show that system (4.23) is asymptotically stable. To this end, we choose a new Lyapunov functional candidate as

$$V(k) = V_1(k) + V_2(k) + V_3(k), \tag{4.29}$$

where

$$\begin{aligned}V_1(k) &= x^T(k) E^T P_1 E x(k), \\ V_2(k) &= \sum_{i=k-\tau}^{k-1} \Upsilon^T(i) Q \Upsilon(i),\end{aligned}$$

$$V_3(k) = \tau \sum_{i=-\tau}^{-1} \sum_{j=k+i}^{k-1} \eta^T(j) Z \eta(j),$$

and

$$\Upsilon(i) = \begin{bmatrix} x(i) \\ x(i-\tau) \\ x(i-2\tau) \\ \vdots \\ x(i-\tau m+\tau) \end{bmatrix}, \quad \eta(j) = Ex(j+1) - Ex(j).$$

Taking the forward difference of the functional in (4.29) along the solution of system (4.23), and defining

$$\xi(k) = \begin{bmatrix} \Upsilon(k) \\ x(k-m\tau) \\ \eta(k) \end{bmatrix},$$

we have

$$\begin{aligned}
\Delta V_1(k) &= x^T(k+1)E^T P_1 Ex(k+1) - x^T(k)E^T P_1 Ex(k) \\
&= \left[Ex(k+1) - Ex(k)\right]^T P_1 \left[Ex(k+1) - Ex(k)\right] \\
&\quad + 2x(k)^T E^T P_1 \left[Ex(k+1) - Ex(k)\right] \\
&\quad + 2\Big[x(k)^T P_2^T + \left[Ex(k+1) - Ex(k)\right]^T P_3^T + x(k-\tau)^T P_4^T\Big] \\
&\quad \times \left[(A-E)x(k) - \left[Ex(k+1) - Ex(k)\right] + A_d x(k-m\tau)\right] \\
&\quad + 2\Big[x(k)^T S_1 R^T + x(k-\tau)^T S_2 R^T \\
&\quad + \left[Ex(k+1) - Ex(k)\right]^T S_3 R^T\Big]\left[Ex(k+1) - Ex(k)\right] \\
&= \xi^T(k) W_P^T P_1 W_P \xi(k) + 2\xi^T(k)(W_{P1}^T E^T P_1 W_P \\
&\quad + P^T W_{P2} + SR^T W_P)\xi(k), \qquad (4.30) \\
\Delta V_2(k) &= \Upsilon^T(k) Q \Upsilon(k) - \Upsilon^T(k-\tau) Q \Upsilon(k-\tau) \\
&= \xi^T(k) W_Q^T \tilde{Q} W_Q \xi(k), \qquad (4.31) \\
\Delta V_3(k) &= \tau^2 \eta^T(k) Z \eta(k) - \tau \sum_{i=k-\tau}^{k-1} \eta^T(i) Z \eta(i).
\end{aligned}$$

By Lemma 4.5, we have

$$\Delta V_3(k) \leq \tau^2\eta^T(k)Z\eta(k) - [x(k)-x(k-\tau)]^T E^T ZE[x(k)-x(k-\tau)]. \tag{4.32}$$

By connecting (4.30)–(4.32), we obtain

$$\begin{aligned} \Delta V(k) &\leq \xi(k)^T W_P^T P_1 W_P \xi(k) + 2\xi(k)^T (W_{P1}^T E^T P_1 W_P + P^T W_{P2} \\ &+ SR^T W_P)\xi(k) + \xi(k)^T W_Q^T \tilde{Q} W_Q \xi(k) + \xi(k)^T W_P^T \tau^2 Z W_P \xi(k) \\ &- \xi(k)^T W_Z^T E^T ZEW_Z \xi(k) \\ &= \xi(k)^T \Theta \xi(k). \end{aligned} \tag{4.33}$$

Thus (4.26) implies $\Delta V(k) < -\epsilon \|x(k)\|^2$, where ϵ is a positive scalar. Then we have

$$-V(0) \leq V(k+1) - V(0) = \sum_{i=0}^{k} \Delta V(i) \leq -\epsilon \sum_{i=0}^{k} \|x(i)\|^2 \leq 0,$$

which implies

$$0 \leq \sum_{i=0}^{k} \|x(i)\|^2 \leq \frac{V(0)}{\epsilon}.$$

Hence, the series $\sum_{i=0}^{\infty} \|x(i)\|^2$ converge, and we have $\lim_{i\to\infty} x(i) = 0$. Then from Definition 4.13, we conclude that the system is asymptotically stable and this completes the proof. □

Remark 4.18. For state-space systems with state delay, that is, $\operatorname{rank} E = n$, then $R = 0$ as a result of $E^T R = 0$. Therefore the term $SR^T W_P$ in (4.26) disappears, and Theorem 4.17 is specialized to the following corollary:

Corollary 4.19. *Given positive integers m, τ, the system in (4.23) with* $\operatorname{rank} E = n$ *is asymptotically stable if there exist matrices $P_1 > 0$, $Q > 0$, $Z > 0$, Y_1, Y_2, T_1, P_2, P_3, and P_4, such that*

$$\breve{\Theta} < 0,$$

where

$$\begin{aligned} \breve{\Theta} &= W_P^T(P_1 + \tau^2 Z)W_P + W_Q^T \tilde{Q} W_Q - W_Z^T E^T ZEW_Z \\ &+ sym(W_{P1}^T E^T P_1 W_P + P^T W_{P2}). \end{aligned}$$

When $m = 1$, in deriving the result of Theorem 4.17, we give up the delay-partitioning approach. In such a case, we have the following result:

Corollary 4.20. *The system in (4.23) is admissible if there exist matrices $P_1 > 0$, $Q > 0$, $Z > 0$, S_1, S_2, S_3, P_2, P_3, and P_4, such that*

$$\begin{bmatrix} \hat{\Theta}_{11} & \hat{\Theta}_{12} & \hat{\Theta}_{13} \\ \star & \hat{\Theta}_{22} & \hat{\Theta}_{23} \\ \star & \star & \hat{\Theta}_{33} \end{bmatrix} < 0, \quad (4.34)$$

where R, $\hat{\Theta}_{11}$, $\hat{\Theta}_{13}$, $\hat{\Theta}_{33}$ are defined in (4.27), and

$$\begin{aligned} \hat{\Theta}_{12} &= P_2^T A_d + (A - E)^T P_4 + E^T Z E, \\ \hat{\Theta}_{22} &= P_4^T A_d + A_d^T P_4 - E^T Z E - Q, \\ \hat{\Theta}_{23} &= A_d^T P_3 - P_4^T + S_2 R^T. \end{aligned}$$

Proof. When $m = 1$, the matrices in (4.26) become

$$\begin{aligned} \tau &= d,\ S = \begin{bmatrix} S_1^T & S_2^T & S_3^T \end{bmatrix}^T,\ W_P = \begin{bmatrix} 0_{n,2n} & I_n \end{bmatrix}, \\ W_{P1} &= \begin{bmatrix} I_n & 0_{n,2n} \end{bmatrix},\ W_{P2} = \begin{bmatrix} A - E & A_d & -I_n \end{bmatrix}, \\ W_Q &= \left[\begin{array}{ccc} \multicolumn{2}{c}{I_n} & 0_{n,2n} \\ \hline 0_{n,n} & I_n & 0_{n,n} \end{array}\right],\ P = \begin{bmatrix} P_2 & P_4 & P_3 \end{bmatrix}, \\ W_Z &= \begin{bmatrix} I_n & -I_n & 0_{n,n} \end{bmatrix},\ Q_{11} = Q,\ \tilde{Q} = \begin{bmatrix} Q & 0_{n,n} \\ 0_{n,n} & -Q \end{bmatrix}. \end{aligned}$$

After some simple manipulations, (4.34) can be obtained and this completes the proof. □

Remark 4.21. The number of LMI decision variables in Theorem 4.17 is $\frac{n}{2}[(m^2 + 14)n - 6r + m + 2]$. Thus the computational burden, including the number of the variables and the computation time will significantly increase with the partitioning number m becoming bigger. However, the LMI toolbox in MATLAB® can still be used to efficiently solve the LMIs.

4.2.2.2 Stability analysis: uncertain case

In this subsection, the robust stability analysis for the uncertain singular system in (4.22) with $w(k) = 0$ is considered; that is, we consider the following uncertain system:

$$
\begin{cases}
Ex(k+1) = (A + \Delta A(k))x(k) + (A_d + \Delta A_d(k))x(k-d) \\
\qquad\qquad\quad +(B + \Delta B(k))u(k), \\
x(k) = \phi(k), \ k \in [-\bar{d},\ 0].
\end{cases} \tag{4.35}
$$

For system (4.35) with time-varying structured uncertainties, we have the following theorem:

Theorem 4.22. *Given positive integers m, τ, the time-delay system in (4.35) with $u(k) = 0$ is admissible for all parameter uncertainties if there exist matrices $P_1 > 0$, $Q > 0$, $Z > 0$, S_1, S_2, S_3, P_2, P_3, P_4, and a scalar $\varepsilon > 0$, such that*

$$
\begin{bmatrix} \Theta + \varepsilon \Xi^T \Xi & P^T M \\ \star & -\varepsilon I \end{bmatrix} < 0, \tag{4.36}
$$

where Θ is defined in (4.26), and

$$
\Xi = \begin{bmatrix} N_1 & 0_{b,(m-1)n} & N_2 & 0_{b,n} \end{bmatrix}.
$$

Proof. Based on Theorem 4.17, by replacing A and A_d in (4.26) with $A + MF(k)N_1$ and $A_d + MF(k)N_2$, respectively, the stability criterion for the uncertain system can be rewritten as

$$
\Theta + \mathrm{sym}(P^T MF(k)\Xi) < 0. \tag{4.37}
$$

Applying Schur complement to (4.36), we obtain

$$
\Theta + \varepsilon \Xi^T \Xi + \varepsilon^{-1} P^T M M^T P < 0,
$$

which by Lemma 4.16 implies (4.37), and the proof is completed. □

Now we are in the position to show that the proposed result will demonstrate its superiority in terms of reduced conservatism with m increasing.

Proposition 4.23. *Suppose that τ_m and d_m are the maximal τ and the maximal delay obtained by Theorem 4.17 for a given number of partitions m. Then for any positive integer β such that $\frac{m}{m+\beta}\tau_m$ is an integer, we have $\frac{m}{m+\beta}\tau_m \le \tau_{m+\beta}$, and thus $d_m \le d_{m+\beta}$.*

Proof. From Theorem 4.17, we know that the following inequality holds for given partitioning number m and the integer τ_m:

$$\begin{aligned}
\Delta V(k) \leq\ & x^T(k+1)E^TP_1Ex(k+1) - x^T(k)E^TP_1Ex(k) + \tau_m^2\eta^T(k)Z\eta(k) \\
& -[x(k) - x(k-\tau_m)]^TE^TZE[x(k) - x(k-\tau_m)] \\
& + \begin{bmatrix} x(k) \\ x(k-\tau_m) \\ x(k-2\tau_m) \\ \vdots \\ x\left(k-(m-1)\tau_m\right) \end{bmatrix}^T Q \begin{bmatrix} x(k) \\ x(k-\tau_m) \\ x(k-2\tau_m) \\ \vdots \\ x\left(k-(m-1)\tau_m\right) \end{bmatrix} \\
& - \begin{bmatrix} x(k-\tau_m) \\ x(k-2\tau_m) \\ x(k-3\tau_m) \\ \vdots \\ x\left(k-m\tau_m\right) \end{bmatrix}^T Q \begin{bmatrix} x(k-\tau_m) \\ x(k-2\tau_m) \\ x(k-3\tau_m) \\ \vdots \\ x\left(k-m\tau_m\right) \end{bmatrix} \\
<\ & 0.
\end{aligned} \tag{4.38}$$

Take $\tau = \frac{m}{m+\beta}\tau_m < \tau_m$. Since τ_m is the maximal value of τ satisfying (4.38), even if we replace τ_m in (4.38) by $\frac{m}{m+\beta}\tau_m$, we still have

$$\begin{aligned}
\Delta \tilde{V}(k) \leq\ & x^T(k+1)E^TP_1Ex(k+1) \\
& -x^T(k)E^TP_1Ex(k) + (\frac{m}{m+\beta}\tau_m)^2\eta^T(k)Z\eta(k) \\
& -[x(k) - x(k-\frac{m}{m+\beta}\tau_m)]^TE^TZE[x(k) - x(k-\frac{m}{m+\beta}\tau_m)] \\
& + \begin{bmatrix} x(k) \\ x(k-\frac{m}{m+\beta}\tau_m) \\ x(k-2\frac{m}{m+\beta}\tau_m) \\ \vdots \\ x\left(k-(m-1)\frac{m}{m+\beta}\tau_m\right) \end{bmatrix}^T Q \begin{bmatrix} x(k) \\ x(k-\frac{m}{m+\beta}\tau_m) \\ x(k-2\frac{m}{m+\beta}\tau_m) \\ \vdots \\ x\left(k-(m-1)\frac{m}{m+\beta}\tau_m\right) \end{bmatrix} \\
& - \begin{bmatrix} x(k-\frac{m}{m+\beta}\tau_m) \\ x(k-2\frac{m}{m+\beta}\tau_m) \\ x(k-3\frac{m}{m+\beta}\tau_m) \\ \vdots \\ x\left(k-m\frac{m}{m+\beta}\tau_m\right) \end{bmatrix}^T Q \begin{bmatrix} x(k-\frac{m}{m+\beta}\tau_m) \\ x(k-2\frac{m}{m+\beta}\tau_m) \\ x(k-3\frac{m}{m+\beta}\tau_m) \\ \vdots \\ x\left(k-m\frac{m}{m+\beta}\tau_m\right) \end{bmatrix}
\end{aligned}$$

$$
\begin{aligned}
&= x^T(k+1)E^T P_1 Ex(k+1) - x^T(k)E^T P_1 Ex(k) \\
&+ (\frac{m}{m+\beta}\tau_m)^2 \eta^T(k) Z \eta(k) \\
&- [x(k) - x(k - \frac{m}{m+\beta}\tau_m)]^T E^T ZE[x(k) - x(k - \frac{m}{m+\beta}\tau_m)] \\
&+ \Upsilon_1(K)^T \grave{Q} \Upsilon_1(K) - \Upsilon_2(K)^T \grave{Q} \Upsilon_2(K) \\
&< 0,
\end{aligned}
$$

where

$$
\Upsilon_1(k) = \begin{bmatrix} x(k) \\ x(k - \frac{m}{m+\beta}\tau_m) \\ x(k - 2\frac{m}{m+\beta}\tau_m) \\ \vdots \\ x\left(k - (m-1)\frac{m}{m+\beta}\tau_m\right) \\ \vdots \\ x\left(k - (m+\beta-1)\frac{m}{m+\beta}\tau_m\right) \end{bmatrix}, \quad \Upsilon_2(k) = \begin{bmatrix} x(k - \frac{m}{m+\beta}\tau_m) \\ x(k - 2\frac{m}{m+\beta}\tau_m) \\ x(k - 3\frac{m}{m+\beta}\tau_m) \\ \vdots \\ x\left(k - m\frac{m}{m+\beta}\tau_m\right) \\ \vdots \\ x\left(k - m\tau_m\right) \end{bmatrix},
$$

$$
\grave{Q} = \begin{bmatrix} Q & 0_{mn,\beta n} \\ \star & 0_{\beta n,\beta n} \end{bmatrix}.
$$

We can always find a small enough scalar $\delta > 0$ such that

$$
\begin{aligned}
\Delta \tilde{V}(k) \leq\ & x^T(k+1)E^T P_1 Ex(k+1) - x^T(k)E^T P_1 Ex(k) \\
& + (\frac{m}{m+\beta}\tau_m)^2 \eta^T(k) Z \eta(k) \\
& - [x(k) - x(k - \frac{m}{m+\beta}\tau_m)]^T E^T ZE[x(k) - x(k - \frac{m}{m+\beta}\tau_m)] \\
& + \Upsilon_1(K)^T \acute{Q} \Upsilon_1(K) - \Upsilon_2(K)^T \acute{Q} \Upsilon_2(K) \\
<\ & 0,
\end{aligned} \tag{4.39}
$$

where

$$
\acute{Q} = \begin{bmatrix} Q & 0_{mn,\beta n} \\ \star & \delta I \end{bmatrix} > 0,
$$

which implies that the inequality in (4.26) still holds for given partitioning number $m+\beta$ and the integer τ. We can obtain

$$
\tau = \frac{m}{m+\beta}\tau_m \leq \tau_{m+\beta}. \tag{4.40}
$$

Then by multiplying (4.40) by $m+\beta$, we have $d_m \le d_{m+\beta}$, and this proposition is proved. □

4.2.3 Stabilization

In this section, we devote our attention to design a robust state feedback controller for system (4.22) with $w(k)=0$, such that the closed-loop system is admissible for all uncertainties. Based on Theorem 4.17, we have the result that follows. Before giving the main result, we give the closed-loop system of nominal singular discrete-time system:

$$Ex(k+1) = (A+BK)x(k) + A_d x(k-d). \tag{4.41}$$

Theorem 4.24. *Given scalars λ_1, λ_2, λ_3, and positive integers m, τ, there exists a state-feedback controller in the form of (4.24) such that the closed-loop system in (4.25) is admissible if there exist matrices $P_1>0$, $Q>0$, $Z>0$, S_1, S_2, S_3, J, X, and a scalar $\varepsilon>0$, such that*

$$\begin{bmatrix} \Psi + \varepsilon\Gamma\Gamma^T & \Phi^T \\ \star & -\varepsilon I \end{bmatrix} < 0, \tag{4.42}$$

where $R \in \mathbb{R}^{n\times(n-r)}$ is any full-column rank matrix satisfying $E^T R = 0$ and

$$\begin{aligned}
\Psi &= W_P^T(P_1+\tau^2 Z)W_P + W_Q^T\tilde{Q}W_Q - W_Z^T EZE^T W_Z \\
&\quad + sym(W_{P1}^T E P_1 W_P + W_E^T \Lambda + SR^T W_P), \\
W_E &= \begin{bmatrix} \lambda_1 I_n & \lambda_3 I_n & 0_{n,(m-1)n} & \lambda_2 I_n \end{bmatrix}, \\
\Lambda &= [\; J^T(A-E)^T + X^T B^T \quad 0_{n,(m-1)n} \quad J^T A_d^T \quad -J^T \;], \\
\Phi &= \begin{bmatrix} \lambda_1(N_1 J + N_3 X) & \lambda_3(N_1 J + N_3 X) & 0_{b,(m-1)n} & \lambda_2(N_1 J + N_3 X) \\ \lambda_1 N_2 J & \lambda_3 N_2 J & 0_{b,(m-1)n} & \lambda_2 N_2 J \end{bmatrix}, \\
\Gamma^T &= \begin{bmatrix} M^T & 0_{l,(m-1)n} & 0_{l,n} & 0_{l,n} \\ 0_{l,n} & 0_{l,(m-1)n} & M^T & 0_{l,n} \end{bmatrix}.
\end{aligned}$$

Moreover, if the above condition is feasible, a desired controller gain matrix in the form of (4.24) is given by

$$K = XJ^{-1}.$$

Proof. It is easy to see that

$$\det(zE - (A+BK)) \;=\; \det(zE^T - (A+BK)^T),$$

$$\deg(\det(zE-(A+BK))) \;=\; \deg(\det(zE^T-(A+BK)^T)),$$

and that $\det(zE-(A+BK)-z^{-d}A_d)=0$ and $\det(zE^T-(A+BK)^T-z^{-d}A_d^T)=0$ have the same solution set. With respect to the regularity, causality, and stability of a system, we obtain that the system in (4.41) is equivalent to the following system based on Definition 4.1 and Lemma 4.12:

$$E^T\delta(k+1)=(A+BK)^T\delta(k)+A_d^T\delta(k-d). \tag{4.43}$$

Substituting E, A, and A_d with E^T, $(A+BK)^T$, and A_d^T in (4.26), respectively, we have the following inequality:

$$\Psi_1<0,$$

where

$$\begin{aligned}\Psi_1 \;=\;& W_P^T(P_1+\tau^2Z)W_P+W_Q^T\tilde{Q}W_Q-W_Z^TEZE^TW_Z\\ &+\mathrm{sym}(W_{P1}^TEP_1W_P+SR^TW_P+\begin{bmatrix} P_2 & P_4 & 0_{n,(m-1)n} & P_3\end{bmatrix}^T\\ &\times\begin{bmatrix}(A-E+BK)^T & 0_{n,(m-1)n} & A_d^T & -I_n\end{bmatrix}).\end{aligned}$$

Then, denoting $P_2=\lambda_1J$, $P_3=\lambda_2J$, $P_4=\lambda_3J$ and $X=KJ$, we obtain

$$\Psi<0. \tag{4.44}$$

Replacing A, A_d, and B by $A+MF(k)N_1$, $A_d+MF(k)N_2$, and $B+MF(k)N_3$ in (4.44), we have

$$\Psi+\Psi_2<0, \tag{4.45}$$

where

$$\begin{aligned}\Psi_2 \;&=\; \mathrm{sym}\left(\begin{bmatrix} MF(k)(N_1J+N_3X)\\ 0_{(m-1)n,n}\\ MF(k)N_2J\\ 0_{n,n}\end{bmatrix}\begin{bmatrix}\gamma_1I_n & \gamma_3I_n & 0_{n,(m-1)n} & \gamma_2I_n\end{bmatrix}\right)\\ &=\; \Gamma\bar{F}\Phi+\Phi^T\bar{F}^T\Gamma^T,\\ \bar{F} \;&=\; \begin{bmatrix} F(k) & 0_{l,b}\\ 0_{l,b} & F(k)\end{bmatrix}.\end{aligned}$$

By applying Schur complement, (4.42) is equivalent to

$$\Psi + \varepsilon \Gamma \Gamma^T + \varepsilon^{-1} \Phi^T \Phi < 0. \tag{4.46}$$

Based on Lemma 4.16, (4.46) implies (4.45) holds.

Thus the theorem is proved. □

Remark 4.25. To establish the stability conditions for the singular systems with constant delay, the considered systems are converted to delay-free systems by the state augmentation approach in [20,212]. However, the order of the transformed systems is high if the delay is large, and the method becomes difficult to apply for unknown delay, or for time-varying delay cases.

Remark 4.26. It is the utilization of the delay-partitioning technique that constitutes the major difference when compared with existing results about singular systems in the literature, from which the reduced conservatism can benefit. The reduction of conservatism is more prominent when the partitioning number m increases, which has been proved in [60] for continuous-time systems. In addition, the delay-partitioning technique has also been applied to stability analysis of continuous systems with multiple delay components in [31], neutral delay systems in [32], and delayed complex network in [177].

Remark 4.27. Although this section deals with the constant delay case, it can be readily extended to discrete-time singular systems with time-varying delay $d(k)$ satisfying $1 \le d_m \le d(k) \le d_M$ in terms of the following Lyapunov functional candidate:

$$\Lambda(k) = \Lambda_1(k) + \Lambda_2(k) + \Lambda_3(k) + \Lambda_4(k),$$

where

$$\begin{aligned}
\Lambda_1(k) &= x^T(k) E^T P_1 E x(k), \\
\Lambda_2(k) &= \sum_{i=k-\tau}^{k-1} \Upsilon^T(i) Q_1 \Upsilon(i) + \sum_{i=k-d_M}^{k-1} x^T(i) Q_2 x(i), \\
\Lambda_3(k) &= \sum_{i=-\tau}^{-1} \sum_{j=k+i}^{k-1} \eta^T(j) Z_1 \eta(j) + \sum_{i=-d_M}^{-m\tau-1} \sum_{j=k+i}^{k-1} \eta^T(j) Z_2 \eta(j), \\
\Lambda_4(k) &= \sum_{i=-d_M+1}^{-m\tau+1} \sum_{j=k-1+i}^{k-1} x^T(j) R x(j).
\end{aligned}$$

Then, the robust stability and stabilization problems for uncertain discrete-time singular systems with time-varying delay can be addressed by following similar lines as developed in this section.

4.2.4 Robust H_∞ control

In this subsection, we consider the admissibility with an H_∞ disturbance attenuation for system (4.22). Based on this condition, we will design a robust H_∞ state feedback controller for system (4.22) such that the closed-loop system is admissible with an H_∞ disturbance attenuation.

4.2.4.1 H_∞ performance analysis

Before presenting the main results, we establish a new version of delay-dependent bounded real lemma for the system in (4.22) with $u(k) = 0$, that is, we consider the following nominal time-delay system:

$$\begin{cases} Ex(k+1) = Ax(k) + A_d x(k-d) + B_w w(k), \\ z(k) = Cx(k), \\ x(k) = \phi(k), \ k \in [-\bar{d}, 0]. \end{cases} \tag{4.47}$$

Theorem 4.28. *Given positive integers m, τ, and a prescribed scalar $\gamma > 0$, the system in (4.47) is admissible with an H_∞ disturbance attenuation γ, if there exist matrices $P_1 > 0$, $Q > 0$, $Z > 0$, S_1, S_2, S_3, P_2, P_3, and P_4, such that*

$$\bar{\Theta} < 0, \tag{4.48}$$

where $R \in \mathbb{R}^{n\times(n-r)}$ is any full-column rank matrix satisfying $E^T R = 0$ and

$$\begin{aligned} \bar{\Theta} &= \bar{W}_P^T(P_1 + \tau^2 Z)\bar{W}_P + \bar{W}_Q^T \tilde{Q} \bar{W}_Q + W_C^T W_C - \gamma^2 W_W^T W_W \\ &\quad - \bar{W}_Z^T E^T Z E \bar{W}_Z + sym(\bar{W}_{P1}^T E^T P_1 \bar{W}_P + \bar{P}^T \bar{W}_{P2} + \bar{S} R^T \bar{W}_P), \\ \tilde{Q} &= \begin{bmatrix} Q & 0_{mn,mn} \\ 0_{mn,mn} & -Q \end{bmatrix}, \ \bar{S} = \begin{bmatrix} S_1^T & S_2^T & 0_{n-r,(m-1)n} & S_3^T & 0_{n-r,p} \end{bmatrix}^T, \\ \bar{W}_P &= \begin{bmatrix} 0_{n,(m+1)n} & I_n & 0_{n,p} \end{bmatrix}, \ \bar{W}_{P1} = \begin{bmatrix} I_n & 0_{n,(m+1)n} & 0_{n,p} \end{bmatrix}, \\ \bar{W}_{P2} &= \begin{bmatrix} A - E & 0_{n,(m-1)n} & A_d & -I_n & B_w \end{bmatrix}, \\ \bar{W}_Z &= \begin{bmatrix} I_n & -I_n & 0_{n,mn+p} \end{bmatrix}, \ \bar{W}_Q = \left[\begin{array}{ccc} \multicolumn{2}{c}{I_{mn}} & 0_{mn,2n+p} \\ \hline 0_{mn,n} & I_{mn} & 0_{mn,n+p} \end{array} \right], \end{aligned}$$

$$\bar{P} = \begin{bmatrix} P_2 & P_4 & 0_{n,(m-1)n} & P_3 & 0_{n,p} \end{bmatrix}, \quad W_C = \begin{bmatrix} C & 0_{s,(m+1)n+p} \end{bmatrix},$$
$$W_W = \begin{bmatrix} 0_{q,(m+2)n} & I_p \end{bmatrix}.$$

Proof. First, (4.48) implies (4.26), and thus the system in (4.47) is admissible. Next, we shall establish the H_∞ performance of the system in (4.47) under zero initial condition. To this end, we introduce the following index:

$$J_{zw} = \sum_{k=0}^{N} [z^T(k)z(k) - \gamma^2 w(k)w(k)].$$

Applying the same technique as in the proof of Theorem 4.17 and noting zero initial condition, it can be shown that for any nonzero $w \in l_2[0, \infty)$,

$$\begin{aligned} J_{zw} &= \sum_{k=0}^{N} [z^T(k)z(k) - \gamma^2 w^T(k)w(k) + V(k+1) - V(k)] - V(N+1) \\ &\leq \sum_{k=0}^{N} [z^T(k)z(k) - \gamma^2 w^T(k)w(k) + \Delta V(k)] \\ &\leq \sum_{k=0}^{N} \bar{\xi}^T(k)\bar{\Theta}\bar{\xi}(k), \end{aligned}$$

where

$$\bar{\xi}(k) = \begin{bmatrix} x^T(k) & x^T(k-\tau) & \cdots & x^T(k-m\tau) & \eta^T(k) & w^T(k) \end{bmatrix}^T.$$

From (4.48), we have $J_{zw} < 0$ for all nonzero $w \in l_2[0, \infty)$ such that $\|z\|_2 < \gamma \|w\|_2$, and this completes the proof. □

Based on Theorem 4.28, we investigate the problem of delay-dependent robust H_∞ performance analysis for discrete-time uncertain singular time-delay system:

$$\begin{cases} Ex(k+1) = (A + \Delta A(k))x(k) + (A_d + \Delta A_d(k))x(k-d) + B_w w(k), \\ z(k) = Cx(k), \\ x(k) = \phi(k),\ k \in [-\bar{d}, 0]. \end{cases} \tag{4.49}$$

Theorem 4.29. *Given positive integers m, τ, and a prescribed scalar $\gamma > 0$, the system in (4.49) is admissible with an H_∞ disturbance attenuation γ, if there exist* *matrices* $P_1 > 0$, $Q > 0$, $Z > 0$, S_1, S_2, S_3, P_2, P_3, P_4, *and a scalar* $\varepsilon > 0$, *such that*

$$\begin{bmatrix} \bar{\Theta} + \varepsilon\bar{\Xi}^T\bar{\Xi} & \bar{P}^T M \\ \star & -\varepsilon I \end{bmatrix} < 0, \tag{4.50}$$

where $\bar{\Theta}$, $\bar{P}$ *are defined in (4.48) and*

$$\bar{\Xi} = \begin{bmatrix} N_1 & 0_{b,(m-1)n} & N_2 & 0_{b,n+q} \end{bmatrix}.$$

Proof. By replacing A and A_d in (4.48) with $A + MF(k)N_1$ and $A_d + MF(k)N_2$, respectively, the admissibility criterion for the uncertain system can be rewritten as

$$\bar{\Theta} + \mathrm{sym}(\bar{P}^T MF(k)\bar{\Xi}) < 0. \tag{4.51}$$

Applying Schur complement to (4.50), we obtain

$$\bar{\Theta} + \varepsilon\bar{\Xi}^T\bar{\Xi} + \varepsilon^{-1}\bar{P}^T MM^T\bar{P} < 0,$$

which, by Lemma 4.16, is equivalent to (4.51), and the proof is completed. □

4.2.4.2 H_∞ controller design

In this subsection, we design a robust H_∞ state feedback controller in the form of (4.24) for system (4.22) such that the closed-loop system is admissible with H_∞ performance. For notational simplicity, we give the following closed-loop system:

$$\begin{cases} Ex(k+1) = (A + BK)x(k) + A_d x(k-d) + B_w w(k), \\ z(k) = (C + DK)x(k), \\ x(k) = \phi(k), \ k \in [-\bar{d}, 0]. \end{cases} \tag{4.52}$$

Based on Theorem 4.28, we can obtain the following result:

Theorem 4.30. *Given scalars* λ_1, λ_2, λ_3, *positive integers* m, τ, *and a prescribed scalar* $\gamma > 0$, *there exists a state-feedback controller in the form of (4.24) such that the closed-loop system in (4.22) is admissible with an* H_∞ *disturbance attenuation* γ *if there exist matrices* $P_1 > 0$, $Q > 0$, $Z > 0$, S_1, S_2, S_3, J, X, *and a scalar* $\varepsilon > 0$, *such that*

$$\begin{bmatrix} \tilde{\Psi} + \varepsilon\tilde{\Gamma}\tilde{\Gamma}^T & \tilde{\Phi}^T \\ \star & -\varepsilon I \end{bmatrix} < 0, \tag{4.53}$$

where $R \in \mathbb{R}^{n\times(n-r)}$ *is any full-column rank matrix satisfying* $E^T R = 0$ *and*

$$\begin{aligned}
\tilde{\Psi} &= \tilde{W}_P^T(P_1+\tau^2 Z)\tilde{W}_P + \tilde{W}_Q^T\tilde{Q}\tilde{W}_Q + \tilde{W}_C^T\tilde{W}_C - \gamma^2\tilde{W}_W^T\tilde{W}_W \\
&\quad - \tilde{W}_Z^T EZE^T\tilde{W}_Z + sym(\tilde{W}_{P1}^T EP_1\tilde{W}_P + \tilde{W}_E^T\tilde{\Lambda} + \tilde{S}R^T\tilde{W}_P), \\
\tilde{W}_E &= \begin{bmatrix} \lambda_1 I_n & \lambda_3 I_n & 0_{n,(m-1)n} & \lambda_2 I_n & 0_{n,s} \end{bmatrix}, \\
\tilde{W}_{P1} &= \begin{bmatrix} I_n & 0_{n,(m+1)n} & 0_{n,s} \end{bmatrix}, \\
\tilde{\Lambda} &= \begin{bmatrix} S_1^T & S_2^T & 0_{n-r,(m-1)n} & S_3^T & 0_{n-r,s} \end{bmatrix}^T, \\
\tilde{W}_P &= \begin{bmatrix} 0_{n,(m+1)n} & I_n & 0_{n,s} \end{bmatrix}, \\
\tilde{S} &= \begin{bmatrix} S_1^T & S_2^T & 0_{n-r,(m-1)n} & S_3^T & 0_{n-r,s} \end{bmatrix}^T, \\
\tilde{W}_Q &= \left[\begin{array}{ccc} I_{mn} & 0_{mn,2n+s} & \\ \hline 0_{mn,n} & I_{mn} & 0_{mn,n+s} \end{array}\right], \quad \tilde{W}_Z = \begin{bmatrix} I_n & -I_n & 0_{n,mn+s} \end{bmatrix}, \\
\tilde{W}_C &= \begin{bmatrix} B_w^T & 0_{p,(m+1)n+s} \end{bmatrix}, \quad \tilde{W}_W = \begin{bmatrix} 0_{s,(m+2)n} & I_s \end{bmatrix}, \\
\tilde{\Phi} &= \begin{bmatrix} \lambda_1(N_1 J + N_3 X) & \lambda_3(N_1 J + N_3 X) & 0_{b,(m-1)n} & \lambda_2(N_1 J + N_3 X) & 0_{e,s} \\ \lambda_1 N_2 J & \lambda_3 N_2 J & 0_{b,(m-1)n} & \lambda_2 N_2 J & 0_{e,s} \end{bmatrix}, \\
\tilde{\Gamma}^T &= \begin{bmatrix} M^T & 0_{l,(m-1)n} & 0_{l,n} & 0_{l,n+s} \\ 0_{l,n} & 0_{l,(m-1)n} & M^T & 0_{l,n+s} \end{bmatrix}.
\end{aligned}$$

Moreover, if the above condition is feasible, a desired controller gain matrix in the form of (4.24) is given by

$$K = XJ^{-1}.$$

Proof. It is easy to see that

$$\begin{aligned}
\det(zE-(A+BK)) &= \det(zE^T-(A+BK)^T), \\
\deg(\det(zE-(A+BK))) &= \deg(\det(zE^T-(A+BK)^T)),
\end{aligned}$$

and there are the same solutions to $\det(zE-(A+BK)-z^{-d}A_d)=0$ as well as $\det(zE^T-(A+BK)^T-z^{-d}A_d^T)=0$. Furthermore, the H_∞ norm of the system in (4.52) is given by

$$\|H(z)\|_\infty = \sup_{w\in[0,2\pi)} \sigma_{\max}((C+DK)(e^{jw}E-(A+BK+e^{-jwd}A_d))^{-1}B_w),$$

which is equivalent to

$$\|H(z)\|_\infty = \sup_{w\in[0,2\pi)} \sigma_{\max}(B_w^T(e^{jw}E^T-(A+BK+e^{-jwd}A_d)^T)^{-1}(C+DK)^T).$$

With respect to the regularity, causality, stability, and H_∞ performance problems of a system, we can obtain that the system in (4.52) is equivalent to the following system based on Definition 4.1, Definition 4.14 and Lemma 4.15:

$$\begin{cases} E^T\delta(k+1) = (A+BK)^T\delta(k) + A_d^T\delta(k-d) + (C+DK)^T\varphi(k), \\ \varkappa(k) = B_w^T\delta(k). \end{cases}$$

Substituting E, A, A_d, B_w, p and C with E^T, $(A+BK)^T$, A_d^T, $(C+DK)^T$, s and B_w^T in (4.48), respectively, we have the following inequality

$$\Psi_1 < 0,$$

where

$$\begin{aligned} \Psi_1 \;=\;& \tilde{W}_P^T(P_1+\tau Z)\tilde{W}_P + \tilde{W}_Q^T\tilde{Q}\tilde{W}_Q + \tilde{W}_C^T\tilde{W}_C - \gamma^2\tilde{W}_W^T\tilde{W}_W \\ & -\tilde{W}_Z^T EZE^T\tilde{W}_Z + \mathrm{sym}(\tilde{W}_{P1}^T EP_1\tilde{W}_P + \tilde{S}R^T\tilde{W}_P \\ & +\begin{bmatrix} P_2 & P_4 & 0_{n,(m-1)n} & P_3 & 0_{n,s} \end{bmatrix}^T \\ & \times\begin{bmatrix} (A-E+BK)^T & 0_{n,(m-1)n} & A_d^T & -I_n & (C+DK)^T \end{bmatrix}). \end{aligned}$$

Then, by denoting $P_2 = \lambda_1 J$, $P_3 = \lambda_2 J$, $P_4 = \lambda_3 J$, and $X = KJ$, $\tilde{\Psi} < 0$ is readily obtained. Following a similar line of argument as in the proof of Theorem 4.24, Theorem 4.30 can be established. □

4.2.5 Illustrative examples

In this subsection, we use numerical examples to illustrate the advantages of the developed results and the applicability of the proposed controller design methods. We first demonstrate the improvement by considering a nominal singular system in Example 4.2.

Example 4.2. Consider the singular system

$$\begin{bmatrix} 3.5 & 0 \\ 0 & 0 \end{bmatrix} x(k+1) = \begin{bmatrix} a_{11} & 0 \\ 0 & -3 \end{bmatrix} x(k) + \begin{bmatrix} -1.3 & 1.5 \\ 0 & 0.5 \end{bmatrix} x(k-d).$$

Our purpose is to determine the allowable time-delay upper bounds $\bar{d}$ such that the system is admissible. To compare our results with those in [72, 172,191,249], we consider $a_{11} = 2.3$, or $a_{11} = 2.4$. The maximum delay

Table 4.2 Comparisons of maximum allowed delay $\bar{d}$.

a_{11}	2.3	2.4
$\bar{d}$ [72]	10	7
$\bar{d}$ [172]	11	7
$\bar{d}$ [191]	11	7
$\bar{d}$ [249]	11	7
$\bar{d}$ (Theorem 4.17)	11 ($m=1$, $\tau=11$)	7 ($m=1$, $\tau=7$)
$\bar{d}$ (Theorem 4.17)	12 ($m=3$, $\tau=4$)	9 ($m=3$, $\tau=3$)
$\bar{d}$ (Theorem 4.17)	14 ($m=7$, $\tau=2$)	10 ($m=5$, $\tau=2$)
$\bar{d}$ (Exact solution)	14	10

bounds with $a_{11}=2.3$ such that the above system is admissible are found by using the methods of [72,172,191,249], to be 10, 11, 11, 11, respectively. However, the maximum bound obtained by using Theorem 4.17 is 14 for $m=7$, $\tau=2$. Table 4.2 gives more detailed comparison results on the maximum allowed bounds for $\bar{d}$ via the methods in [72,172,191,249] and Theorem 4.17 in this section. The results in Table 4.2 clearly show that Theorem 4.17 in this section outperforms those in [72,172,191,249] in terms of conservatism.

On the other hand, by using Lemma 4.15, the exact solutions of the maximal delay are listed under the given parameters. From Table 4.2, we can see that the results obtained is the same as the exact solution, so there is no room for further improvement.

In this example, we obtain the same maximal delay as that in [191]. However, the number of variables involved in Corollary 4.20 is $\frac{n}{2}(15n-6r+3)$, whereas the method in [191] requires $\frac{n}{2}(21n-6r+3)$. Thus the presented approach is more favorable computationally.

Next, the advantages of our results will be shown by considering an uncertain discrete-time singular delay system in Example 4.3.

Example 4.3. Consider the uncertain discrete-time singular system with the following parameters (borrowed from [191]):

$$\begin{bmatrix} 2 & 0 \\ 0 & 0 \end{bmatrix} x(k+1) = \begin{bmatrix} 0.9977+0.1\alpha & 1.1972 \\ 0.1001 & -1.9 \end{bmatrix} x(k) + \begin{bmatrix} -1.1972 & 1.5772 \\ 0 & 0.9757+0.1\alpha \end{bmatrix} x(k-d).$$

Table 4.3 Allowable maximum absolute value of α obtained by different methods.

$\bar{d}$	2	3	4
$\bar{\alpha}$ [72]	1.9464	0.8033	0.1563
$\bar{\alpha}$ [191]	2.1359	1.0325	0.2853
$\bar{\alpha}$ (Theorem 4.22)	2.1359 ($m=1$, $\tau=2$)	1.0325 ($m=1$, $\tau=3$)	0.2853 ($m=1$, $\tau=4$)
$\bar{\alpha}$ (Theorem 4.22)	2.6847 ($m=2$, $\tau=1$)	1.6734 ($m=3$, $\tau=1$)	0.8972 ($m=4$, $\tau=1$)

Table 4.4 Allowable maximum time delays obtained by different methods.

$\bar{d}$ [72]	$\bar{d}$ [191]	$\bar{d}$ (Corollary 4.19) $m=1$	$\bar{d}$ (Corollary 4.19) $m=5$	$\bar{d}$ (Corollary 4.19) $m=7$
41	42	42 ($\tau=42$)	55 ($\tau=11$)	56 ($\tau=8$)

The purpose is to determine the upper bounds for the absolute value of uncertain parameter α, that is, $\bar{\alpha}$. To illustrate the benefits of our results, Table 4.3 gives the comparison results on $\bar{\alpha}$.

These comparison results show that the result in Theorem 4.22 for delay singular systems with uncertainties in this section is less conservative than those in [72,191].

From the above two examples, it can be seen that better results have been obtained in this section for discrete-time singular systems with time delay. Moreover, it is worth pointing out that our result is also effective for standard state-space delay systems, which is shown in Example 4.4.

Example 4.4. Consider the following standard state-space system:

$$x(k+1)=\begin{bmatrix}0.8 & 0\\ 0 & 0.91\end{bmatrix}x(k)+\begin{bmatrix}-0.1 & 0\\ -0.1 & -0.1\end{bmatrix}x(k-d).$$

By solving the feasibility problem of the LMIs in Corollary 4.19, we conclude that this system is stable for any constant time-delay d satisfying $d \leq 56$ (with $m=7$, $\tau=8$). However, the methods of [72] and [191] fail to yield feasible solutions when $d > 41$ and $d > 42$, see computational results shown in Table 4.4.

In Example 4.5, the applicability of the proposed controller design methods will be demonstrated.

Example 4.5. Consider the uncertain singular system in (4.22) with the following parameters:

$$\begin{aligned} E &= \begin{bmatrix} 1 & 0 \\ 0 & 0 \end{bmatrix}, \ A = \begin{bmatrix} 1.7 & 2 \\ 1 & 2 \end{bmatrix}, \ A_d = \begin{bmatrix} 1.5 & 1 \\ 1 & 0.05 \end{bmatrix}, \\ B &= \begin{bmatrix} -2 & 3 \\ 0 & -2 \end{bmatrix}, \ M = \begin{bmatrix} 0.2 \\ 0.2 \end{bmatrix}, \ N_1 = N_2 = \begin{bmatrix} 0.2 & 0.2 \end{bmatrix}, \\ N_3 &= \begin{bmatrix} 0.01 & 0.02 \end{bmatrix}, \ F(k) = \sin(k). \end{aligned}$$

In this example, we choose

$$\lambda_1 = 0.6, \ \lambda_2 = 1, \ \lambda_3 = -0.35, \ R = \begin{bmatrix} 0 & 1 \end{bmatrix}^T, \ d = 3,$$

and obtain the solution as follows by solving the LMIs in (4.42):

$$\begin{aligned} X &= \begin{bmatrix} -396.3955 & 752.8114 \\ -138.9248 & 406.9560 \end{bmatrix}, \\ J &= \begin{bmatrix} 181.7492 & -314.1157 \\ -314.3584 & 352.1828 \end{bmatrix}. \end{aligned}$$

Therefore, by Theorem 4.24, an admissible state-feedback control law can be obtained as

$$u(k) = \begin{bmatrix} -2.7939 & -0.3544 \\ -2.2744 & -0.8730 \end{bmatrix} x(k).$$

Fig. 4.1 and Fig. 4.2 give the simulation results of two states with and without the state-feedback control, respectively. From Fig. 4.1 and Fig. 4.2, we can see that the open-loop system is unstable and the closed-loop system is stable.

Example 4.6. Consider the following system:

$$\begin{aligned} E &= \begin{bmatrix} 3 & -0.5 \\ 0 & 0 \end{bmatrix}, \ A = \begin{bmatrix} 0.8 & 0 \\ 0 & -1.6 \end{bmatrix}, \ A_d = \begin{bmatrix} -2 & 1 \\ 0 & 1 \end{bmatrix}, \\ B_w &= \begin{bmatrix} 0.3 \\ 0.3 \end{bmatrix}, \ C = \begin{bmatrix} 0.5 & 0.5 \end{bmatrix}. \end{aligned}$$

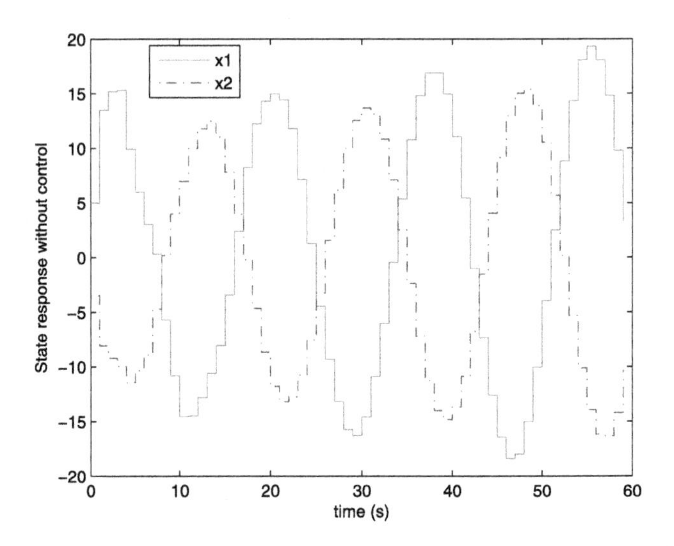

Figure 4.1 The state trajectories of the open-loop system.

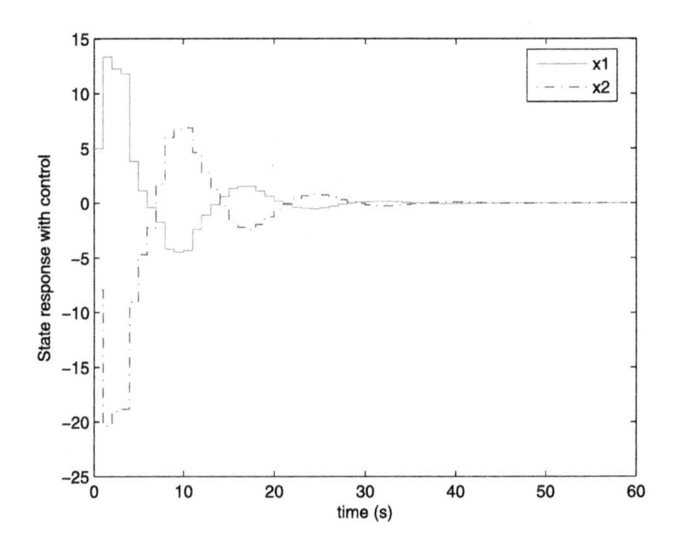

Figure 4.2 The state trajectories of the closed-loop system.

For a given $\gamma > 0$, the maximum allowed delay $\bar{d}$ satisfying the LMI in Theorem 4.28 can be calculated by solving a quasi-convex optimization problem. Similarly, for a given $d > 0$, the minimum allowed γ satisfying the LMI in Theorem 4.28 can also be computed by solving a quasiconvex optimization problem. Tables 4.5 and 4.6 give the comparison results on

Table 4.5 Comparisons of maximum allowed delay d given $\gamma > 0$.

γ	**0.66**	**0.65**	**0.64**
$\bar{d}$ **[196]**	**27**	**13**	**10**
$\bar{d}$ (Theorem 4.28)	27 ($m=1$, $\tau=27$)	13 ($m=1$, $\tau=13$)	11 ($m=1$, $\tau=11$)
$\bar{d}$ (Theorem 4.28)	38 ($m=2$, $\tau=19$)	18 ($m=2$, $\tau=9$)	14 ($m=2$, $\tau=7$)
$\bar{d}$ (Theorem 4.28)	39 ($m=3$, $\tau=13$)	18 ($m=3$, $\tau=6$)	15 ($m=3$, $\tau=5$)
$\bar{d}$ (Theorem 4.28)	40 ($m=4$, $\tau=20$)	20 ($m=4$, $\tau=5$)	15 ($m=5$, $\tau=3$)

Table 4.6 Comparisons of minimum allowed γ given $d > 0$.

d	**6**	**12**	**18**	**24**
γ^* **[196]**	**0.5977**	**0.6457**	**0.6555**	**0.6590**
γ^* (Theorem 4.28), $m=1$	0.5977 ($\tau=6$)	0.6457 ($\tau=12$)	0.6555 ($\tau=18$)	0.6590 ($\tau=24$)
γ^* (Theorem 4.28), $m=2$	0.5493 ($\tau=3$)	0.6290 ($\tau=6$)	0.6477 ($\tau=9$)	0.6546 ($\tau=12$)
γ^* (Theorem 4.28), $m=3$	0.5402 ($\tau=2$)	0.6251 ($\tau=4$)	0.6458 ($\tau=6$)	0.6534 ($\tau=8$)

the maximum allowed delay d for given $\gamma > 0$, and the minimum allowed γ for given $d > 0$, respectively, via the methods in [196] and Theorem 4.28 in this section. It can be seen that these comparisons show that our results for delay systems without uncertainties (in this section) is less conservative than those in [196].

Example 4.7. Consider the linear uncertain discrete singular delay system in (4.22) with parameters as follows:

$$
\begin{aligned}
E &= \begin{bmatrix} 2 & 0 \\ 0 & 0 \end{bmatrix}, \; A = \begin{bmatrix} 1.47 & 0 \\ 0.5 & -1.5 \end{bmatrix}, \; A_d = \begin{bmatrix} -1 & 0.5 \\ 0 & 0.5 \end{bmatrix}, \\
B &= \begin{bmatrix} 0.36 \\ 0.5 \end{bmatrix}, \; B_w = \begin{bmatrix} 0.3 \\ 0.3 \end{bmatrix}, \; C = \begin{bmatrix} 0.5 & 0.5 \end{bmatrix}, \\
M &= \begin{bmatrix} -0.02 & 0.02 \\ -0.02 & -0.02 \end{bmatrix}, \; N_1 = \begin{bmatrix} 0.025 & 0.01 \\ 0.1 & 0.025 \end{bmatrix}, \; D = 0.3, \\
N_2 &= \begin{bmatrix} 0.02 & 0.01 \\ 0.01 & 0.02 \end{bmatrix}, \; N_3 = \begin{bmatrix} 0.3 \\ -0.4 \end{bmatrix}, \; d = 6, \\
\lambda_1 &= 0.2, \; \lambda_2 = 0.1, \; \lambda_3 = -0.1, \; R = \begin{bmatrix} 0 & 1 \end{bmatrix}^T.
\end{aligned}
$$

The purpose is to design a state-feedback controller in the form of (4.24) such that the system in (4.22) is asymptotically stable with a guaranteed H_∞ performance γ. In this example, the minimum H_∞ performance $\gamma^* = 0.48$ is obtained by solving LMI (4.53), and the feasible solutions are

$$\begin{aligned} J &= \begin{bmatrix} 2.6517 & 2.9523 \\ 2.1264 & 7.0467 \end{bmatrix}, \\ X &= \begin{bmatrix} -7.1433 & -11.0540 \end{bmatrix}. \end{aligned}$$

Therefore, by Theorem 4.30, the corresponding stabilizing state-feedback controller can be obtained as

$$K = \begin{bmatrix} -2.1625 & -0.6627 \end{bmatrix}.$$

Next, to illustrate the disturbance attenuation performance, assume zero initial condition and the external disturbance

$$w(k) = \begin{cases} 3, & 0 \le k \le 10 \\ -3, & 11 \le k \le 20 \\ 0, & \text{elsewhere.} \end{cases}$$

Fig. 4.3 shows the signals $w(k)$ and $z(k)$. It is found that $||z||_2 = 2.3628$ and $||w||_2 = 13.4164$, which yields $\gamma = 0.1761$ (below the minimum $\gamma^* = 0.48$),

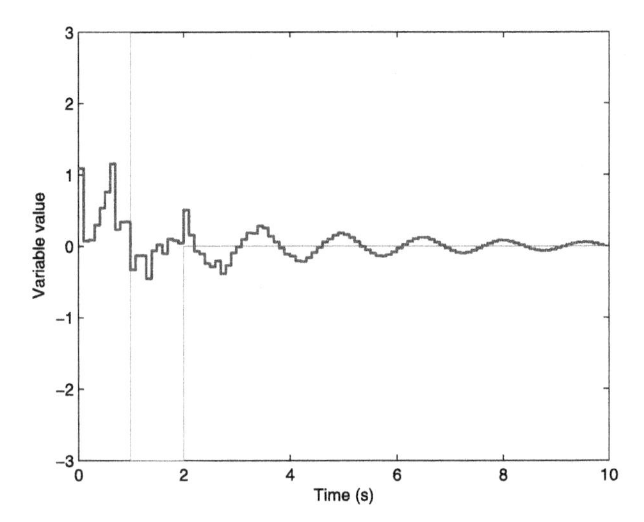

Figure 4.3 Signals $w(k)$ and $z(k)$.

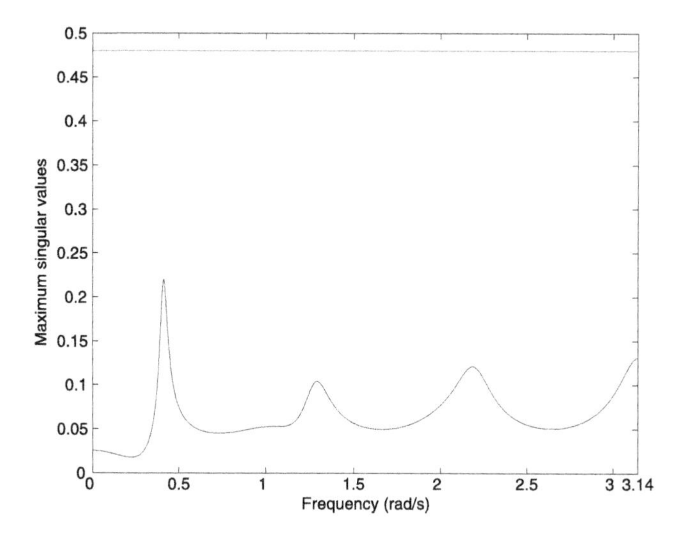

Figure 4.4 Maximum singular values over frequency range $[0, \pi)$.

showing the effectiveness of the controller design. For illustration, we take one of the worst-case perturbation such that $F(k) = I$. The maximum singular value plot of the linear uncertain discrete singular delay system in (4.22) is shown in Fig. 4.4 for the frequency range $[0, \pi)$. The effectiveness of the guaranteed H_∞ disturbance is apparent.

4.2.6 Conclusion

In this chapter, improved functionals based on the delay partitioning technique have been introduced to derive improved results for robust stability and stabilization problems of linear uncertain discrete-time singular systems with state delay, which guarantees the closed-loop system is admissible. Moreover, the proposed new results have been utilized to investigate robust H_∞ control problem which assures the resulting closed-loop system is admissible with an H_∞ disturbance attenuation. Less conservative and easily verifiable conditions have been formulated in terms of strict LMIs involving no decomposition of the system matrices. It is also proved that the conservatism of the results is non-increasing with the reduction of the partition size. Numerical examples have been given to demonstrate the advantages and the merits of the proposed results. Extending the method proposed here to tackle the robust control, guaranteed cost and variable structure control for singular system with time-varying delay will be interesting topics for future research.

CHAPTER 5

Delay-dependent dissipativity analysis and synthesis of singular delay systems

The problem of delay-dependent dissipativity analysis and synthesis for singular delay system has been addressed in this chapter. Firstly, based on developed inequality and improved reciprocally convex combination approach, the problem of dissipative control for discrete-time singular delay system is studied. Secondly, using the delay partitioning technique, the state-feedback controller is designed. And the sufficient conditions of dissipativity synthesis for continuous-time singular system with time delay is constructed. Finally, by introducing some variables to decouple the Lyapunov matrices and the filtering error system matrices, we consider the problem of robust reliable dissipative filtering for discrete-time singular systems with polytopic uncertainties, time-varying delays, and sensor failures.

5.1 Dissipativity analysis for discrete singular systems with time-varying delay

In this section, the issue of dissipativity analysis for discrete singular systems with time-varying delay is investigated. By using a recently developed inequality, which is less conservative than the Jensen inequality, and the improved reciprocally convex combination approach, sufficient criteria are established to guarantee the admissibility and dissipativity of the considered system. Moreover, H_∞ performance characterization and passivity analysis are carried out. Numerical examples are presented to illustrate the effectiveness of the proposed method.

5.1.1 Problem formulation

Consider discrete-time singular systems with time-varying delay described by

$$\begin{aligned} Ex(k+1) &= Ax(k) + A_d x(k-d(k)) + B_\omega \omega(k), \\ z(k) &= Lx(k) + L_d x(k-d(k)) + G_\omega \omega(k), \\ x(k) &= \phi(k), \ \ k \in [-d_2, 0], \end{aligned} \tag{5.1}$$

Analysis and Synthesis of Singular Systems
https://doi.org/10.1016/B978-0-12-823739-7.00012-4

where $x(k) \in \mathbb{R}^n$ is the state vector; $\omega(k) \in \mathbb{R}^l$ represents a set of exogenous inputs, which includes disturbances to be rejected; $z(k) \in \mathbb{R}^q$ is the control output; $d(k)$ is a time-varying delay satisfying $0 < d_1 \leq d(k) \leq d_2$, where d_1 and d_2 are prescribed positive integers representing the lower and upper bounds of the time delay, respectively. $\phi(k)$ is the compatible initial condition. The matrix $E \in \mathbb{R}^{n\times n}$ may be singular, and it is assumed that $\text{rank}(E) = r \leq n$. A, A_d, B_ω, C, C_d, D, L, L_d, and G_ω are known real constant matrices with appropriate dimensions.

The lemmas and definitions that follow will be used in the proof of the primary results in this section.

Denote $y(k) = x(k+1) - x(k)$ and a new inequality is derived in the following lemma:

Lemma 5.1. [136] *For a given positive definite matrix R and three given nonnegative integers a, b, k satisfying $a \leq b \leq k$, denote*

$$\chi(k,a,b) = \begin{cases} \frac{1}{b-a}\left[2\sum_{s=k-b}^{k-a-1} x(s) + x(k-a) + x(k-b)\right], & a < b \\ 2x(k-a), & a = b. \end{cases}$$

Then, we have

$$-(b-a)\sum_{s=k-b}^{k-a-1} y^T(s)Ry(s) \leq -\begin{bmatrix} \Theta_0 \\ \Theta_1 \end{bmatrix}^T \begin{bmatrix} R & 0 \\ 0 & 3R \end{bmatrix} \begin{bmatrix} \Theta_0 \\ \Theta_1 \end{bmatrix}, \tag{5.2}$$

where

$$\Theta_0 = x(k-a) - x(k-b),$$
$$\Theta_1 = x(k-a) + x(k-b) - \chi(k,a,b).$$

Remark 5.2. There is a difference between Lemma 3 in [136] and Lemma 5.1 in this section. After checking, when $a < b$, it is found that the sign of $x(k-b)$ in $\chi(k,a,b)$ should be "+" instead of "−" in [136]. Moreover, it is pointed out that there are minor typographical errors in the Lyapunov function. V_3 is written as

$$\begin{aligned} V_3 =& \tau_m \sum_{s=-\tau_m}^{-1}\sum_{v=k+s}^{k-1} y^T(v)S_1y(v) + (\tau_a - \tau_m)\sum_{s=-\tau_a}^{\tau_m-1}\sum_{v=k+s}^{k-1} y^T(v)S_2y(v) \\ &+ (\tau_M - \tau_a)\sum_{s=-\tau_M}^{\tau_a-1}\sum_{v=k+s}^{k-1} y^T(v)S_3y(v). \end{aligned}$$

However, to get the forward difference of V_3 in (25) of [136], it should be written as

$$\begin{aligned} V_3 =& \tau_m \sum_{s=-\tau_m}^{-1} \sum_{v=k+s}^{k-1} y^T(v) S_1 y(v) + (\tau_a - \tau_m) \sum_{s=-\tau_a+1}^{-\tau_m} \sum_{v=k+s-1}^{k-1} y^T(v) S_2 y(v) \\ &+ (\tau_M - \tau_a) \sum_{s=-\tau_M+1}^{-\tau_a} \sum_{v=k+s-1}^{k-1} y^T(v) S_3 y(v). \end{aligned}$$

Lemma 5.3. [140] *Let n, m be two positive integers and two matrices R_1 in $\mathbb{S}_n^+$ and R_2 in $\mathbb{S}_m^+$. The improved reciprocally convex combination guarantees that, if there exists a matrix X in $\mathbb{R}^{n\times m}$ such that $\begin{bmatrix} R_1 & X \\ X^T & R_2 \end{bmatrix} \geq 0$, then the following inequality holds for any scalar α in the interval $(0, 1)$:*

$$\begin{bmatrix} \frac{1}{\alpha} R_1 & 0 \\ 0 & \frac{1}{1-\alpha} R_2 \end{bmatrix} \geq \begin{bmatrix} R_1 & X \\ X^T & R_2 \end{bmatrix}. \tag{5.3}$$

The nominal discrete singular system with time-varying delay of system (5.1) can be written as

$$\begin{aligned} Ex(k+1) =& Ax(k) + A_d x(k - d(k)), \\ x(k) =& \phi(k), \; k \in [-d_2, 0]. \end{aligned} \tag{5.4}$$

Throughout the section, the following definitions will be adopted:

Definition 5.4. [101] Given some scalar $\alpha > 0$, matrices $\mathcal{Q}$, $\mathcal{S}$ and $\mathcal{R}$ with $\mathcal{Q} \leq 0$ and $\mathcal{R}$ real symmetric, singular system (5.1) is called strictly $(\mathcal{Q}, \mathcal{S}, \mathcal{R})$-$\alpha$-dissipative if for any $\tau \geq 0$, under zero initial state, the following condition is satisfied:

$$\langle z, \mathcal{Q} z \rangle_\tau + 2 \langle z, \mathcal{S} \omega \rangle_\tau + \langle \omega, \mathcal{R} \omega \rangle_\tau \geq \alpha \langle \omega, \omega \rangle_\tau . \tag{5.5}$$

The aim of this section is to study issue of α-dissipativity analysis for singular system (5.1). By utilizing the discrete Wirtinger inequality and the improved reciprocally convex combination approach, we will derive some sufficient criteria in terms of LMIs so that singular system (5.1) is admissible and strictly $(\mathcal{Q}, \mathcal{S}, \mathcal{R})$-$\alpha$-dissipative.

5.1.2 Main results

In this section, the admissibility and strict $(\mathcal{Q}, \mathcal{S}, \mathcal{R})$-$\alpha$-dissipativity of the discrete singular system with time-varying delay are addressed by adopting the discrete Wirtinger inequality and the improved reciprocally convex combination approach. To simplify the matrices and vector notations, the following notations will be used in our development:

$$
\begin{aligned}
\mu_1(k) &= \chi(k, 0, d_1), \mu_2(k) = \chi(k, d_1, d(k)), \mu_3(k) = \chi(k, d(k), d_2), \\
\xi^T(k) &= \Big[\ x^T(k) \quad x^T(k-d_1) \quad x^T(k-d(k)) \quad x^T(k-d_2) \\
&\qquad \mu_1^T(k)E^T \quad \mu_2^T(k)E^T \quad \mu_3^T(k)E^T \quad \omega^T(k)\ \Big], \\
e_i &= \left[\ 0_{n\times(i-1)n} \quad I_n \quad 0_{n\times(7-i)n} \quad 0_{n\times l}\ \right], \ i = 1, 2, ..., 7, \\
e_8 &= \left[\ 0_{l\times 7n} \quad I_l\ \right], \\
\rho_1 &= \frac{1}{2}\left(d_1 e_5 - Ee_1 - Ee_2\right), \\
\rho_2 &= Ae_1 + A_d e_3 + B_\omega e_8, \\
\rho_3 &= \rho_2 - Ee_1, \\
\rho_4^T &= \left[\ \rho_3^T \quad (Ee_1 - Ee_2)^T \quad (Ee_2 - Ee_4)^T\ \right], \\
\rho_5 &= Le_1 + L_d e_3 + G_\omega e_8, \\
\phi_1(k) &= \frac{1}{2}\left((d(k) - d_1)e_6 - Ee_2 - Ee_3\right), \\
\phi_2(k) &= \frac{1}{2}((d_2 - d(k))e_7 - Ee_3 - Ee_4), \\
\phi_3(k) &= \phi_1(k) + \phi_2(k) \\
&= \frac{1}{2}\left((d(k) - d_1)e_6 + (d_2 - d(k))e_7 - Ee_2 - 2Ee_3 - Ee_4\right), \\
\phi_4^T(k) &= \left[\ (Ee_1)^T \quad \rho_1^T \quad \phi_3^T(k)\ \right], \\
\Pi^T &= \Big[\ (Ee_1 - Ee_2)^T \quad (Ee_1 + Ee_2 - e_5)^T \quad (Ee_2 - Ee_3)^T \\
&\qquad (Ee_2 + Ee_3 - e_6)^T \quad (Ee_3 - Ee_4)^T \quad (Ee_3 + Ee_4 - e_7)^T\ \Big].
\end{aligned}
$$

Theorem 5.5. *For given integers d_1 and d_2 satisfying $0 < d_1 \le d(k) \le d_2$, let scalar $\alpha > 0$, matrices $\mathcal{Q}$, $\mathcal{S}$ and $\mathcal{R}$ be given with $\mathcal{Q}$ and $\mathcal{R}$ real symmetric and $\mathcal{Q} \le 0$. Then system (5.1) is admissible and strictly $(\mathcal{Q}, \mathcal{S}, \mathcal{R})$-$\alpha$-dissipative if there exist matrices $P > 0$, $Q_i > 0$ $(i = 1, 2, 3)$, $R_j > 0$ $(j = 1, 2)$ and matrices X, W*

such that the following set of LMIs hold:

$$\Xi + \mathrm{sym}(\rho_4^T P \phi_{41}) < 0, \quad (5.6)$$

$$\Xi + \mathrm{sym}(\rho_4^T P \phi_{42}) < 0, \quad (5.7)$$

where $R \in \mathbb{R}^{n\times(n-r)}$ *is any matrix with full column rank and satisfies* $E^T R = 0$, *and*

$$\begin{aligned}
\Xi &= \Xi_1 - \Xi_2, \\
\Xi_1 &= \rho_4^T P \rho_4 + e_1^T(Q_1 + (\tilde{d}+1)Q_3)e_1 + e_2^T(-Q_1 + Q_2)e_2 - e_3^T Q_3 e_3 \\
&\quad - e_4^T Q_2 e_4 + \rho_3^T(d_1^2 R_1 + \tilde{d}^2 R_2)\rho_3 - \Pi^T \Phi \Pi + \mathrm{sym}(e_1^T W R^T \rho_2), \\
\Xi_2 &= \rho_5^T \mathcal{Q} \rho_5 + \mathrm{sym}(\rho_5^T \mathcal{S} e_8) + e_8^T \mathcal{R} e_8, \\
\phi_{41}^T &= \begin{bmatrix} (Ee_1)^T & \rho_1^T & \frac{1}{2}(\tilde{d}e_7 - Ee_2 - 2Ee_3 - Ee_4)^T \end{bmatrix}, \\
\phi_{42}^T &= \begin{bmatrix} (Ee_1)^T & \rho_1^T & \frac{1}{2}(\tilde{d}e_6 - Ee_2 - 2Ee_3 - Ee_4)^T \end{bmatrix},
\end{aligned}$$

$$P = \begin{bmatrix} P_{11} & P_{12} & P_{13} \\ \star & P_{22} & P_{23} \\ \star & \star & P_{33} \end{bmatrix}, \ \tilde{d} = d_2 - d_1,$$

$$\Phi = \begin{bmatrix} \tilde{R}_1 & 0 & 0 \\ 0 & \tilde{R}_2 & X \\ 0 & X^T & \tilde{R}_2 \end{bmatrix}, \ \tilde{R}_i = \begin{bmatrix} R_i & 0 \\ 0 & 3R_i \end{bmatrix}, \ i = 1,2.$$

Proof. Firstly, we will prove that the system (5.1) is regular and causal. Due to $\mathrm{rank}(E) = r$, we choose two nonsingular matrices M and N such that

$$MEN = \begin{bmatrix} I & 0 \\ 0 & 0 \end{bmatrix}. \quad (5.8)$$

Set

$$MAN = \begin{bmatrix} A_1 & A_2 \\ A_3 & A_4 \end{bmatrix}, \ N^T W = \begin{bmatrix} W_1 \\ W_2 \end{bmatrix}, \ M^{-T} R = \begin{bmatrix} 0 \\ I \end{bmatrix} F, \quad (5.9)$$

where $F \in \mathbb{R}^{(n-r)\times(n-r)}$ is nonsingular.

Expand $\Xi + \mathrm{sym}(\rho_4^T P \phi_{41})$ as

$$\Xi + \mathrm{sym}(\rho_4^T P \phi_{41}) = \begin{bmatrix} \Xi_{11} & \bullet \\ \bullet & \bullet \end{bmatrix},$$

where $\bullet$ means that the elements of the matrix are not relevant in our discussion, and

$$\begin{aligned}\Xi_{11} =&(A-E)^T P_{11}(A-E) + (A-E)^T P_{12}E + E^T P_{12}^T(A-E) + Q_1 \\ &+ (\widetilde{d}+1)Q_3 - 4E^T R_1 E + (A-E)^T(d_1^2 R_1 + \widetilde{d}^2 R_2)(A-E) \\ &+ WR^T A + A^T R W^T + (A-E)^T P_{11} E + E^T P_{11}(A-E) - L^T \mathcal{Q} L \\ &- \frac{1}{2}(A-E)^T P_{12} E - \frac{1}{2} E^T P_{12}^T (A-E) + E^T P_{12} E + E^T P_{12}^T E.\end{aligned}$$

Due to $\Xi + \mathrm{sym}(\rho_4^T P \phi_{41}) < 0$, thus $\Xi_{11} < 0$, together with $P > 0$, $Q_1 > 0$, $Q_3 > 0$, $R_i > 0$, $S_i > 0$ $(i = 1, 2)$, and $-\mathcal{Q} \geq 0$, we have

$$\begin{aligned}\Omega =&(A-E)^T P_{12} E + E^T P_{12}^T (A-E) - 4E^T R_1 E + WR^T A + A^T R W^T \\ &+ (A-E)^T P_{11} E + E^T P_{11}(A-E) - \frac{1}{2}(A-E)^T P_{12} E - \frac{1}{2} E^T P_{12}^T (A-E) \\ &+ E^T P_{12} E + E^T P_{12}^T E < 0.\end{aligned}$$

Premultiplying and postmultiplying $\Omega < 0$ by N^T and N, respectively, substituting (5.8) and (5.9) into the previous inequalities gives

$$\begin{bmatrix} \bullet & \bullet \\ \bullet & W_2 F^T A_4 + A_4^T F W_2^T \end{bmatrix} < 0.$$

From the above inequality, it is easy to see that $W_2 F^T A_4 + A_4^T F W_2^T < 0$, which implies A_4 is nonsingular. Hence, the pair (E, A) is regular and causal.

To prove the stability of system (5.1), we design a Lyapunov function as

$$V(k) = V_1(k) + V_2(k) + V_3(k) + V_4(k),$$

where

$$\begin{aligned}V_1(k) =&\varepsilon^T(k) P \varepsilon(k), \\ V_2(k) =& \sum_{i=k-d_1}^{k-1} x^T(i) Q_1 x(i) + \sum_{i=k-d_2}^{k-d_1-1} x^T(i) Q_2 x(i) + \sum_{i=k-d(k)}^{k-1} x^T(i) Q_3 x(i) \\ &+ \sum_{i=-d_2+1}^{-d_1} \sum_{j=k+i}^{k-1} x^T(j) Q_3 x(j),\end{aligned}$$

$$V_3(k) = d_1 \sum_{i=-d_1+1}^{0} \sum_{j=k+i}^{k-1} y^T(j) E^T R_1 E y(j) + \tilde{d} \sum_{i=-d_2+1}^{-d_1} \sum_{j=k+i}^{k-1} y^T(j) E^T R_2 E y(j)$$

with

$$\varepsilon(k) = \begin{bmatrix} Ex(k) \\ \sum_{i=k-d_1}^{k-1} Ex(i) \\ \sum_{i=k-d_2}^{k-d_1-1} Ex(i) \end{bmatrix}.$$

By denoting the forward difference of $V(k)$ as $\Delta V(k) = V(k+1) - V(k)$ and calculating it along the solution of system (5.1), we have

$$\Delta V_1(k) = \Delta\varepsilon^T(k) P \Delta\varepsilon(k) + 2\Delta\varepsilon^T(k) P \varepsilon(k),$$

where

$$\Delta\varepsilon(k) = \begin{bmatrix} (A-E)x(k) + A_d x(k-d_1) \\ Ex(k) - Ex(k-d_1) \\ Ex(k-d_1) - Ex(k-d_2) \end{bmatrix},$$

so

$$\Delta V_1(k) = \xi^T(k)(\rho_4^T P \rho_4 + 2\rho_4^T P \phi_4(k))\xi(k) \tag{5.10}$$

The estimation of the forward difference of $V_2(k)$ is

$$\begin{aligned}
\Delta V_2(k) =& x^T(k) Q_1 x(k) - x^T(k-d_1) Q_1 x(k-d_1) + x^T(k-d_1) Q_2 x(k-d_1) \\
& - x^T(k-d_2) Q_2 x(k-d_2) + x^T(k) Q_3 x(k) + \sum_{i=k+1-d(k+1)}^{k-1} x^T(i) Q_3 x(i) \\
& - \sum_{i=k+1-d(k)}^{k-1} x^T(i) Q_3 x(i) - x^T(k-d(k)) Q_3 x(k-d(k)) \\
& + \tilde{d} x^T(k) Q_3 x(k) - \sum_{i=k-d_2+1}^{k-d_1} x^T(i) Q_3 x(i) \\
=& x^T(k)(Q_1 + (\tilde{d}+1) Q_3) x(k) + x^T(k-d_1)(-Q_1 + Q_2) x(k-d_1) \\
& - x^T(k-d_2) Q_2 x(k-d_2) - x^T(k-d(k)) Q_3 x(k-d(k)) \\
& + \sum_{i=k+1-d_1}^{k-1} x^T(i) Q_3 x(i) + \sum_{i=k+1-d(k+1)}^{k-d_1} x^T(i) Q_3 x(i) \\
& - \sum_{i=k+1-d(k)}^{k-1} x^T(i) Q_3 x(i) - \sum_{i=k-d_2+1}^{k-d_1} x^T(i) Q_3 x(i)
\end{aligned}$$

$$\leq \xi^T(k)(e_1^T(Q_1+(\tilde{d}+1)Q_3)e_1+e_2^T(-Q_1+Q_2)e_2 \\ -e_3^TQ_3e_3-e_4^TQ_2e_4)\xi(k). \tag{5.11}$$

According to Lemma 5.1 and Lemma 5.3, when $d_1 < d(k) < d_2$, the forward difference of $V_3(k)$ is calculated as

$$\begin{aligned}
\Delta V_3(k) =& y^T(k)(d_1^2E^TR_1E+\tilde{d}^2E^TR_2E)y(k)-d_1\sum_{i=k-d_1+1}^{k}y^T(i)E^TR_1Ey(i) \\
&-\tilde{d}\sum_{i=k-d(k)+1}^{k-d_1}y^T(i)E^TR_2Ey(i)-\tilde{d}\sum_{i=k-d_2+1}^{k-d(k)}y^T(i)E^TR_2Ey(i) \\
\leq& y^T(k)E^T(d_1^2R_1+\tilde{d}^2R_2)Ey(k) \\
&-\xi^T(k)\Pi^T\begin{bmatrix} \widetilde{R}_1 & 0 & 0 \\ 0 & \frac{\tilde{d}}{d(k)-d_1}\widetilde{R}_2 & 0 \\ 0 & 0 & \frac{\tilde{d}}{d_2-d(k)}\widetilde{R}_2 \end{bmatrix}\Pi\xi(k) \\
\leq& \xi^T(k)\left\{\rho_3^T(d_1^2R_1+\tilde{d}^2R_2)\rho_3-\Pi^T\Phi\Pi\right\}\xi(k).
\end{aligned} \tag{5.12}$$

Notice that when $d(k)=d_1$, we have $\xi^T(k)\begin{bmatrix} E(e_2-e_3) & Ee_2+Ee_3-e_6 \end{bmatrix} =0$, so the inequality (5.12) still stands. So does it when $d(k)=d_2$.

In addition, it is clear that

$$2x^T(k)WR^TEx(k+1)\equiv 0. \tag{5.13}$$

According to the inequalities from (5.10) to (5.13), we can obtain

$$\begin{aligned}
\Delta V(k) &= \Delta V_1(k)+\Delta V_2(k)+\Delta V_3(k)+2x^T(k)WR^TEx(k+1) \\
&\leq \xi^T(k)(\Xi_1+\mathrm{sym}(\rho_4^TP\phi_4(k)))\xi(k).
\end{aligned}$$

Define

$$J(k)=z^T(k)\mathcal{Q}z(k)+2z^T(k)\mathcal{S}\omega(k)+\omega^T(k)\mathcal{R}\omega(k).$$

Then

$$\Delta V(k)-J(k)\leq \xi^T(k)(\Xi+\mathrm{sym}(\rho_4^TP\phi_4(k)))\xi(k).$$

Due to the convexity of $\Xi+\mathrm{sym}(\rho_4^TP\phi_4(k))$ with respect to $d(k)$, conditions (5.6) and (5.7) can guarantee $\Xi+\mathrm{sym}(\rho_4^TP\phi_4(k)<0$, so $\Delta V(k)-J(k)<0$, which implies the strict $(\mathcal{Q},\mathcal{S},\mathcal{R})$-$\alpha$-dissipativity. When $\omega(k)\equiv 0$, from $\Delta V(k)-J(k)\leq 0$ and $\mathcal{Q}\leq 0$, we have $\Delta V(k)\leq -c\left\|x(k)\right\|^2$ for some

$c > 0$, so the discrete singular system (5.1) is stable. Then, the singular system (5.1) is admissible and strictly $(\mathcal{Q}, \mathcal{S}, \mathcal{R})$-$\alpha$-dissipative in the sense of Definition 4.1, Definition 4.2, and Definition 5.4. Then the proof is completed. □

To facilitate comparison of our result with the existing ones, considering the stability of the nominal system (5.4) and following the same lines as that in proof of Theorem 5.5, it is easy to acquire the admissibility criterion of the singular system (5.4).

Corollary 5.6. *For given integers d_1 and d_2 satisfying $0 < d_1 \leq d(k) \leq d_2$, system (5.4) is admissible if there exist matrices $P > 0$, $Q_i > 0$ $(i = 1, 2, 3)$, $R_j > 0$ $(j = 1, 2)$, and matrices X, W such that the following set of LMIs hold:*

$$\begin{aligned}
&\overline{\Xi}_1 + \mathrm{sym}(\overline{\rho}_4^T P \overline{\phi}_{41}) < 0, \\
&\overline{\Xi}_1 + \mathrm{sym}(\overline{\rho}_4^T P \overline{\phi}_{42}) < 0,
\end{aligned}$$

where $R \in \mathbb{R}^{n\times(n-r)}$ is any matrix with full column rank and satisfies $E^T R = 0$, and

$$\begin{aligned}
\overline{\Xi}_1 &= \overline{\rho}_4^T P \overline{\rho}_4 + \overline{e}_1^T (Q_1 + (\widetilde{d}+1) Q_3) \overline{e}_1 + \overline{e}_2^T(-Q_1 + Q_2)\overline{e}_2 - \overline{e}_3^T Q_3 \overline{e}_3 \\
&\quad - \overline{e}_4^T Q_2 \overline{e}_4 + \overline{\rho}_3^T (d_1^2 R_1 + \widetilde{d}^2 R_2)\overline{\rho}_3 - \Pi^T \Phi \Pi + \mathrm{sym}(\overline{e}_1^T W R^T \overline{\rho}_2) \\
\overline{\rho}_1 &= \frac{1}{2}\left(d_1 \overline{e}_5 - E\overline{e}_1 - E\overline{e}_2\right), \ \overline{\rho}_2 = A\overline{e}_1 + A_d \overline{e}_3, \ \overline{\rho}_3 = \overline{\rho}_2 - E\overline{e}_1 \\
\overline{\rho}_4^T &= \begin{bmatrix} \overline{\rho}_3^T & (E\overline{e}_1 - E\overline{e}_2)^T & (E\overline{e}_2 - E\overline{e}_4)^T \end{bmatrix} \\
\overline{\phi}_{41}^T &= \begin{bmatrix} (E\overline{e}_1)^T & \overline{\rho}_1^T & \frac{1}{2}(\widetilde{d}\overline{e}_7 - E\overline{e}_2 - 2E\overline{e}_3 - E\overline{e}_4)^T \end{bmatrix} \\
\overline{\phi}_{42}^T &= \begin{bmatrix} (E\overline{e}_1)^T & \overline{\rho}_1^T & \frac{1}{2}(\widetilde{d}\overline{e}_6 - E\overline{e}_2 - 2E\overline{e}_3 - E\overline{e}_4)^T \end{bmatrix} \\
\overline{e}_i &= \begin{bmatrix} 0_{n\times(i-1)n} & I_n & 0_{n\times(7-i)n} \end{bmatrix}, \ i = 1, 2, ..., 7
\end{aligned}$$

5.1.3 Numerical examples

Example 5.1. Considering the singular system (5.4) with the following parameters:

$$E = \begin{bmatrix} 3.5 & 0 \\ 0 & 0 \end{bmatrix}, \ A = \begin{bmatrix} 2.3 & 0 \\ 0 & -3 \end{bmatrix}, \ A_d = \begin{bmatrix} -1.3 & 1.5 \\ 0 & 0.5 \end{bmatrix}.$$

To guarantee the admissibility of the above system, we calculate the allowable maximum of $d(k)$. Set $d(k)$ as a constant time delay d. By using z-transformation method, one can get the characteristic polynomial as

$\det(zE - A - A_d z^{-d}) = (3.5z - 2.3 + 1.3z^{-d})(3 - 0.5z^{-d})$. The characteristic values must satisfy $|z| < 1$ to ensure the above system admissible, which yields that the maximum value of d is 15. The allowable maximum values of d_2 that guarantees system (5.4) to be admissible, with various d_1 by applying Corollary 5.6 and some other methods are shown in Table 5.1. From the results, it is clear to see that the criterion in this section has reduced conservatism significantly for the admissibility condition of Example 5.1.

Example 5.2. In this example, the generality of the dissipativity is demonstrated, which unifies H_∞ performance and passivity performance. Consider singular system (5.1) with the parameters

$$E = \begin{bmatrix} 1 & 0 \\ 0 & 0 \end{bmatrix}, \ A = \begin{bmatrix} 0.8 & 0 \\ 0.05 & 0.9 \end{bmatrix}, \ A_d = \begin{bmatrix} -0.1 & 0 \\ -0.2 & -0.1 \end{bmatrix},$$

$$B_\omega = \begin{bmatrix} 0.1 \\ 0.5 \end{bmatrix}, \ L = \begin{bmatrix} -0.1 & 0.4 \end{bmatrix}, \ L_d = \begin{bmatrix} -0.2 & -0.1 \end{bmatrix}, \ G_\omega = 0.1.$$

- H_∞ performance: Let $\mathcal{Q} = -I_q$, $\mathcal{S} = 0_{q\times l}$, $\mathcal{R} - \alpha I_l = \gamma^2 I_l$. The dissipativity reduces to standard H_∞ performance. By adopting Theorem 5.5, the allowable minimum H_∞ disturbance attenuation γ can be acquired for stationary $d_1 = 1$ and varied d_2. The relationship between γ and d_2 is illustrated in Table 5.2, which demonstrates that the minimal value of γ increases as the value of d_2 grows. Compared with Corollary 1 of paper [252] and Corollary 1 in [27], it is obvious that the method proposed in this paper has a better performance to bear the tolerance according to the lower disturbance attenuation level γ.
- Passivity: When $q = l$ and $\mathcal{Q} = 0_q$, $\mathcal{S} = I_q$, $\mathcal{R} - \alpha I_q = \gamma I_q$, it becomes passivity performance. Setting $d_1 = 3$, the different values of minimum γ are shown in Table 5.3, according to Theorem 5.5, with various d_2. It is clear to see that the minimum value of γ also becomes larger when d_2 increases.
- Dissipativity: Set

$$\mathcal{Q} = -0.4, \ \mathcal{S} = 1.1, \ \mathcal{R} = 3.$$

If the LMIs (5.6) and (5.7) are feasible, singular system (5.1) is admissible and strictly $(\mathcal{Q}, \mathcal{S}, \mathcal{R})$-$\alpha$-dissipative, where α denotes the level of dissipativity. A higher α means that singular system (5.1) has a better performance to tolerate uncertainties and disturbances. From Table 5.4,

Table 5.1 Comparison of the allowable maximum values of d_2 for various d_1.

d_1	1	2	3	4	5	6	7	8
Theorem 2 [33]	3	3	4	infeasible	infeasible	infeasible	infeasible	infeasible
Theorem 2 [36]	3	3	4	5	infeasible	infeasible	infeasible	infeasible
Theorem 1 [188]	3	4	4	5	6	7	infeasible	infeasible
Theorem 1 [46]	4	4	5	5	6	7	infeasible	infeasible
Corollary 5.6	10	7	5	6	6	7	8	9

Table 5.2 Allowable minimum γ for different d_2.

d_2	3	6	9	12	15
Corollary 4.1 [252]	0.1945	0.2614	0.4465	3.763	infeasible
Corollary 1 [27]	0.1945	0.2379	0.3044	0.4335	0.8182
Theorem 5.5	0.1907	0.2313	0.2628	0.2865	0.3107

Table 5.3 Different minimal values of γ for different d_2.

d_2	6	11	16	21	26
Theorem 5.5	0.3892	0.4304	0.4569	0.4754	0.4904

Table 5.4 Allowable maximum dissipativity level α for different d_2.

d_2	5	10	15	20	25
Theorem 5.5	2.558	2.508	2.476	2.454	2.433

we can see that for fixed $d_1 = 2$, when the upper bound of delay d_2 becomes larger, the dissipativity level becomes smaller.

Example 5.3. In this example, Corollary 5.6 is applied to check the admissibility of a class of economic system, which can be modeled by singular system [120]. We use the Leontief model given in [4],

$$Ex(k+1) = (I - U + E)x(k) - Wu(k).$$

The physical meaning of $x(k)$ is the production level in the sectors at time k. $Ux(k)$ represents the amount required as direct input for the current production, and $u(k)$ denotes the amount of production for the current demand, respectively. The element e_{ij} of matrix E denotes the amount of stock of commodity i, as a capital good, that sector j must have on hand for each unit of production. Since not every sector produces significant capital goods, it is common for some rows of the matrix E to contain only zero elements such that matrix E is singular. Therefore the Leontief model is a singular system. By using the system parameters as those in [4], the singular system is given as follows:

$$\begin{aligned} \begin{bmatrix} 1 & 1 \\ 0 & 0 \end{bmatrix} x(k+1) &= \begin{bmatrix} -0.05 & 0.6 \\ -0.7695 & -1.83 \end{bmatrix} x(k) \\ &+ \begin{bmatrix} 0.1676 & 0.117 \\ -0.5112 & -0.3569 \end{bmatrix} x(k-d(k)). \end{aligned}$$

The lower bound and upper bound of time-delay $d(k)$ are chosen as $d_1 = 1$ and $d_2 = 4$, respectively. By solving the LMIs in Corollary 5.6, the condition is feasible, which implies that the singular system is admissible under the variation of delay $d(k)$ between 1 and 4.

5.1.4 Conclusion

The problem of admissibility and strict $(\mathcal{Q}, \mathcal{S}, \mathcal{R})$-$\alpha$-dissipativity analysis for discrete singular system with time-varying delay has been studied in this section. Through combining a discrete Wirtinger inequality and the improved reciprocally convex combination approach, a sufficient criterion in terms of strict LMIs has been proposed for the admissibility and dissipativity analysis of the considered systems. Compared with existing results, the results of this section show less conservative performance. The obtained results also address the analysis of H_∞ performance, passivity, and strict $(\mathcal{Q}, \mathcal{S}, \mathcal{R})$-$\alpha$-dissipativity of the discrete singular system with time-varying delay in a unified framework. The effectiveness and advantages of the proposed approach have been shown by numerical examples. The dissipative controller design of the considered system will be addressed in our future work.

5.2 Dissipativity analysis and dissipative control of singular time-delay systems

In this section, the problem of delay-dependent α-dissipative control is investigated for continuous-time singular systems with time delay. Using the delay-partitioning technique, sufficient conditions are established in terms of LMIs, which guarantees a singular system to be admissible and strictly (Q, S, R)-α-dissipative. Based on the criteria, a design algorithm for a state feedback controller is proposed. In addition to delay dependence, the obtained results are also dependent on the parameter α. Moreover, the results of H_∞ control and passive control are also derived from the dissipative control results. The results developed in this section are less conservative than existing ones in the literature, which are illustrated by several examples.

5.2.1 Problem formulation

Consider a class of linear continuous-time singular systems described by

$$\begin{cases} E\dot{x}(t) &= Ax(t) + A_d x(t-h) + Bu(t) + B_w w(t) \\ z(t) &= Cx(t) + C_d x(t-h) + Du(t) + D_w w(t) \\ x(t) &= \varphi(t), \forall t \in [-h, 0], \end{cases} \tag{5.14}$$

where $x(t) \in \mathbb{R}^n$ is the state vector; $u(t) \in \mathbb{R}^p$ is the control input; $w(t) \in \mathbb{R}^l$ represents a set of exogenous inputs, which includes disturbances to be rejected, and $z(k) \in \mathbb{R}^q$ is the controlled output; h represents the system delay satisfying $0 < h \leq \bar{h}$; A, A_d, B, B_w, C, C_d, D, and D_w denote constant matrices with appropriate dimensions. In contrast with standard linear systems with $E = I$, the matrix $E \in \mathbb{R}^{n \times n}$ has $\mathrm{rank}(E) = p \leq n$.

Consider the following memoryless state-feedback controller for system (5.14):

$$u(t) = Kx(t). \tag{5.15}$$

By applying controller (5.15) to system (5.14), the closed-loop system can be described by

$$\begin{cases} E\dot{x}(t) &= (A + BK)x(t) + A_d x(t-h) + B_w w(t) \\ z(t) &= (C + DK)x(t) + C_d x(t-h) + D_w w(t) \\ x(t) &= \varphi(t), \forall t \in [-h, 0]. \end{cases} \tag{5.16}$$

Before moving on, we give some definitions and lemmas concerning the following nominal unforced counterpart of the system in (5.14) with $w(t) = 0$:

$$\begin{cases} E\dot{x}(t) &= Ax(t) + A_d x(t-h) \\ z(t) &= Cx(t) + C_d x(t-h) \\ x(t) &= \varphi(t), \forall t \in [-h, 0]. \end{cases} \tag{5.17}$$

Lemma 5.7. [216] *Suppose the pair* (E, A) *is regular and impulse free, then the solution to system (5.17) is impulse free and unique on* $[0, \infty)$.

In view of this, we introduce the following definition for singular delay system (5.17):

Definition 5.8. 1) The singular delay system in (5.17) is said to be regular and impulse free if the pair (E, A) is regular and impulse free.

2) The singular system in (5.17) is said to be asymptotically stable if, for any $\varepsilon > 0$, there exists a scalar $\delta(\varepsilon) > 0$, such that for any compatible initial

conditions x_0 satisfying $\| x_0 \| \leq \delta(\varepsilon)$, the solution $x(t)$ of (5.17) satisfies $\| x(t) \| \leq \varepsilon$ for $t \geq 0$; furthermore, $x(t) \rightarrow 0$, when $t \rightarrow \infty$.

3) The singular system in (5.17) is said to be admissible, or the pair (E, A) is admissible, if the system is regular, impulse free, and asymptotically stable.

Definition 5.9. [101] Given some scalar $\alpha > 0$, matrices Q, R, and S with Q and R real symmetric, systems (5.14) with $u(t) = 0$ is called strictly (Q, S, R)-α-dissipative if, for any $\tau \geq 0$, under zero initial state, the following condition is satisfied:

$$\langle z, Qz \rangle_\tau + 2\langle z, Sw \rangle_\tau + \langle w, Rw \rangle_\tau \geq \alpha \langle w, w \rangle_\tau. \tag{5.18}$$

As in [202], we assume that $Q \leq 0$. Then we can get

$$-Q = Q_-^T Q_-$$

for some Q_-.

Our main objective is to study the problem of α-dissipative control for singular system (5.14). More specially, we are concerned with the following two problems:

1. Establish a sufficient condition in terms of LMIs such that singular system (5.14) is admissible and strictly (Q, S, R)-α-dissipative.
2. Design a state-feedback controller in the form of (5.15) such that the closed-loop systems in (5.16) is admissible and strictly (Q, S, R)-α-dissipative.

5.2.2 Dissipative analysis

In this section, the improved sufficient condition is derived firstly by employing the delay-partitioning technology, which guarantees that the unforced system of (5.14) is admissible and strictly (Q, S, R)-α-dissipative.

The result of strict dissipative analysis for system (5.14) is presented in the following theorem:

Theorem 5.10. *Let scalar $\alpha > 0$, matrices Q, S, and R be given with Q and R real symmetric and $Q \leq 0$. Then the system in (5.14) with $u(t) = 0$ is admissible and strictly (Q, S, R)-α-dissipative if there exist matrices P, Y, U, $V > 0$, and $W > 0$ such that the following set of LMIs hold:*

$$E^T P = P^T E \geq 0, \tag{5.19}$$

$$\Omega < 0, \tag{5.20}$$

where

$$\begin{aligned}
\Omega &= \text{sym}(W_X^T P^T W_D + \frac{h}{m} W_X^T Y^T W_V - W_X^T Y^T W_E + \frac{h}{m} W_d^T U^T W_V \\
&\quad - W_d^T U^T W_E - W_S^T S W_Q) + \frac{h}{m} W_D^T V W_D - \frac{h}{m} W_V^T V W_V \\
&\quad + W_W^T \bar{W} W_W - W_Q^T Q W_Q - W_S^T (R - \alpha I) W_S \\
W_X &= \begin{bmatrix} I_n & 0_{n,(m+1)n+l} \end{bmatrix}, \ W_E = \begin{bmatrix} E & -E & 0_{n,mn+l} \end{bmatrix}, \\
W_Q &= \begin{bmatrix} C & 0_{q,(m-1)n} & C_d & 0_{q,n} & D_w \end{bmatrix}, \\
W_D &= \begin{bmatrix} A & 0_{n,(m-1)n} & A_d & 0_{n,n} & B_w \end{bmatrix}, \\
W_S &= \begin{bmatrix} 0_{l,(m+2)n} & I_l \end{bmatrix}, \ W_W = \left[\begin{array}{ccc} \multicolumn{3}{c}{I_{mn} \quad 0_{mn,2n+l}} \\ \hline 0_{mn,n} & I_{mn} & 0_{mn,n+l} \end{array} \right], \\
W_d &= \begin{bmatrix} 0_{n,n} & I_n & 0_{n,mn+l} \end{bmatrix}, \ W_V = \begin{bmatrix} 0_{n,(m+1)n} & I_n & 0_{n,l} \end{bmatrix}, \\
\bar{W} &= \begin{bmatrix} W & 0_{mn,mn} \\ \star & -W \end{bmatrix}.
\end{aligned}$$

Proof. The proof contains two parts. The admissibility of the singular system is tackled in the first part, and the second part deals with the α-dissipative analysis of the singular system. For the first part, we consider system (5.17) and choose two nonsingular matrices M and N such that

$$E = M \begin{bmatrix} I & 0 \\ \star & 0 \end{bmatrix} N, \ A = M \begin{bmatrix} \bar{A}_1 & \bar{A}_2 \\ \bar{A}_3 & \bar{A}_4 \end{bmatrix} N.$$

Denote

$$M^T P N^{-1} = \begin{bmatrix} \bar{P}_1 & \bar{P}_2 \\ \bar{P}_3 & \bar{P}_4 \end{bmatrix}, \ M^T Y N^{-1} = \begin{bmatrix} \bar{Y}_1 & \bar{Y}_2 \\ \bar{Y}_3 & \bar{Y}_4 \end{bmatrix}.$$

Furthermore, the following inequality is obtained from (5.20):

$$\begin{bmatrix} \text{sym}(A^T P - E^T Y) + W & \bullet & \bullet & \bullet \\ \star & \bullet & \bullet & \bullet \\ \star & \star & \bullet & \bullet \\ \star & \star & \star & \bullet \end{bmatrix}$$

$$
+\frac{h}{m}\begin{bmatrix} A^T \\ 0_{(m-1)\times n,n} \\ A_d^T \\ B_w \end{bmatrix} V\begin{bmatrix} A & 0_{n,(m-1)\times n} & A_d & B_w \end{bmatrix}
-\begin{bmatrix} C^T \\ 0_{(m-1)\times n,q} \\ C_d^T \\ D_w \end{bmatrix} Q\begin{bmatrix} C & 0_{q,(m-1)\times n} & C_d & D_w \end{bmatrix} < 0. \tag{5.21}
$$

Due to $-Q \geq 0$ and $V > 0$, $\text{sym}(A^T P - E^T Y) + W < 0$ is derived from (5.21). Then following a similar analysis method of [214], we can obtain $\bar{A}_4$ is nonsingular. Therefore following from [25] and Definition 5.8, we conclude that the system in (5.17) is regular and impulse-free. For the stability property, let us choose the following Lyapunov functional:

$$
\begin{aligned}
V(x(t)) &= x(t)^T P^T E x(t) + \int_{t-\frac{h}{m}}^{t} \Upsilon(s)^T W \Upsilon(s) ds \\
&\quad + \int_{-\frac{h}{m}}^{0} \int_{t+\theta}^{t} (E\dot{x}(s))^T V E\dot{x}(s) ds d\theta,
\end{aligned} \tag{5.22}
$$

where

$$
\Upsilon(t) = \begin{bmatrix} x(t) \\ x(t-\frac{h}{m}) \\ \vdots \\ x(t-\frac{(m-1)h}{m}) \end{bmatrix}.
$$

Evaluating the derivative of $V(x(t))$ along the solutions of system (5.17), we obtain

$$
\begin{aligned}
\dot{V}(x(t)) &= 2x(t)^T P^T E\dot{x}(t) + \frac{h}{m}(E\dot{x}(t))^T V(E\dot{x}(t)) \\
&\quad + \Upsilon(t)^T W \Upsilon(t) - \Upsilon(t-\frac{h}{m})^T W \Upsilon(t-\frac{h}{m}) \\
&\quad - \int_{t-\frac{h}{m}}^{t} (E\dot{x}(s))^T V(E\dot{x}(s)) ds \\
&\quad + 2x(t)^T Y^T E[\int_{t-\frac{h}{m}}^{t} \dot{x}(s) ds - x(t) + x(t-\frac{h}{m})] \\
&\quad + 2x(t-\frac{h}{m})^T U^T E[\int_{t-\frac{h}{m}}^{t} \dot{x}(s) ds - x(t) + x(t-\frac{h}{m})] \\
&= \frac{m}{h}\int_{t-\frac{h}{m}}^{t} \zeta_1(t,s)^T \Omega_1 \zeta_1(t,s) ds,
\end{aligned}
$$

where

$$\begin{aligned}
\Omega_1 &= \operatorname{sym}(W_{X1}^T P^T W_{D1} + \frac{h}{m} W_{X1}^T Y^T W_{V1} - W_{X1}^T Y^T W_{E1} \\
&\quad + \frac{h}{m} W_{d1}^T U^T W_{V1} - W_{d1}^T U^T W_{E1}) + \frac{h}{m} W_{D1}^T V W_{D1} \\
&\quad - \frac{h}{m} W_{V1}^T V W_{V1} + W_{W1}^T \bar{W} W_{W1}, \\
W_{X1} &= \begin{bmatrix} I_n & 0_{n,(m+1)n} \end{bmatrix}, \ W_{E1} = \begin{bmatrix} E & -E & 0_{n,mn} \end{bmatrix}, \\
W_{D1} &= \begin{bmatrix} A & 0_{n,(m-1)n} & A_d & 0_{n,n} \end{bmatrix}, \ W_{W1} = \left[\begin{array}{ccc} \multicolumn{3}{c}{I_{mn} \quad 0_{mn,2n}} \\ \hline 0_{mn,n} & I_{mn} & 0_{mn,n} \end{array} \right], \\
W_{d1} &= \begin{bmatrix} 0_{n,n} & I_n & 0_{n,mn} \end{bmatrix}, \ W_{V1} = \begin{bmatrix} 0_{n,(m+1)n} & I_n \end{bmatrix}, \ \zeta_1(t,s) = \begin{bmatrix} \Upsilon(t) \\ x(t-h) \\ E\dot{x}(s) \end{bmatrix}.
\end{aligned}$$

It can be seen that $\Omega < 0$ implies $\Omega_1 < 0$. Then

$$\dot{V}(x(t)) < 0 \tag{5.23}$$

is derived.

On the other hand, from (5.20), we have

$$\begin{aligned}
\Omega_2 &= \begin{bmatrix} \Lambda_{11} & \Lambda_{12} & 0_{n,(m-2)n} & P^T A_d \\ \star & \Lambda_{22} & 0_{n,n} & 0_{n,n} \\ \star & \star & 0_{(m-2)n,(m-2)n} & 0_{(m-2)n,n} \\ \star & \star & \star & 0_{n,n} \end{bmatrix} + \begin{bmatrix} W & 0_{mn,n} \\ \star & 0_{n,n} \end{bmatrix} \\
&\quad - \begin{bmatrix} 0_{n,n} & 0_{n,mn} \\ \star & W \end{bmatrix} < 0,
\end{aligned} \tag{5.24}$$

where

$$\begin{aligned}
\Lambda_{11} &= P^T A + A^T P - Y^T E - E^T Y, \\
\Lambda_{12} &= Y^T E - E^T U, \\
\Lambda_{22} &= U^T E + E^T U.
\end{aligned}$$

Pre- and postmultiplying (5.24) by $[I, \ldots, I] \in \mathbb{R}^{n \times mn}$ and its transpose, respectively, the following inequality is obtained

$$P^T (A + A_d) + (A + A_d)^T P < 0,$$

which implies P is nonsingular. Due to the regularity and the absence of impulses of the pair (E, A), there always exist two nonsingular matrices $\tilde{M}$ and $\tilde{N}$ such that

$$E = \tilde{M} \begin{bmatrix} I_p & 0_{p,n-p} \\ 0 & 0_{n-p,n-p} \end{bmatrix} \tilde{N}, \ A = \tilde{M} \begin{bmatrix} \tilde{A}_1 & 0_{p,n-p} \\ 0 & I_{n-p,n-p} \end{bmatrix} \tilde{N}. \tag{5.25}$$

Define

$$\tilde{P} = \tilde{M}^T P \tilde{N}^{-1} = \begin{bmatrix} \tilde{P}_1 & \tilde{P}_2 \\ \tilde{P}_3 & \tilde{P}_4 \end{bmatrix}, \ \tilde{U} = \tilde{M}^T U \tilde{N}^{-1} = \begin{bmatrix} \tilde{U}_1 & \tilde{U}_2 \\ \tilde{U}_3 & \tilde{U}_4 \end{bmatrix}, \tag{5.26}$$

$$\tilde{Y} = \tilde{M}^T Y \tilde{N}^{-1} = \begin{bmatrix} \tilde{Y}_1 & \tilde{Y}_2 \\ \tilde{Y}_3 & \tilde{Y}_4 \end{bmatrix}, \ \tilde{A}_d = \tilde{M}^{-1} A_d \tilde{N}^{-1} = \begin{bmatrix} \tilde{A}_{d1} & \tilde{A}_{d2} \\ \tilde{A}_{d3} & \tilde{A}_{d4} \end{bmatrix}, \tag{5.27}$$

$$W = \begin{bmatrix} W_{11} & \cdots & W_{1m} \\ * & \ddots & \vdots \\ * & * & W_{mm} \end{bmatrix}, \ \tilde{W}_{jj} = \tilde{N}^{-T} W_{jj} \tilde{N}^{-1} = \begin{bmatrix} \tilde{W}_{jj1} & \tilde{W}_{jj2} \\ \tilde{W}_{jj2}^T & \tilde{W}_{jj4} \end{bmatrix}, \tag{5.28}$$

where the partition is compatible with that of A in (5.25). Then, by (5.19), it can be shown that $\tilde{P}_2 = 0$ and $\tilde{P}_1 > 0$. Now, pre- and postmultiplying (5.24) by

$$\mathrm{diag}(\tilde{N}^{-T}, \ldots, \tilde{N}^{-T}) \in R^{(m+1)n \times (m+1)n}$$

and its transpose, respectively, we obtain

$$\begin{bmatrix} \bullet & \bullet & \bullet & \bullet & \bullet & \bullet & \bullet & \bullet & \bullet & \bullet \\ \star & \tilde{P}_4^T + \tilde{P}_4 + \tilde{W}_{114} & \bullet & \bullet & \bullet & \bullet & \bullet & \bullet & \bullet & \tilde{P}_4^T \tilde{A}_{d4} \\ \star & \star & \bullet & \bullet & \bullet & \bullet & \bullet & \bullet & \bullet & \bullet \\ \star & \star & \star & \tilde{W}_{224} - \tilde{W}_{114} & \bullet & \bullet & \bullet & \bullet & \bullet & \bullet \\ \star & \star & \star & \star & \ddots & \vdots & \vdots & \vdots & \vdots & \vdots \\ \star & \star & \star & \star & \cdots & \ddots & \vdots & \vdots & \vdots & \vdots \\ \star & \star & \star & \star & \cdots & \cdots & \bullet & \bullet & \bullet & \bullet \\ \star & \star & \star & \star & \cdots & \cdots & \star & \tilde{W}_{mm4} - \tilde{W}_{(m-1)(m-1)4} & \bullet & \bullet \\ \star & \star & \star & \star & \cdots & \cdots & \star & \star & \bullet & \bullet \\ \star & \star & \star & \star & \cdots & \cdots & \star & \star & \star & -\tilde{W}_{mm4} \end{bmatrix} < 0, \tag{5.29}$$

which implies

$$\begin{bmatrix} \tilde{P}_4^T + \tilde{P}_4 + \tilde{W}_{114} & \tilde{P}_4^T \tilde{A}_{d4} \\ \star & -\tilde{W}_{mm4} \end{bmatrix} < 0, \tag{5.30}$$

$$\tilde{W}_{224} - \tilde{W}_{114} < 0, \ \tilde{W}_{334} - \tilde{W}_{224} < 0, \ \cdots, \ \tilde{W}_{mm4} - \tilde{W}_{(m-1)(m-1)4} < 0. \tag{5.31}$$

Pre- and postmultiplying (5.30) by $\begin{bmatrix} -\tilde{A}_{d4}^T & I \end{bmatrix}$ and its transpose, we have

$$\tilde{A}_{d4}^T \tilde{W}_{114} \tilde{A}_{d4} - \tilde{W}_{mm4} < 0,$$

that is,

$$\tilde{A}_{d4}^T \tilde{W}_{114} \tilde{A}_{d4} - \tilde{W}_{114} + (\tilde{W}_{114} - \tilde{W}_{mm4}) < 0. \tag{5.32}$$

Since (5.31) implies $\tilde{W}_{114} - \tilde{W}_{mm4} > 0$, and we have from (5.32) that

$$\tilde{A}_{d4}^T \tilde{W}_{114} \tilde{A}_{d4} - \tilde{W}_{114} < 0,$$

therefore

$$\rho(\tilde{A}_{d4}) < 1.$$

Noting this and noting (5.22) and (5.23), and following a similar line to that in the proof of Theorem 1 in [214], we can deduce that the singular time-delay system in (5.17) is stable.

Let us prove that the system is strictly dissipative. For this purpose, considering (5.14) with $u(t) = 0$ and (5.23), we obtain

$$\begin{aligned}
& \dot{V}(x(t)) - z(t)^T Q z(t) - 2w(t)^T S z(t) - w(t)^T (R - \alpha I) w(t) \\
= \ & 2x(t)^T P^T [A x(t) + A_d x(t-h) + B_w w(t)] + \frac{h}{m} (E\dot{x}(t))^T V (E\dot{x}(t)) \\
& + \Upsilon(t)^T W \Upsilon(t) - \Upsilon(t - \frac{h}{m})^T W \Upsilon(t - \frac{h}{m}) \\
& - \int_{t-\frac{h}{m}}^{t} (E\dot{x}(s))^T V (E\dot{x}(s)) ds \\
& + 2x(t)^T Y^T E[\int_{t-\frac{h}{m}}^{t} \dot{x}(s) ds - x(t) + x(t - \frac{h}{m})] \\
& + 2x(t - \frac{h}{m})^T U^T E[\int_{t-\frac{h}{m}}^{t} \dot{x}(s) ds - x(t) + x(t - \frac{h}{m})] \\
& - 2w(t)^T S[C_z x(t) + C_d x(t-h) + D_w w(t)] - w(t)^T (R - \alpha I) w(t) \\
& - [C_z x(t) + C_d x(t-h) + D_w w(t)]^T Q [C_z x(t) + C_d x(t-h) + D_w w(t)] \\
= \ & \frac{m}{h} \int_{t-\frac{h}{m}}^{t} \zeta(t,s) \Omega \zeta(t,s) ds,
\end{aligned} \tag{5.33}$$

where

$$\zeta(t,s) = \begin{bmatrix} \Upsilon(t) \\ x(t-h) \\ E\dot{x}(s) \\ w(t) \end{bmatrix}.$$

Then, since $\Omega < 0$, we have

$$\dot{V}(x(t)) - z(t)^T Qz(t) - 2w(t)^T Sz(t) - w(t)^T (R - \alpha I)w(t) < 0. \qquad (5.34)$$

Integrating (5.34) over the range $[0, \tau]$, we obtain

$$V(x(T)) + \alpha w(t)^T w(t) < \int_0^{\tau} z(t)^T Qz(t) + 2w(t)^T Sz(t) + w(t)^T Rw(t)dt,$$

which implies that (5.18) is satisfied. This completes the proof. □

Remark 5.11. When $h = 0$, the delay term $A_d x(t-h)$ disappears. In this case, if one sets V, W, U, and Y all identically equal to zero, and introduces the identity $w(t)^T Z^T (Ax(t) + B_w w(t)) = 0$ with $Z \in \mathbb{R}^{n\times l}$ satisfying $E^T Z = 0$ in the derivative of the Lyapunov function, then we have the condition in (5.20) replaced with

$$\hat{\Omega} < 0,$$

where

$$\begin{aligned} \hat{\Omega} &= \mathrm{sym}(\breve{W}_X^T P^T \breve{W}_D - \breve{W}_S^T S \breve{W}_Q + \breve{W}_S^T Z^T \breve{W}_D) \\ &\quad - \breve{W}_Q^T Q \breve{W}_Q - \breve{W}_S^T (R - \alpha I) \breve{W}_S, \\ \breve{W}_X &= \begin{bmatrix} I_n & 0_{n,l} \end{bmatrix}, \ \breve{W}_Q = \begin{bmatrix} C & D_w \end{bmatrix}, \\ \breve{W}_D &= \begin{bmatrix} A & B_w \end{bmatrix}, \ \breve{W}_S = \begin{bmatrix} 0_{l,n} & I_l \end{bmatrix}. \end{aligned}$$

When α is chosen to be sufficiently small, the condition obtained corresponds to the necessary and sufficient condition for dissipativity of linear time-invariant singular systems (Corollary 7 in [127]).

Remark 5.12. The main technique utilized in this section is the delay-partitioning idea, in which the time delay is divided into m equal partitions. It can be shown (see Proposition 5.14) that the condition in Theorem 5.10 will reduce its conservatism if the number of partitions increases. When considering the stability analysis problem of state-space delay systems, a proof of reduced conservatism is provided in Proposition 2 in [60]. Benefiting from this approach, many results for other problems have been improved. To mention a few, the stability of recurrent neural networks with time-invariant delay is studied in [30]; the stabilization results for discrete singular delay systems are given in [44]. For time-varying delay set, the problem of delay-dependent stability analysis for discrete-time systems with time-varying delay is investigated in [132]; the stabilization result of Markovian jump systems with time delay is presented in [38].

Remark 5.13. The number of LMI decision variables in Theorem 5.10 is $\frac{n}{2}[(m^2+7)n+m+1]$. This shows that the computational complexity increases quadratically with the partitioning size as well as with the state dimension of the system.

Proposition 5.14. *Suppose that h_m is the maximal delay obtained by Theorem 5.10 for a given number of partitions m. Then we have $h_m \leq h_{m+1}$.*

Proof. From Theorem 5.10, we know that the following inequality holds for given partitioning number m and the delay h_m:

$$
\begin{aligned}
&\dot{V}(x(t)) - z(t)^T Qz(t) - 2w(t)^T Sz(t) - w(t)^T(R-\alpha I)w(t)\\
=&2x(t)^T P^T E\dot{x}(t) + \frac{h_m}{m}(E\dot{x}(t))^T V(E\dot{x}(t)) - \int_{t-\frac{h_m}{m}}^{t} (E\dot{x}(s))^T V(E\dot{x}(s))ds\\
&- z(t)^T Qz(t) - 2w(t)^T Sz(t) - w(t)^T(R-\alpha I)w(t)\\
&+ 2x(t)^T Y^T E[\int_{t-\frac{h_m}{m}}^{t} \dot{x}(s)ds - x(t) + x(t-\frac{h_m}{m})]\\
&+ 2x(t-\frac{h_m}{m})^T U^T E[\int_{t-\frac{h_m}{m}}^{t} \dot{x}(s)ds - x(t) + x(t-\frac{h_m}{m})]\\
&+ \begin{bmatrix} x(t) \\ x(t-\frac{h_m}{m}) \\ \vdots \\ x(t-\frac{(m-1)h_m}{m}) \end{bmatrix}^T W \begin{bmatrix} x(t) \\ x(t-\frac{h_m}{m}) \\ \vdots \\ x(t-\frac{(m-1)h_m}{m}) \end{bmatrix}\\
&- \begin{bmatrix} x(t-\frac{h_m}{m}) \\ x(t-\frac{2h_m}{m}) \\ \vdots \\ x(t-h_m) \end{bmatrix}^T W \begin{bmatrix} x(t-\frac{h_m}{m}) \\ x(t-\frac{2h_m}{m}) \\ \vdots \\ x(t-h_m) \end{bmatrix}\\
&< 0.
\end{aligned}
\tag{5.35}
$$

Take $h = \frac{m}{m+1}h_m < h_m$. Even if we replace h_m in (5.35) by $\frac{m}{m+1}h_m$, we still have

$$
\begin{aligned}
&\dot{V}(x(t)) - z(t)^T Qz(t) - 2w(t)^T Sz(t) - w(t)^T(R-\alpha I)w(t)\\
=&2x(t)^T P^T E\dot{x}(t) + \frac{h_m}{m+1}(E\dot{x}(t))^T V(E\dot{x}(t)) - \int_{t-\frac{h_m}{m+1}}^{t} (E\dot{x}(s))^T V(E\dot{x}(s))ds\\
&- z(t)^T Qz(t) - 2w(t)^T Sz(t) - w(t)^T(R-\alpha I)w(t)\\
&+ 2x(t)^T Y^T E[\int_{t-\frac{h_m}{m+1}}^{t} \dot{x}(s)ds - x(t) + x(t-\frac{h_m}{m+1})]\\
&+ 2x(t-\frac{h_m}{m+1})^T U^T E[\int_{t-\frac{h_m}{m+1}}^{t} \dot{x}(s)ds - x(t) + x(t-\frac{h_m}{m+1})]\\
&+ \begin{bmatrix} x(t) \\ x(t-\frac{h_m}{m+1}) \\ \vdots \\ x(t-\frac{(m-1)h_m}{m+1}) \\ x(t-\frac{mh_m}{m+1}) \end{bmatrix}^T \hat{W} \begin{bmatrix} x(t) \\ x(t-\frac{h_m}{m+1}) \\ \vdots \\ x(t-\frac{(m-1)h_m}{m+1}) \\ x(t-\frac{mh_m}{m+1}) \end{bmatrix}\\
&- \begin{bmatrix} x(t-\frac{h_m}{m+1}) \\ x(t-\frac{2h_m}{m+1}) \\ \vdots \\ x(t-\frac{mh_m}{m+1}) \\ x(t-h_m) \end{bmatrix}^T \hat{W} \begin{bmatrix} x(t-\frac{h_m}{m}) \\ x(t-\frac{2h_m}{m}) \\ \vdots \\ x(t-\frac{mh_m}{m+1}) \\ x(t-h_m) \end{bmatrix}\\
<&0,
\end{aligned}
\tag{5.36}
$$

where

$$
\hat{W} = \begin{bmatrix} W & 0_{mn,n} \\ \star & 0_{n,n} \end{bmatrix}.
$$

Then, we can always find a small enough scalar $\delta > 0$ such that (5.36) still holds when $\hat{W} = \begin{bmatrix} W & 0_{mn,n} \\ \star & \delta I \end{bmatrix}$, which implies that the inequality in (5.20) holds for given partitioning number $m+1$. We can obtain $h_m \leq h_{m+1}$, and this proposition is proved. □

Remark 5.15. The dissipative result obtained in Theorem 5.10 is not only dependent on time-delay, but also dependent on scalar α, which can represent the dissipative margin. Moreover, we can find that α increases with the growth of partitioning size m.

Remark 5.16. As in special cases, the above problem of strictly (Q, S, R)-α-dissipative reduces to the problem of positive realness when $Q=0$, $S=I$ and $R=\alpha I$, whereas for a scalar $\gamma>0$, the H_∞ performance optimizing problem corresponds to $Q=-I$, $S=0$ and $R=(\alpha+\gamma^2)I$.

Based on Remark 5.16, the following corollaries give LMIs conditions for strict positive realness and H_∞ performance analysis:

Corollary 5.17. *The system in (5.14) with $u(t)=0$ is admissible and strictly positive real if there exist matrices P, Y, U, $V>0$, and $W>0$ such that the following set of LMIs hold:*

$$E^T P = P^T E \geq 0, \tag{5.37}$$

$$\Omega_P < 0, \tag{5.38}$$

where

$$\begin{aligned} \Omega_P &= sym(W_X^T P^T W_D + \frac{h}{m} W_X^T Y^T W_V - W_X^T Y^T W_E + \frac{h}{m} W_d^T U^T W_V \\ &- W_d^T U^T W_E - W_S^T W_Q) + \frac{h}{m} W_D^T V W_D - \frac{h}{m} W_V^T V W_V \\ &+ W_W^T \bar{W} W_W. \end{aligned}$$

Corollary 5.18. *For a given $\gamma>0$, the system in (5.14) with $u(t)=0$ is admissible with disturbance rejection of level γ if there exist matrices P, Y, U, $V>0$, and $W>0$ such that the following set of LMIs hold:*

$$E^T P = P^T E \geq 0, \tag{5.39}$$

$$\Omega_\infty < 0, \tag{5.40}$$

where

$$\begin{aligned} \Omega_\infty &= sym(W_X^T P^T W_D + \frac{h}{m} W_X^T Y^T W_V - W_X^T Y^T W_E + \frac{h}{m} W_d^T U^T W_V \\ &\quad - W_d^T U^T W_E) + \frac{h}{m} W_D^T V W_D - \frac{h}{m} W_V^T V W_V \\ &\quad + W_W^T \bar{W} W_W + W_Q^T W_Q - \gamma^2 W_S^T W_S. \end{aligned}$$

The result obtained in Corollary 5.18 coincides with the following corollary as a special case when $m=1$:

Corollary 5.19. *Given scalars $\gamma > 0$ and $h > 0$, then the unforced singular system of (5.14) is admissible with disturbance rejection of level γ if there exist matrices P, Y, U, and matrices $V > 0$ and $W > 0$ such that the following set of LMIs hold:*

$$E^T P = P^T E \geq 0, \tag{5.41}$$

$$\begin{bmatrix} J_1 & J_2 & hY^T & P^T B_w & hA^T V & C^T \\ \star & J_3 & hU^T & 0 & hA_d^T V & C_d^T \\ \star & \star & -hV & 0 & 0 & 0 \\ \star & \star & \star & -\gamma^2 I & hB_d^T V & D_w \\ \star & \star & \star & \star & -hV & 0 \\ \star & \star & \star & \star & \star & -I \end{bmatrix} < 0, \tag{5.42}$$

where

$$\begin{aligned} J_1 &= P^T A + A^T P + W - Y^T E + E^T Y, \\ J_2 &= P^T A_d + Y^T E - E^T U, \\ J_3 &= U^T E + E^T U - W. \end{aligned}$$

Remark 5.20. Corollary 5.19 is equivalent to Theorem 1 in [214] when the system considered therein takes the same form as (5.14). Furthermore, it can be easily shown that Corollary 5.17 will be less conservative than Corollary 5.19 when $m > 1$.

To compare our results with that in [62], the admissibility criterion of singular delay system is obtained from Corollary 5.18 as follows:

Corollary 5.21. *For a given $h > 0$, the singular delay system*

$$E\dot{x}(t) = Ax(t) + A_d x(t-h) \tag{5.43}$$

is admissible if there exist matrices P, Y, U, $V > 0$, and $W > 0$ such that the following set of LMIs hold:

$$E^T P = P^T E \geq 0, \tag{5.44}$$

$$\Omega_A < 0, \tag{5.45}$$

where

$$\Omega_A = sym(W_{X0}^T P^T W_{D0} + \frac{h}{m} W_{X0}^T Y^T W_{V0} - W_{X0}^T Y^T W_{E0}$$

$$+\frac{h}{m}W_{d0}^TU^TW_{V0} - W_{d0}^TU^TW_{E0}) + \frac{h}{m}W_{D0}^TVW_{D0}$$
$$-\frac{h}{m}W_{V0}^TVW_{V0} + W_{W0}^T\bar{W}W_{W0},$$

and W_{X0}, W_{D0}, W_{V0}, W_{E0}, W_{d0}, W_{W0} *are obtained from* W_X, W_D, W_V, W_E, W_d, W_W *defined in Theorem 5.10 by setting* $l=0$.

Remark 5.22. Corollary 5.21 provides a delay-dependent condition for singular delay system (5.43) to be admissible. Although only discrete delay is considered in system (5.43), the obtained condition can be extended to the case of systems with both discrete and distributed delays. For the system

$$E\dot{x}(t) = Ax(t) + A_dx(t-h) + A_h\int_{t-d}^{t}x(s)ds,$$

we can choose a new Lyapunov functional as follows:

$$\begin{aligned} V(x(t)) &= x(t)^TE^TPx(t) + \int_{t-\frac{h}{m}}^{t}\lambda(\theta)^TZ_1\lambda(\theta)d\theta \\ &+\int_{-\frac{h}{m}}^{0}\int_{t+s}^{t}\dot{x}(\theta)^TE^TZ_2E\dot{x}(\theta)d\theta ds \\ &+\int_{t-\frac{d}{r}}^{t}\eta(\theta)^TQ\eta(\theta)d\theta + \int_{-\frac{d}{r}}^{0}\int_{t+\theta}^{t}x(s)^TZx(s)dsd\theta \\ &+\int_{-\frac{d}{r}}^{0}\int_{\theta}^{0}\int_{t+\lambda}^{t}\dot{x}(s)^TE^TRE\dot{x}(s)dsd\lambda d\theta, \end{aligned}$$

where

$$\lambda(\theta) = \begin{bmatrix} x(\theta) \\ x(\theta-\frac{h}{m}) \\ \vdots \\ x(\theta-\frac{(m-1)h}{m}) \end{bmatrix}, \quad \eta(\theta) = \begin{bmatrix} \int_{\theta-\frac{d}{r}}^{\theta}x(s)ds \\ \vdots \\ \int_{\theta-\frac{(r-1)d}{r}}^{\theta-\frac{(r-2)d}{r}}x(s)ds \\ \int_{\theta-d}^{\theta-\frac{(r-1)d}{r}}x(s)ds \end{bmatrix};$$

m and r denote the different number of partitions for discrete delay h and distributed delay d, respectively.

Remark 5.23. For singular system with multiple delays such as

$$E\dot{x}(t) = Ax(t) + \sum_{i=1}^{q}A_{di}x(t-h_i),$$

we can choose a new Lyapunov functional as follows:

$$\begin{aligned} V(x(t)) &= x(t)^T E^T P x(t) + \sum_{i=1}^{q} \int_{t-\frac{h_i}{m_i}}^{t} \lambda_i(\theta)^T Q_i \lambda_i(\theta) d\theta \\ &+ \sum_{i=1}^{q} \int_{-\frac{h_i}{m_i}}^{0} \int_{t+s}^{t} \dot{x}(\theta)^T E^T Z_i E \dot{x}(\theta) d\theta ds, \end{aligned}$$

where

$$\lambda_i(\theta) = \begin{bmatrix} x(\theta) \\ x(\theta - \frac{h_i}{m_i}) \\ \vdots \\ x(\theta - \frac{(m_i-1)h_i}{m_i}) \end{bmatrix},$$

and m_i denotes the different number of partitions for different delays h_i, $i = 1, \ldots, q$.

5.2.3 State-feedback dissipative control

In this section, attention will be devoted to design the state-feedback controller in the form of (5.15) such that closed-loop system (5.16) is admissible and strictly (Q, S, R)-α-dissipative. Based on the results of Theorem 5.10, the dissipative stabilization results for singular system (5.14) is presented in the following theorem:

Theorem 5.24. *Let scalar $\alpha > 0$, matrices Q, S, and R be given with Q and R real symmetric and $Q \leq 0$. Then the system in (5.14) is admissible and strictly (Q, S, R)-α-dissipative if there exist matrices $\mathcal{P}$, $\mathcal{Z}$, $\mathcal{U}$, $\mathcal{Y}$, $\mathcal{V} > 0$, and $\mathcal{W} > 0$ such that the following set of LMIs hold:*

$$E\mathcal{P} = \mathcal{P}^T E^T \geq 0, \tag{5.46}$$

$$\begin{bmatrix} \tilde{\Omega} & \frac{h}{m}\tilde{W}_D^T & \tilde{W}_Q^T Q_-^T \\ \star & -\frac{h}{m}\mathcal{V} & 0 \\ \star & \star & -I \end{bmatrix} < 0, \tag{5.47}$$

where

$$\begin{aligned} \tilde{\Omega} &= sym(W_X^T \tilde{W}_D + \frac{h}{m} W_X^T \mathcal{Y}^T W_V - W_X^T \mathcal{Y}^T \tilde{W}_E + \frac{h}{m} W_d^T \mathcal{U}^T W_V \\ &\quad - W_d^T \mathcal{U}^T \tilde{W}_E - W_S^T S \tilde{W}_Q) - \frac{h}{m} W_V^T (\mathcal{P} + \mathcal{P}^T - \mathcal{V}) W_V \\ &\quad + W_W^T \tilde{W} W_W - W_S^T (R - \alpha I) W_S, \end{aligned}$$

$$\begin{aligned}
\tilde{W}_Q &= \begin{bmatrix} C\mathcal{P}+D\mathcal{Z} & 0_{q,(m-1)n} & C_d\mathcal{P} & 0_{q,n} & D_w \end{bmatrix}, \\
\tilde{W}_E &= \begin{bmatrix} E^T & -E^T & 0_{n,mn+l} \end{bmatrix}, \\
\tilde{W}_D &= \begin{bmatrix} A\mathcal{P}+B\mathcal{Z} & 0_{n,(m-1)n} & A_d\mathcal{P} & 0_{n,n} & B_w \end{bmatrix}, \; \tilde{W} = \begin{bmatrix} \mathcal{W} & 0_{mn,mn} \\ \star & -\mathcal{W} \end{bmatrix}.
\end{aligned}$$

When the above conditions are satisfied, an admissible and (Q, S, R)-α-dissipative controller is given by

$$K = \mathcal{Z}\mathcal{P}^{-1}. \tag{5.48}$$

Proof. Substitute A and C in (5.20) with $A+BK$ and $C+DK$, respectively. Then applying the Schur complement lemma, we have

$$\begin{bmatrix} \bar{\Omega} & \frac{h}{m}\bar{W}_D^T V \\ \star & -\frac{h}{m}V \end{bmatrix} < 0, \tag{5.49}$$

where

$$\begin{aligned}
\bar{\Omega} &= \mathrm{sym}(W_X^T P^T \bar{W}_D + \frac{h}{m} W_X^T Y^T W_V - W_X^T Y^T W_E \\
&\quad + \frac{h}{m} W_d^T U^T W_V - W_d^T U^T W_E - W_S^T S \bar{W}_Q) - \frac{h}{m} W_V^T V W_V \\
&\quad + W_W^T \bar{W} W_W - \bar{W}_Q^T Q \bar{W}_Q - W_S^T (R-\alpha I) W_S \\
\bar{W}_D &= \begin{bmatrix} A+BK & 0_{n,(m-1)n} & A_d & 0_{n,n} & B_w \end{bmatrix}, \\
\bar{W}_Q &= \begin{bmatrix} C+DK & 0_{q,(m-1)n} & C_d & 0_{q,n} & D_w \end{bmatrix}.
\end{aligned}$$

Then we introduce the new variable $\mathcal{P} = P^{-1}$, $\mathcal{V} = V^{-1}$, and let

$$\begin{aligned}
T_1 &= \mathrm{diag}(\mathcal{P}, \mathcal{P}, \mathcal{P}, \ldots, \mathcal{P}) \in \mathbb{R}^{mn \times mn}, \\
T_2 &= \mathrm{diag}(T_1, \mathcal{P}^T, I_l, \mathcal{V}^T) \in \mathbb{R}^{((m+3)n+l) \times ((m+3)n+l)}.
\end{aligned}$$

Pre- and postmultiplying to (5.49) by T_2^T and T_2, yields

$$\begin{bmatrix} \breve{\Omega} - \tilde{W}_Q^T Q \tilde{W}_Q & \frac{h}{m}\tilde{W}_D^T \\ \star & -\frac{h}{m}\mathcal{V} \end{bmatrix} < 0 \tag{5.50}$$

by denoting $\mathcal{Z} = K\mathcal{P}$, $\mathcal{U} = \mathcal{P}U\mathcal{P}$, $\mathcal{Y} = \mathcal{P}Y\mathcal{P}$ and $\mathcal{W} = T_1^T W T_1$, where

$$\breve{\Omega} = \mathrm{sym}(W_X^T \tilde{W}_D + \frac{h}{m} W_X^T \mathcal{Y}^T W_V - W_X^T \mathcal{Y}^T W_E + \frac{h}{m} W_d^T \mathcal{U}^T W_V$$

$$-W_d^T\mathcal{U}^T W_E - W_S^T S\tilde{W}_Q) - \frac{h}{m}W_V^T\mathcal{P}V\mathcal{P}^T W_V$$
$$+W_W^T\tilde{W}W_W - W_S^T(R-\alpha I)W_S.$$

Note

$$-\mathcal{P}V\mathcal{P}^T \leq \mathcal{V} - \mathcal{P} - \mathcal{P}^T = -(\mathcal{P}+\mathcal{P}^T - \mathcal{V}).$$

Therefore (5.50) can be derived from

$$\begin{bmatrix} \tilde{\Omega} - \tilde{W}_Q^T Q \tilde{W}_Q & \frac{h}{m}\tilde{W}_D^T \\ \star & -\frac{h}{m}\mathcal{V} \end{bmatrix} < 0. \tag{5.51}$$

By using the Schur complement lemma, (5.51) equals to (5.47). The LMI in (5.19) under the congruent transformation $\mathcal{P}^T$ becomes (5.46). This completes the proof. □

Similar to Corollary 5.17 and Corollary 5.18, based on Theorem 5.24, the controller design methods for strict positive realness and H_∞ control are given in Corollary 5.25 and Corollary 5.26, respectively.

Corollary 5.25. *The system in (5.14) is admissible and strict positive realness if there exist matrices $\mathcal{P}$, $\mathcal{Z}$, $\mathcal{U}$, $\mathcal{Y}$, $\mathcal{V} > 0$, and $\mathcal{W} > 0$ such that the following set of LMIs hold:*

$$E\mathcal{P} = \mathcal{P}^T E^T \geq 0, \tag{5.52}$$

$$\begin{bmatrix} \tilde{\Omega}_P & \frac{h}{m}\tilde{W}_D^T \\ \star & -\frac{h}{m}\mathcal{V} \end{bmatrix} < 0. \tag{5.53}$$

When the above conditions are satisfied, a desired state-feedback controller is given by

$$K = \mathcal{Z}\mathcal{P}^{-1}, \tag{5.54}$$

where

$$\begin{aligned} \tilde{\Omega}_P \; = \; & sym(W_X^T\tilde{W}_D + \frac{h}{m}W_X^T\mathcal{Y}^T W_V - W_X^T\mathcal{Y}^T\tilde{W}_E + \frac{h}{m}W_d^T\mathcal{U}^T W_V \\ & -W_d^T\mathcal{U}^T\tilde{W}_E - W_S^T\tilde{W}_Q) - \frac{h}{m}W_V^T(\mathcal{P}+\mathcal{P}^T - \mathcal{V})W_V \\ & +W_W^T\tilde{W}W_W. \end{aligned}$$

Corollary 5.26. *For a given $\gamma > 0$, the system in (5.14) is admissible with disturbance rejection of level γ if there exist matrices $\mathcal{P}$, $\mathcal{Z}$, $\mathcal{U}$, $\mathcal{Y}$, $\mathcal{V} > 0$, and $\mathcal{W} > 0$ such that the following set of LMIs hold:*

$$E\mathcal{P} = \mathcal{P}^T E^T \geq 0, \tag{5.55}$$

$$\begin{bmatrix} \tilde{\Omega}_\infty & \frac{h}{m}\tilde{W}_D^T & \tilde{W}_Q^T \\ \star & -\frac{h}{m}\mathcal{V} & 0 \\ \star & \star & -I \end{bmatrix} < 0. \tag{5.56}$$

When the above conditions are satisfied, a desired state-feedback controller is given by

$$K = \mathcal{Z}\mathcal{P}^{-1}, \tag{5.57}$$

where

$$\begin{aligned} \tilde{\Omega}_\infty &= sym(W_X^T \tilde{W}_D + \frac{h}{m} W_X^T \mathcal{Y}^T W_V - W_X^T \mathcal{Y}^T \tilde{W}_E \\ &\quad + \frac{h}{m} W_d^T \mathcal{U}^T W_V - W_d^T \mathcal{U}^T \tilde{W}_E) - \frac{h}{m} W_V^T (\mathcal{P} + \mathcal{P}^T - \mathcal{V}) W_V \\ &\quad + W_W^T \tilde{W} W_W - \gamma^2 W_S^T W_S. \end{aligned}$$

Remark 5.27. There are nonstrict inequality constraints in our results, such as $E^T P = P^T E$ in (5.19) and $E\mathcal{P} = \mathcal{P}^T E^T$ in (5.46), which will lead to numerical problems when solving such nonstrict LMIs. For this case, the methods in [192], [216], and [223] can be employed by setting $P = Z_1 E + \Phi_1 H_1$ and $\mathcal{P} = Z_2 E^T + \Phi_2 H_2$, where $Z_1 > 0$, $Z_2 > 0$, $H_1, H_2 \in \mathbb{R}^{(n-p)\times n}$, and $\Phi_1, \Phi_2 \in \mathbb{R}^{n\times(n-p)}$ with $rank(\Phi_1) = rank(\Phi_2) = n - p$ such that $E^T \Phi_1 = 0$ and $E\Phi_2 = 0$.

5.2.4 Illustrative examples

In this section, some examples are provided to illustrate the applicability and reduced conservatism of the proposed approach.

Example 5.4. Consider a singular time-delay system in (5.14) with following parameters:

$$\begin{aligned} E &= \begin{bmatrix} 1 & 1 \\ 0 & 0 \end{bmatrix}, \quad A = \begin{bmatrix} -2 & 0 \\ 0 & -3 \end{bmatrix}, \quad A_d = \begin{bmatrix} -1.5 & 0.5 \\ 0.4 & -0.3 \end{bmatrix}, \\ C &= \begin{bmatrix} -1 & 0.4 \\ 1 & 1 \end{bmatrix}, \quad C_d = 0, \quad B_w = \begin{bmatrix} 1 & 0 \\ 2.5 & 0 \end{bmatrix}, \quad D_w = \begin{bmatrix} 1 & 1.5 \\ 3 & 2 \end{bmatrix}. \end{aligned}$$

Table 5.5 Allowable maximum time-delay h obtained by different methods.

Methods	[143]	[123]	$m=1$	$m=2$	$m=4$	$m=10$
h	delay-independent	infeasible	0.8501	0.9816	1.0152	1.0245

Table 5.6 Allowable maximum scalar α obtained by different methods.

Methods	[123]	$m=1$	$m=2$	$m=4$	$m=10$
α	infeasible	0.3859	0.4462	0.4569	0.4597

The purpose is to find the maximal allowable time-delay h such that the singular system is admissible and strictly (Q, S, R)-α-dissipative. To this end, we choose

$$S=\begin{bmatrix}1.1 & 3\\ 0.5 & 2\end{bmatrix},\quad Q=\begin{bmatrix}-0.4 & 0\\ 0 & -1\end{bmatrix},\quad R=\begin{bmatrix}3 & 0\\ 0 & 1\end{bmatrix},\quad \alpha=0.2.$$

Then, the maximal allowable time-delay h satisfying (5.19) and (5.20) can be calculated by using standard software. Table 5.5 presents the comparison, which shows that the result in Theorem 5.10 is less conservative than those in [143] and [123]. Furthermore, the conservatism is reduced as the number of partitions increases. On the other hand, to illustrate the dissipative results are also dependent on α, we give the different maximal allowed α in Table 5.6 with $h=0.7$ for different partitioning size. From Table 5.6, we can see that the more we partition the time-delay, the larger α is.

Example 5.5. To compare the delay-dependent BRL in Corollary 5.18 with existing results, we consider a singular system described by [214]:

$$\begin{aligned}\begin{bmatrix}1 & 0\\ 0 & 0\end{bmatrix}\dot{x}(t) &= \begin{bmatrix}-0.3012 & 0.1257\\ 0.2351 & -2.5652\end{bmatrix}x(t)\\ &\quad+\begin{bmatrix}-0.5124 & 0.9648\\ 0.1023 & 0.8197\end{bmatrix}x(t-h)+\begin{bmatrix}0.2102\\ -0.8152\end{bmatrix}w(t)\\ z(t) &= \begin{bmatrix}1.2321 & 0.3185\end{bmatrix}x(t).\end{aligned}$$

For a given $\gamma>0$, we can calculate the maximum allowable time-delay h satisfying the LMIs in (5.39) and (5.40) by solving a quasiconvex optimization problem. Some comparison results between [6], [194], [196], [214], [246] and Corollary 5.18 in this secttion are presented in Table 5.7, which illustrates the reduced conservatism of our results.

Table 5.7 Allowable maximum time-delay h obtained by different methods.

γ	2	2.5	3	3.5
[6]	infeasible	infeasible	infeasible	infeasible
[194]	2.6921	2.9586	3.1710	3.3443
[196]	2.5920	2.9090	3.1522	3.3408
[214]	2.5920	2.9090	3.1522	3.3408
[246]	2.2465	2.5736	2.8348	3.0505
Corollary 5.18				
$m=1$	2.5920	2.9090	3.1522	3.3408
$m=2$	3.0673	3.3954	3.6582	3.8737
$m=4$	3.1635	3.5083	3.7847	3.8737

Example 5.6. To illustrate the applicability of our controller design method, we consider the system in (5.14) with parameters as follows:

$$
\begin{aligned}
E &= \begin{bmatrix} 2 & 1 \\ 2 & 1 \end{bmatrix}, \quad A=\begin{bmatrix} 0 & 1 \\ -1 & 1 \end{bmatrix}, \quad A_d=\begin{bmatrix} 0.5 & 0 \\ 0 & 0.5 \end{bmatrix}, \\
B_w &= \begin{bmatrix} 1 \\ 0.5 \end{bmatrix}, \quad B=\begin{bmatrix} -0.5 \\ 0.5 \end{bmatrix}, \quad C=\begin{bmatrix} 0.8 & 0.8 \end{bmatrix}, \\
D_w &= 0.1, \quad D=0.1, \quad C_d=\begin{bmatrix} -0.2 & 0.5 \end{bmatrix}.
\end{aligned}
$$

The purpose is to design a state feedback controller such that the closed-loop singular system in (5.16) is admissible and strictly (Q, S, R)-α-dissipative. We select

$$S=1, \quad Q=-1, \quad R=10.$$

For given time-delay $h=1$, a state feedback controller can be calculated by solving the LMIs in (5.46) and (5.47). To illustrate the reduced conservatism of our result, Table 5.8 is given to compare the value of α for different m.

Moreover, Fig. 5.1 and Fig. 5.2 give the simulation results of two states with $w(t)=e^{-t}\sin t$. Fig. 5.1 depicts the state response for open-loop system, whereas Fig. 5.2 give the state trajectories for the closed-loop system with $K=\begin{bmatrix} 1.4248 & 1.4248 \end{bmatrix}$. From Fig. 5.1 and Fig. 5.2, we can see that the open-loop system is unstable and the closed-loop system is stable.

Example 5.7. To compare Corollary 5.21 obtained by employing the delay-partitioning approach with Theorem 8 in [62], obtained by using

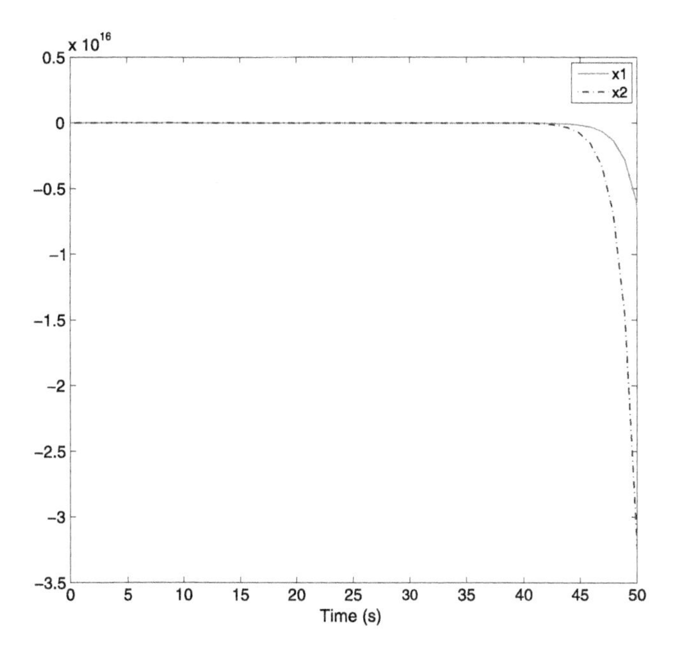

Figure 5.1 State trajectories of open-loop system.

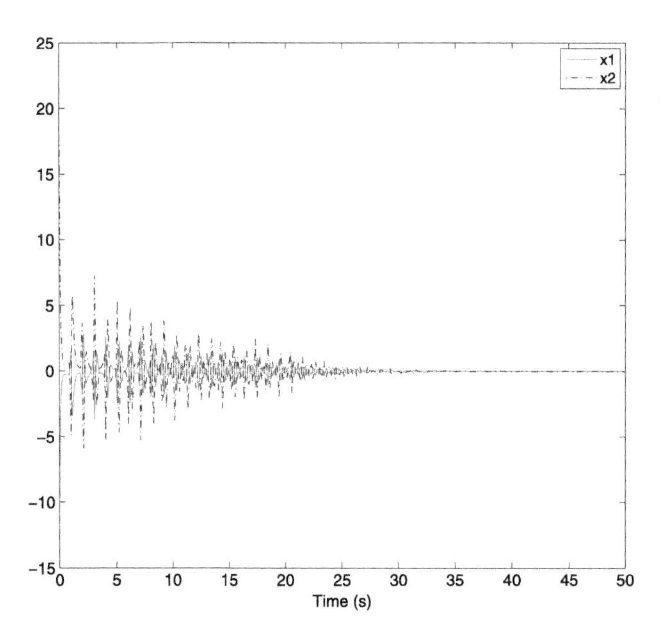

Figure 5.2 State trajectories of closed-loop system.

Table 5.8 Allowable maximum scalar α obtained for different m.

m	α	K
1	8.0508	$\begin{bmatrix} 10.8299 & 5.4150 \end{bmatrix}$
2	9.6380	$\begin{bmatrix} 6.5408 & 3.2704 \end{bmatrix}$
3	9.8477	$\begin{bmatrix} 6.6569 & 3.3284 \end{bmatrix}$
4	9.9255	$\begin{bmatrix} 6.9973 & 3.4987 \end{bmatrix}$

Table 5.9 Comparison of maximum allowable time-delay h.

Corollary 5.21	$m=1$	$m=2$	$m=10$	$m=24$
h	0.3111	0.3154	0.3166	0.3167

Theorem 8 in [62]	$N=1$	$N=2$	$N=3$
h	0.3166	0.3167	0.3167

discretized Lyapunov method, we consider a singular system described by

$$\begin{bmatrix} 1 & 0 \\ 0 & 0 \end{bmatrix} \dot{x}(t) = \begin{bmatrix} 0.5 & 0 \\ 0 & -1 \end{bmatrix} x(t) + \begin{bmatrix} -1 & -1 \\ 1 & 0.5 \end{bmatrix} x(t-h).$$

Let

$$\bar{x}(t) = x_1(t), \ \ y(t) = x(t).$$

The system considered in this example can be rewritten as

$$\begin{aligned} \dot{\bar{x}}(t) &= 0.5\bar{x}(t) + \begin{bmatrix} -1 & -1 \end{bmatrix} y(t-h), \\ y(t) &= \begin{bmatrix} 1 \\ 0 \end{bmatrix} \bar{x}(t) + \begin{bmatrix} 0 & 0 \\ 1 & 0.5 \end{bmatrix} y(t-h). \end{aligned}$$

We can calculate the maximum allowable time-delay h satisfying the LMIs in (5.44) and (5.45) by solving a quasiconvex optimization problem. Some comparison results between [62] and Corollary 5.21 in this section are presented in Table 5.9, which illustrates that both the maximum allowable time-delay h obtained by the two different methods are very close to the analytical solution $h_{max} = 0.3167$. However, there are some disadvantages of the discretized Lyapunov method compared with the delay partitioning method:

1. The discretized Lyapunov method is effective in the analysis of delay-dependent stability for time-delay systems, but it is not desirable for investigating synthesis problems due to the cross terms of A^TP and A^TQ_i, $(i=1,\ldots,N)$ [52], [211].
2. The results in [62] obtained by discretized Lyapunov method has not been shown to be monotonic with respect to the delay h. For this reason, it is only useful to test the stability of delay systems for a given time delay and is not easily employed to test the stability of delay systems on a given interval of delay.
3. It is difficult to extend the discretized Lyapunov method to singular delay systems. The results obtained in [62] were not developed for checking the admissibility of singular delay systems; the regularity and impulse free have to be checked separately.

5.2.5 Conclusion

The problem of α-dissipative control for continuous-time singular systems with time-delay has been studied in this section. The delay-dependent conditions in terms of LMIs have been proposed for guaranteeing singular systems admissible and strictly (Q, S, R)-α-dissipative. Based on this result, a state-feedback controller has been designed to guarantee the closed-loop system is admissible and strictly (Q, S, R)-α-dissipative. The results presented in this section are not only dependent on the delay, but also dependent on the dissipative margin α. Moreover, the results on H_∞ control and passive control of singular systems with time-delay are unified in the proposed results. Finally, some examples are given to demonstrate the reduced conservatism and the applicability of our methods.

5.3 Robust reliable dissipative filtering for discrete delay singular systems

This section is concerned with the problem of robust reliable dissipative filtering for uncertain discrete-time singular system with interval time-varying delay and sensor failures. The uncertainty and the sensor failures considered are polytopic uncertainty and varying in a given interval, respectively. The purpose is to design a filter such that the filtering error singular systems is regular, causal, asymptotically stable, and strictly (Q, S, R)-dissipative. By utilizing reciprocally convex approach, firstly, sufficient reliable dissipativity analysis condition is established in terms of LMIs for singular systems with time-varying delay and sensor failures. Based on this criterion, the result is extended to uncertain singular systems with time-varying delay and sensor failures. Moreover, the reliable dissipative filter is designed in terms of LMIs for uncertain singular systems with time-varying delay and sensor failures. Finally, the effectiveness of the filter design method in this section is illustrated by numerical examples.

5.3.1 Problem statement

Consider a class of linear discrete-time singular systems with time-varying delay described by

$$\begin{cases} Ex(k+1) &= Ax(k)+A_d x(k-d(k))+Bw(k) \\ y(k) &= Cx(k)+C_d x(k-d(k))+Dw(k) \\ z(k) &= Lx(k)+L_d x(k-d(k))+Gw(k) \\ x(k) &= \phi(k),\ k=-d_2,-d_2+1,\ldots,0, \end{cases} \tag{5.58}$$

where $x(k)\in\mathbb{R}^n$ is the state vector; $y(k)\in\mathbb{R}^m$ is the measured output; $z(k)\in\mathbb{R}^p$ represents the signal to be estimated; $w(k)\in\mathbb{R}^l$ is assumed to be an arbitrary noise belonging to l_2, and $\phi(k)$ is a known given initial condition sequence; $d(k)$ is a time-varying delay satisfying

$$1\le d_1\le d(k)\le d_2<\infty,\quad k=1,2,\ldots \tag{5.59}$$

The system matrices A, A_d, B, C, C_d, D, L, L_d, and G with appropriate dimension belong to a convex polytopic set

$$\chi := (A,\ A_d,\ B,\ C,\ C_d,\ D,\ L,\ L_d,\ G)\in\Omega, \tag{5.60}$$

where

$$\Omega := \left\{\chi(\lambda)=\sum_{i=1}^{q}\lambda_i\chi_i,\ \sum_{i=1}^{q}\lambda_i=1,\ \lambda_i\ge 0\right\}$$

and $\chi_i := (A_i,\ A_{di},\ B_i,\ C_i,\ C_{di},\ D_i,\ L_i,\ L_{di},\ G_i)\in\Omega$, $i=1,\ldots,q$ denoting the ith vertex of the polyhedral domain Ω. In contrast with standard linear state-space systems with $E=I$, the matrix $E\in\mathbb{R}^{n\times n}$ has $\mathrm{rank}(E)=r\le n$.

The singular system in (5.58) is assumed to be admissible over the whole domain Ω. Our purpose is to design a full order linear filter with sensor failures for the estimate of $z(k)$:

$$\begin{cases} \hat{x}(k+1) &= A_f\hat{x}(k)+B_f y^F(k),\quad \hat{x}(0)=0 \\ \hat{z}(k) &= C_f\hat{x}(k)+D_f y^F(k), \end{cases} \tag{5.61}$$

where A_f, B_f, C_f, and D_f are the filter gains to be determined, and $y^F(k) = [y_1^F(k), \ldots, y_m^F(k)]^T$ denotes the signal from the sensor that may be faulty. The following failure model from [110] is adopted here:

$$y_i^F(k) = \alpha_{si} y_i(k), \ i = 1, 2, \ldots, m,$$

where

$$0 \le \underline{\alpha}_{si} \le \alpha_{si} \le \bar{\alpha}_{si}, i = 1, 2, \ldots, m$$

with $0 \le \underline{\alpha}_{si} \le \bar{\alpha}_{si} \le 1$, in which the variables α_{si} quantify the failures of the sensors.

Then we have

$$y^F(k) = \mathcal{A}_s y(k), \ \mathcal{A}_s = \mathrm{diag}\{\alpha_{s1}, \alpha_{s2}, \ldots, \alpha_{sm}\}.$$

Remark 5.28. In the model mentioned above, when $\underline{\alpha}_{si} = \bar{\alpha}_{si} = 1$, it corresponds to the normal fully operating case, that is, $y_i^F = y_i$; when $\underline{\alpha}_{si} = 0$, then it covers the outage case in [167]; when $\alpha_{ai} \neq 0$ and $\alpha_{ai} \neq 1$, then it corresponds to the case, where the intensity of the feedback signal from sensor may vary.

Denote

$$\begin{aligned} \bar{\mathcal{A}}_s &= \mathrm{diag}\{\bar{\alpha}_{s1}, \bar{\alpha}_{as2}, \ldots, \bar{\alpha}_{sm}\}, \\ \underline{\mathcal{A}}_s &= \mathrm{diag}\{\underline{\alpha}_{s1}, \underline{\alpha}_{s2}, \ldots, \underline{\alpha}_{sm}\}, \\ \mathcal{A}_{s0} &= \mathrm{diag}\{\alpha_{s01}, \alpha_{s02}, \ldots, \alpha_{s0m}\}, \\ \mathcal{B} &= \mathrm{diag}\{\beta_1, \beta_2, \ldots, \beta_m\}, \\ \Delta_s &= \mathrm{diag}\{\Delta_{s1}, \Delta_{s2}, \ldots, \Delta_{sm}\}, \end{aligned}$$

where $\alpha_{s0i} = \frac{\underline{\alpha}_{si} + \bar{\alpha}_{si}}{2}$ and $\beta_i = \frac{\bar{\alpha}_{si} - \underline{\alpha}_{si}}{2}$. Then we have

$$\mathcal{A}_s = \mathcal{A}_{s0} + \Delta_s, \ |\Delta_{si}| \le \beta_i.$$

Let the augmented state vector $\tilde{x}(k) = [x^T(k) \ \ \hat{x}^T(k)]^T$ and $\tilde{z}(k) = z(k) - \hat{z}(k)$. Then the filtering error singular system is described as

$$\begin{cases} \tilde{E}\tilde{x}(k+1) = \tilde{A}\tilde{x}(k) + \tilde{A}_d \Phi \tilde{x}(k - d(k)) + \tilde{B}w(k) \\ \tilde{z}(k) = \tilde{L}\tilde{x}(k) + \tilde{L}_d \Phi \tilde{x}(k - d(k)) + \tilde{G}w(k) \\ \tilde{x}(k) = [\phi^T(k) \ 0]^T, k = -d_2, -d_2 + 1, \ldots, 0, \end{cases} \tag{5.62}$$

where $\Phi = [I \;\; 0]$, and

$$\begin{aligned}\tilde{E} &= \begin{bmatrix} E & 0 \\ 0 & I \end{bmatrix}, \; \tilde{A} = \begin{bmatrix} A & 0 \\ B_f \mathcal{A}_s C & A_f \end{bmatrix}, \; \tilde{A}_d = \begin{bmatrix} A_d \\ B_f \mathcal{A}_s C_d \end{bmatrix}, \; \tilde{B} = \begin{bmatrix} B \\ B_f \mathcal{A}_s D \end{bmatrix}, \\ \tilde{L} &= \begin{bmatrix} L - D_f \mathcal{A}_s C & -C_f \end{bmatrix}, \; \tilde{L}_d = L_d - D_f \mathcal{A}_s C_d, \; \tilde{G} = G - D_f \mathcal{A}_s D.\end{aligned}$$

Before moving on, we give some definitions and lemmas concerning the following system:

$$Ex(k+1) = Ax(k) + A_d x(k - d(k)). \tag{5.63}$$

Definition 5.29. [209]
1. The singular delay system in (5.63) is said to be regular and causal if $\det(sE - A)$ is not identically zero.
2. The singular delay system in (5.63) is said to be causal if $\deg\{\det(sE - A)\} = \text{rank}\, E$.
3. The singular delay system in (5.63) is said to be asymptotically stable if, for any $\varepsilon > 0$, there exists a scalar $\delta(\varepsilon) > 0$ such that for any consistent initial conditions $\phi(k)$ satisfying $\sup_{-d_2 \le k \le -1} \|\phi(k)\| \le \delta(\varepsilon)$, the solution $x(k)$ of (5.63) satisfies $\|x(k)\| \le \varepsilon$ for $k \ge 0$; furthermore, $x(k) \to 0$ when $k \to \infty$.
4. The singular delay system in (5.63) is said to be admissible if the system is regular, causal, and asymptotically stable.

Lemma 5.30. [208] *The discrete-time singular system $Ex(k+1) = Ax(k)$ is admissible if and only if there exist matrices $P > 0$ and Q such that*

$$A^T PA - E^T PE + \text{sym}(A^T S Q^T) < 0,$$

where $S \in \mathbb{R}^{n \times (n-r)}$ is any matrix with full column and satisfies $E^T S = 0$.

Lemma 5.31. [7] *For any matrices U and $V > 0$, the following inequality holds*

$$UV^{-1}U^T \ge U + U^T - V.$$

The purpose of this section is to design a filter in the form of (5.61) such that the filter error system in (5.62) is robustly admissible and strictly (Q, S, R)-dissipative.

5.3.2 Reliable dissipativity analysis

In this subsection, we first give the result of reliable dissipative analysis for system (5.62) with fixed system matrix χ.

Theorem 5.32. *Let matrices Q, S, and R be given with Q and R real symmetric and $Q \leq 0$. Then the system in (5.62) with sensor failure is admissible and strictly (Q, S, R)-dissipative if there exist matrices $P > 0$, $\bar{Q}_i > 0$, $i = 1, 2, 3$, $S_j > 0$, $j = 1, 2$, M, and $\tilde{Z}$ such that the following LMIs hold:*

$$\begin{bmatrix} S_2 & M \\ \star & S_2 \end{bmatrix} \geq 0, \tag{5.64}$$

$$\tilde{\Pi} = \begin{bmatrix} \tilde{\Pi}_{11} & \Phi^T E^T S_1 E & \tilde{Z}^T \tilde{U}^T \tilde{A}_d & 0 & \tilde{\Pi}_{15} & \tilde{\Pi}_{16} S_1 & \tilde{\Pi}_{17} S_2 & \tilde{L}^T Q_-^{\frac{1}{2}} & \tilde{A}^T P \\ \star & \tilde{\Pi}_{22} & \tilde{\Pi}_{23} & E^T M^T E & 0 & 0 & 0 & 0 & 0 \\ \star & \star & \tilde{\Pi}_{33} & \tilde{\Pi}_{34} & -\tilde{L}_d^T S & d_1 \tilde{A}_d^T S_1 & \tilde{d} \tilde{A}_d^T S_2 & \tilde{L}_d^T Q_-^{\frac{1}{2}} & \tilde{A}_d^T P \\ \star & \star & \star & \tilde{\Pi}_{44} & 0 & 0 & 0 & 0 & 0 \\ \star & \star & \star & \star & \tilde{\Pi}_{55} & d_1 \tilde{B}^T S_1 & \tilde{d} \tilde{B}^T S_2 & \tilde{G}^T Q_-^{\frac{1}{2}} & \tilde{B}^T P \\ \star & \star & \star & \star & \star & -S_1 & 0 & 0 & 0 \\ \star & \star & \star & \star & \star & \star & -S_2 & 0 & 0 \\ \star & \star & \star & \star & \star & \star & \star & -I & 0 \\ \star & \star & \star & \star & \star & \star & \star & \star & -P \end{bmatrix} < 0, \tag{5.65}$$

where

$$\begin{aligned} \tilde{\Pi}_{11} &= -\tilde{E}^T P \tilde{E} + Q_1 + Q_2 + (\tilde{d} + 1) Q_3 + \mathrm{sym}(\tilde{A}^T \tilde{U} \tilde{Z}) - \Phi^T E^T S_1 E \Phi, \\ \tilde{\Pi}_{22} &= -\bar{Q}_1 - E^T (S_1 + S_2) E, \ \tilde{\Pi}_{23} = E^T (S_2 - M^T) E, \\ \tilde{\Pi}_{33} &= -\bar{Q}_3 + \mathrm{sym}(E^T (M - S_2) E), \ \tilde{\Pi}_{34} = E^T (S_2 - M^T) E, \\ \tilde{\Pi}_{44} &= -\bar{Q}_2 - E^T S_2 E, \ \tilde{\Pi}_{15} = \tilde{Z}^T \tilde{U}^T \tilde{B} - \tilde{L}^T S, \\ \tilde{\Pi}_{55} &= -\tilde{G}^T S - S^T \tilde{G} - R, \ \tilde{\Pi}_{16} = d_1 \Phi^T (A - E)^T, \ \tilde{\Pi}_{17} = \tilde{d} \Phi^T (A - E)^T, \\ Q_i &= \mathrm{diag}\{\bar{Q}_i, \bar{Q}_i\}, \ i = 1, 2, 3, \ \tilde{d} = d_2 - d_1, \end{aligned}$$

and $\tilde{U} \in \mathbb{R}^{2n \times (2n-r)}$ is any matrix with full column rank satisfying $\tilde{E}^T \tilde{U} = 0$.

Proof. First, we prove that the system in (5.62) is regular and causal. From the LMI in (5.66), we have

$$\begin{bmatrix} \tilde{\Pi}_{11} & \tilde{A}^T P \\ \star & -P \end{bmatrix} < 0,$$

which implies the following inequality holds by using Schur complement equivalence:

$$\tilde{\Pi}_{11} + \tilde{A}^T P \tilde{A} < 0. \tag{5.66}$$

Letting $\tilde{S} = \begin{bmatrix} S_1 & 0 \\ 0 & 0 \end{bmatrix}$ and noting $\Phi^T E^T S_1 E \Phi = \tilde{E}^T \tilde{S} \tilde{E}$, it follows from (5.66) that

$$\tilde{A}^T P \tilde{A} - \tilde{E}^T (P + \tilde{S}) \tilde{E} + \mathrm{sym}(\tilde{A}^T \tilde{U} \tilde{Z}^T) < 0.$$

Combining Lemma 5.30 and Definition 5.29, the regularity and causality of the system in (5.62) are guaranteed. For the stability property, let us define $\tilde{\eta}(k) = \tilde{x}(k+1) - \tilde{x}(k)$ and choose the following Lyapunov functional:

$$V(k) = V_1(k) + V_2(k) + V_3(k) + V_4(k) + V_5(k), \tag{5.67}$$

where

$$\begin{aligned}
V_1(k) &= \tilde{x}^T(k)\tilde{E}^T P \tilde{E}\tilde{x}(k), \\
V_2(k) &= \sum_{j=1}^{2} \sum_{i=k-d_j}^{k-1} \tilde{x}^T(i) Q_i \tilde{x}(i), \\
V_3(k) &= \sum_{i=k-d(k)}^{k-1} \tilde{x}^T(i) Q_3 \tilde{x}(i) + \sum_{j=-d_2+1}^{-d_1} \sum_{i=k+j}^{k-1} \tilde{x}^T(i) Q_3 \tilde{x}(i), \\
V_4(k) &= \sum_{j=-d_1}^{-1} \sum_{i=k+j}^{k-1} d_1 \tilde{\eta}^T(i) \tilde{E}^T \Phi^T S_1 \Phi \tilde{E} \tilde{\eta}(i), \\
V_5(k) &= \sum_{j=-d_2}^{-d_1-1} \sum_{i=k+j}^{k-1} \tilde{d} \tilde{\eta}^T(i) \tilde{E}^T \Phi^T S_2 \Phi \tilde{E} \tilde{\eta}(i).
\end{aligned}$$

Calculate the forward difference of $V(k)$ along the trajectories of filtering error singular system (5.62) with $w(k) = 0$ yields

$$\begin{aligned}
\Delta V_1(k) &= \tilde{x}^T(k+1)\tilde{E}^T P \tilde{E}\tilde{x}(k+1) - \tilde{x}^T(k)\tilde{E}^T P \tilde{E}\tilde{x}(k) \\
&= (\tilde{A}\tilde{x}(k) + \tilde{A}_d \Phi \tilde{x}(k-d(k)))^T P (\tilde{A}\tilde{x}(k) + \tilde{A}_d \Phi \tilde{x}(k-d(k))) \\
&- \tilde{x}^T(k)\tilde{E}^T P \tilde{E}\tilde{x}(k), \qquad (5.68) \\
\Delta V_2(k) &= \sum_{j=1}^{2} \tilde{x}^T(k) Q_j \tilde{x}(k) - \sum_{j=1}^{2} \tilde{x}(k-d_j) Q_j \tilde{x}(k-d_j) \\
&\leq \sum_{j=1}^{2} \tilde{x}^T(k) Q_j \tilde{x}(k) - \sum_{j=1}^{2} \tilde{x}(k-d_j) \Phi^T \bar{Q}_j \Phi \tilde{x}(k-d_j), \qquad (5.69)
\end{aligned}$$

$$
\begin{aligned}
\Delta V_3(k) &= (\tilde{d}+1)\tilde{x}^T(k)Q_3\tilde{x}(k) - \tilde{x}^T(k-d(k))Q_3\tilde{x}(k-d(k)) \\
&\quad + \sum_{i=k+1-d(k+1)}^{k-1} \tilde{x}^T(i)Q_3\tilde{x}(i) - \sum_{i=k+1-d(k)}^{k-1} \tilde{x}^T(i)Q_3\tilde{x}(i) \\
&\quad - \sum_{i=k-d_2+1}^{k-d_1} \tilde{x}^T(i)Q_3\tilde{x}(i) \\
&= (\tilde{d}+1)\tilde{x}^T(k)Q_3\tilde{x}(k) - \tilde{x}^T(k-d(k))Q_3\tilde{x}(k-d(k)) \\
&\quad + \sum_{i=k+1-d_1}^{k-1} \tilde{x}^T(i)Q_3\tilde{x}(i) + \sum_{i=k+1-d(k+1)}^{k-d_1} \tilde{x}^T(i)Q_3\tilde{x}(i) \\
&\quad - \sum_{i=k+1-d(k)}^{k-1} \tilde{x}^T(i)Q_3\tilde{x}(i) - \sum_{i=k-d_2+1}^{k-d_1} \tilde{x}^T(i)Q_3\tilde{x}(i) \\
&\le (\tilde{d}+1)\tilde{x}^T(k)Q_3\tilde{x}(k) - \tilde{x}^T(k-d(k))Q_3\tilde{x}(k-d(k)) \\
&\le (\tilde{d}+1)\tilde{x}^T(k)Q_3\tilde{x}(k) - \tilde{x}^T(k-d(k))\Phi^T\bar{Q}_3\Phi\tilde{x}(k-d(k)).
\end{aligned} \tag{5.70}
$$

By using Lemma 4.5, we have

$$
\begin{aligned}
\Delta V_4(k) &= d_1^2\tilde{\eta}^T(k)\tilde{E}^T\Phi^T S_1\Phi\tilde{E}\tilde{\eta}(k) - d_1\sum_{i=k-d_1}^{k-1}\tilde{\eta}^T(i)\tilde{E}^T\Phi^T S_1\Phi\tilde{E}\tilde{\eta}(i) \\
&\le d_1^2((A-E)\Phi\tilde{x}(k) + A_d\Phi\tilde{x}(k-d(k)))^T S_1((A-E)\Phi\tilde{x}(k) \\
&\quad + A_d\Phi\tilde{x}(k-d(k))) - (E\Phi\tilde{x}(k) - E\Phi\tilde{x}(k-d_1))^T S_1(E\Phi\tilde{x}(k) \\
&\quad - E\Phi\tilde{x}(k-d_1)).
\end{aligned} \tag{5.71}
$$

Since $\begin{bmatrix} S_2 & M \\ \star & S_2 \end{bmatrix} \ge 0$, the following inequality holds:

$$
\begin{bmatrix} \sqrt{\frac{\alpha_1}{\alpha_2}}(Ex(k-d(k)) - Ex(k-d_2)) \\ -\sqrt{\frac{\alpha_2}{\alpha_1}}(Ex(k-d_1) - Ex(k-d(k))) \end{bmatrix}^T \begin{bmatrix} S_2 & M \\ \star & S_2 \end{bmatrix}
\times \begin{bmatrix} \sqrt{\frac{\alpha_1}{\alpha_2}}(Ex(k-d(k)) - Ex(k-d_2)) \\ -\sqrt{\frac{\alpha_2}{\alpha_1}}(Ex(k-d_1) - Ex(k-d(k))) \end{bmatrix} \ge 0,
$$

where $\alpha_1 = \frac{d_2 - d(k)}{\tilde{d}}$, $\alpha_2 = \frac{d(k) - d_1}{\tilde{d}}$. Then employing Lemma 4.4, for $d_1 < d(k) < d_2$, we have

$$
\begin{aligned}
\Delta V_5(k) &= \tilde{d}^2 \tilde{\eta}^T(k) \tilde{E}^T \Phi^T S_2 \Phi \tilde{E} \tilde{\eta}(k) - \tilde{d} \sum_{i=k-d_2}^{k-d(k)-1} \tilde{\eta}^T(i) \tilde{E}^T \Phi^T S_2 \Phi \tilde{E} \tilde{\eta}(i) \\
&- \tilde{d} \sum_{i=k-d(k)}^{k-d_1-1} \tilde{\eta}^T(i) \tilde{E}^T \Phi^T S_2 \Phi \tilde{E} \tilde{\eta}(i) \\
&\leq -\frac{\tilde{d}}{d_2 - d(k)} \left(\sum_{i=k-d_2}^{k-d(k)-1} \Phi \tilde{E} \tilde{\eta}(i) \right)^T S_2 \left(\sum_{i=k-d_2}^{k-d(k)-1} \Phi \tilde{E} \tilde{\eta}(i) \right) \\
&- \frac{\tilde{d}}{d(k) - d_1} \left(\sum_{i=k-d(k)}^{k-d_1-1} \Phi \tilde{E} \tilde{\eta}(i) \right)^T S_2 \left(\sum_{i=k-d(k)}^{k-d_1-1} \Phi \tilde{E} \tilde{\eta}(i) \right) \\
&+ \tilde{d}^2 \tilde{\eta}^T(k) \tilde{E}^T \Phi^T S_2 \Phi \tilde{E} \tilde{\eta}(k) \\
&\leq \tilde{d}^2 \left((A - E) \Phi \tilde{x}(k) + A_d \Phi \tilde{x}(k - d(k)) \right)^T \\
&\times S_2 \left((A - E) \Phi \tilde{x}(k) + A_d \Phi \tilde{x}(k - d(k)) \right) \\
&- \begin{bmatrix} Ex(k - d(k)) - Ex(k - d_2) \\ Ex(k - d_1) - Ex(k - d(k)) \end{bmatrix}^T \begin{bmatrix} S_2 & M \\ \star & S_2 \end{bmatrix} \\
&\times \begin{bmatrix} Ex(k - d(k)) - Ex(k - d_2) \\ Ex(k - d_1) - Ex(k - d(k)) \end{bmatrix}. \qquad (5.72)
\end{aligned}
$$

Note that when $d(k) = d_1$ or $d(k) = d_2$, it yields $Ex(k - d_1) - Ex(k - d(k)) = 0$, or $Ex(k - d(k)) - Ex(k - d_2) = 0$. Hence, the inequality in (5.72) still holds. From $\tilde{E}^T \tilde{U} = 0$, we have

$$
2\tilde{x}(k+1)^T \tilde{E}^T \tilde{U} \tilde{Z} \tilde{x}(k) = 0. \qquad (5.73)
$$

Combining the conditions from (5.68) to (5.73), we have

$$
\Delta V(k) = \zeta^T(k) \Pi_s \zeta(k),
$$

where

$$
\zeta(k) = \begin{bmatrix} \tilde{x}^T(k) & \tilde{x}^T(k - d_1) \Phi^T & \tilde{x}^T(k - d(k)) \Phi^T & \tilde{x}^T(k - d_2) \Phi^T \end{bmatrix}^T,
$$

$$\Pi_s = \begin{bmatrix} \tilde{\Pi}_{11} & \Phi^T E^T S_1 E & \tilde{Z}^T \tilde{U}^T \tilde{A}_d & 0 \\ \star & \tilde{\Pi}_{22} & \tilde{\Pi}_{23} & E^T M^T E \\ \star & \star & \tilde{\Pi}_{33} & \tilde{\Pi}_{34} \\ \star & \star & \star & \tilde{\Pi}_{44} \end{bmatrix}$$
$$+ \begin{bmatrix} \tilde{\Pi}_{16} \\ 0 \\ d_1 \tilde{A}_d^T \\ 0 \end{bmatrix} S_1 \begin{bmatrix} \tilde{\Pi}_{16} \\ 0 \\ d_1 \tilde{A}_d^T \\ 0 \end{bmatrix}^T + \begin{bmatrix} \tilde{\Pi}_{17} \\ 0 \\ \tilde{d} \tilde{A}_d^T \\ 0 \end{bmatrix} S_2 \begin{bmatrix} \tilde{\Pi}_{17} \\ 0 \\ \tilde{d} \tilde{A}_d^T \\ 0 \end{bmatrix}^T$$
$$+ \begin{bmatrix} \tilde{A}^T \\ 0 \\ \tilde{A}_d^T \\ 0 \end{bmatrix} P \begin{bmatrix} \tilde{A}^T \\ 0 \\ \tilde{A}_d^T \\ 0 \end{bmatrix}^T .$$

On the other hand, the following inequality can be obtained from (5.66):

$$\Pi_{s1} = \begin{bmatrix} \tilde{\Pi}_{11} & \Phi^T E^T S_1 E & \tilde{Z}^T \tilde{U}^T \tilde{A}_d & 0 & \tilde{\Pi}_{16} S_1 & \tilde{\Pi}_{17} S_2 & \tilde{A}^T P \\ \star & \tilde{\Pi}_{22} & \tilde{\Pi}_{23} & E^T M^T E & 0 & 0 & 0 \\ \star & \star & \tilde{\Pi}_{33} & \tilde{\Pi}_{34} & d_1 \tilde{A}_d^T S_1 & \tilde{d} \tilde{A}_d^T S_2 & \tilde{A}_d^T P \\ \star & \star & \star & \tilde{\Pi}_{44} & 0 & 0 & 0 \\ \star & \star & \star & \star & -S_1 & 0 & 0 \\ \star & \star & \star & \star & \star & -S_2 & 0 \\ \star & \star & \star & \star & \star & \star & -P \end{bmatrix} < 0, \tag{5.74}$$

which is equivalent to $\Pi_s < 0$. Hence, $\Delta V(k) < 0$, which implies the filter error singular system in (5.62) with $w(k) = 0$ is asymptotically stable. Then combining the regularity and casualty of system (5.62), the admissibility can be guaranteed based on Definition 5.29.

Next, we show the dissipativity of system (5.62). To this end, we define

$$J(T) = \sum_{k=0}^{T} \left[\tilde{z}^T(k) Q \tilde{z}(k) + 2\tilde{z}^T(k) S w(k) + w^T(k) R w(k) \right].$$

Then under the zero initial condition, that is, $x(k) = 0$ for $k = -d_2, -d_2 + 1, \ldots, 0$, it can be shown that for any nonzero $w \in l_2$, let $\Psi = V(T+1) - V(0) - J(T)$,

$$\Psi = \sum_{k=0}^{T} \left[\Delta V(k) - \tilde{z}^T(k) Q \tilde{z}(k) - 2\tilde{z}^T(k) S w(k) - w^T(k) R w(k) \right]$$

$$\begin{aligned}
&= \sum_{k=0}^{T} \xi^T(k)(\bar{\Pi} + \bar{\Pi}_1^T P \bar{\Pi}_1 + d_1^2 \bar{\Pi}_2^T S_1 \bar{\Pi}_2 + d_{12}^2 \bar{\Pi}_2^T S_2 \bar{\Pi}_2 - \bar{\Pi}_3^T Q \bar{\Pi}_3)\xi(k) \\
&< 0,
\end{aligned}$$

where

$$\begin{aligned}
\xi(k) &= \begin{bmatrix} \tilde{x}^T(k) & \tilde{x}^T(k-d_1)\Phi^T & \tilde{x}^T(k-d(k))\Phi^T & \tilde{x}^T(k-d_2)\Phi^T & w(k) \end{bmatrix}^T, \\
\bar{\Pi} &= \begin{bmatrix} \tilde{\Pi}_{11} & \Phi^T E^T S_1 E & \tilde{Z}^T \tilde{U}^T \tilde{A}_d & 0 & \tilde{\Pi}_{15} \\ \star & \tilde{\Pi}_{22} & \tilde{\Pi}_{23} & E^T M^T E & 0 \\ \star & \star & \tilde{\Pi}_{33} & \tilde{\Pi}_{34} & -\tilde{L}_d^T S \\ \star & \star & \star & \tilde{\Pi}_{44} & 0 \\ \star & \star & \star & \star & \tilde{\Pi}_{55} \end{bmatrix}, \\
\bar{\Pi}_1 &= \begin{bmatrix} \tilde{A} & 0 & \tilde{A}_d & 0 & \tilde{B} \end{bmatrix}, \\
\bar{\Pi}_2 &= \begin{bmatrix} (A-E)\Phi & 0 & A_d & 0 & B \end{bmatrix}, \\
\bar{\Pi}_3 &= \begin{bmatrix} \tilde{L} & 0 & \tilde{L}_d & 0 & \tilde{G} \end{bmatrix}.
\end{aligned}$$

By using Schur complement equivalence and noticing $-Q = (Q_-^{\frac{1}{2}})^2$, the inequality in (5.66) is equivalent to $\bar{\Pi} + \bar{\Pi}_1^T P \bar{\Pi}_1 + d_1^2 \bar{\Pi}_2^T S_1 \bar{\Pi}_2 + d_{12}^2 \bar{\Pi}_2^T S_2 \bar{\Pi}_2 - \bar{\Pi}_3^T Q \bar{\Pi}_3 < 0$. Then we have $J(T) > V(T+1)$, and there exists a sufficiently small $\alpha > 0$ such that

$$J(T) \geq \alpha \sum_{k=0}^{T} w^T(k) w(k).$$

By Definition 2.14, the system in (5.62) is strictly (Q, S, R)-dissipative. This completes the proof. □

Remark 5.33. The result in Theorem 5.32 is obtained using the reciprocally convex approach proposed in [140]. The method has led to less conservative stability criterion for continuous-time system with time-varying delay than that in [151]. It is extended to solve the filtering problem for discrete-time singular system with time-varying delay in this section.

Remark 5.34. The reduction of conservatism of the approach used in this section, due to utilizing the reciprocally convex approach, which bounds the integral term $-\sum_{i=k-d_2}^{k-d_1-1}(d_2 - d_1)\eta^T(i) S \eta(i)$ with $d_1 \leq d(k) \leq d_2$, $\eta(i) =$

$x(i+1)-x(i)$ by the term

$$-\begin{bmatrix} x(k-d(k))-x(k-d_1) \\ x(k-d(k))-x(k-d_2) \end{bmatrix}^T \begin{bmatrix} E^TSE & E^TZE \\ \star & E^TSE \end{bmatrix} \begin{bmatrix} x(k-d(k))-x(k-d_1) \\ x(k-d(t))-x(k-d_2) \end{bmatrix}$$

with $\begin{bmatrix} S & Z \\ \star & S \end{bmatrix} \geq 0$. When $Z=0$, it reduces to the Jensen inequality method used in [36], [239]. Moreover, as explained in [140], the reciprocally convex approach directly deals with the inversely weighted convex combination of quadratic terms of integral quantities, rather than approximating the difference between delay bounds using a convex combination approach in [150].

Based on Theorem 5.32, we give the following by-product condition for the admissibility of the delay singular system in (5.63):

Corollary 5.35. *Given integers $1 \leq d_1 \leq d_2$, system (5.63) with time-varying delay $d(k)$ satisfying (5.59) is admissible if there exist matrices $P>0$, $Q_i>0$, $i=1,2,3$, $S_j>0$, $j=1,2$, M, and Z such that the following LMIs hold:*

$$\begin{bmatrix} S_2 & M \\ \star & S_2 \end{bmatrix} \geq 0, \tag{5.75}$$

$$\Pi = \begin{bmatrix} \Pi_{11} & E^TS_1E & Z^TU^TA_d & 0 & \Pi_{16}S_1 & \Pi_{17}S_2 & A^TP \\ \star & \Pi_{22} & \Pi_{23} & E^TM^TE & 0 & 0 & 0 \\ \star & \star & \Pi_{33} & \Pi_{34} & d_1A_d^TS_1 & \tilde{d}A_d^TS_2 & A_d^TP \\ \star & \star & \star & \Pi_{44} & 0 & 0 & 0 \\ \star & \star & \star & \star & -S_1 & 0 & 0 \\ \star & \star & \star & \star & \star & -S_2 & 0 \\ \star & \star & \star & \star & \star & \star & -P \end{bmatrix} < 0, \tag{5.76}$$

where

$$\begin{aligned} \Pi_{11} &= -E^TPE+Q_1+Q_2+(\tilde{d}+1)Q_3+\mathrm{sym}(A^TUZ)-E^TS_1E, \\ \Pi_{22} &= -E^TS_2E-Q_1-E^TS_1E,\ \Pi_{23}=E^TS_2E-E^TM^TE, \\ \Pi_{33} &= -Q_3-2E^TS_2E+\mathrm{sym}(E^TME),\ \Pi_{34}=E^TS_2E-E^TM^TE, \\ \Pi_{44} &= -Q_2-E^TS_2E,\ \tilde{\Pi}_{16}=d_1(A-E)^T,\ \tilde{\Pi}_{17}=\tilde{d}(A-E)^T, \end{aligned}$$

and $U \in \mathbb{R}^{n\times(n-r)}$ is any matrix with full column rank satisfying $E^TU=0$.

Proof. Choose a Lyapunov functional candidate as follows:

$$\tilde{V}(k)=\tilde{V}_1(k)+\tilde{V}_2(k)+\tilde{V}_3(k)+\tilde{V}_4(k)+\tilde{V}_5(k),$$

where

$$\begin{aligned}
\tilde{V}_1(k) &= x(k)E^T PEx(k), \\
\tilde{V}_2(k) &= \sum_{j=1}^{2} \sum_{i=k-d_j}^{k-1} x^T(i) Q_i x(i), \\
\tilde{V}_3(k) &= \sum_{i=k-d(k)}^{k-1} x^T(i) Q_3 x(i) + \sum_{j=-d_2+1}^{-d_1} \sum_{i=k+j}^{k-1} x^T(i) Q_3 x(i), \\
\tilde{V}_4(k) &= \sum_{j=-d_1}^{-1} \sum_{i=k+j}^{k-1} d_1 \eta^T(i) E^T S_1 E \eta(i), \\
\tilde{V}_5(k) &= \sum_{j=-d_2}^{-d_1-1} \sum_{i=k+j}^{k-1} \tilde{d} \eta^T(i) E^T S_2 E \eta(i), \\
\eta(k) &= x(k+1) - x(k).
\end{aligned}$$

Then following a similar line as in the proof of Theorem 5.32 yields Corollary 5.35. □

Similar to Theorem 2 in [56], a less conservative condition of LMI (5.65) is obtained by introducing three slack matrices H_1, H_2, and T, which is presented in the following theorem:

Theorem 5.36. *Let matrices Q, S, and R be given with Q and R real symmetric and $Q \leq 0$. Then the system in (5.62) with sensor failure is admissible and strictly (Q, S, R)-dissipative if there exist matrices $P > 0$, $\bar{Q}_i > 0$, $i = 1, 2, 3$, $S_j > 0$, $H_j, j = 1, 2$, T, M, and $\tilde{Z}$ such that the following LMIs hold:*

$$\begin{bmatrix} S_2 & M \\ \star & S_2 \end{bmatrix} \geq 0, \tag{5.77}$$

$$\begin{bmatrix}
\tilde{\Pi}_{11} & \Phi^T E^T S_1 E & \tilde{Z}^T \tilde{U}^T \tilde{A}_d & 0 & \tilde{\Pi}_{15} & \tilde{\Pi}_{16} H_1^T & \tilde{\Pi}_{17} H_2^T & \tilde{L}^T Q_-^{\frac{1}{2}} & \tilde{A}^T T^T \\
\star & \tilde{\Pi}_{22} & \tilde{\Pi}_{23} & E^T M^T E & 0 & 0 & 0 & 0 & 0 \\
\star & \star & \tilde{\Pi}_{33} & \tilde{\Pi}_{34} & -\tilde{L}_d^T S & d_1 \tilde{A}_d^T H_1^T & \tilde{d} \tilde{A}_d^T H_2^T & \tilde{L}_d^T Q_-^{\frac{1}{2}} & \tilde{A}_d^T T^T \\
\star & \star & \star & \tilde{\Pi}_{44} & 0 & 0 & 0 & 0 & 0 \\
\star & \star & \star & \star & \tilde{\Pi}_{55} & d_1 \tilde{B}^T H_1^T & \tilde{d} \tilde{B}^T H_2^T & \tilde{G}^T Q_-^{\frac{1}{2}} & \tilde{B}^T T^T \\
\star & \star & \star & \star & \star & S_1 - H_1^T - H_1 & 0 & 0 & 0 \\
\star & \star & \star & \star & \star & \star & S_2 - H_2^T - H_2 & 0 & 0 \\
\star & \star & \star & \star & \star & \star & \star & -I & 0 \\
\star & \star & \star & \star & \star & \star & \star & \star & P - T^T - T
\end{bmatrix} < 0, \tag{5.78}$$

where $\tilde{\Pi}_{ii}$, $i = 1, \ldots, 5$, $\tilde{\Pi}_{15}$, $\tilde{\Pi}_{16}$, $\tilde{\Pi}_{17}$, $\tilde{\Pi}_{23}$, $\tilde{\Pi}_{34}$, and $\tilde{U}$ are defined in (5.66).

Proof. If (5.65) holds, then there exist $H_j = H_j^T = S_j, j = 1, 2$, and $T = T^T = P$ such that (5.78) holds. On the other hand if (5.78) holds, we have the following inequality based on Lemma 5.31:

$$\begin{bmatrix} \tilde{\Pi}_{11} & \Phi^T E^T S_1 E & \tilde{Z}^T \tilde{U}^T \tilde{A}_d & 0 & \tilde{\Pi}_{15} & \tilde{\Pi}_{16} H_1^T & \tilde{\Pi}_{17} H_2^T & \tilde{L}^T Q_-^{\frac{1}{2}} & \tilde{A}^T T^T \\ \star & \tilde{\Pi}_{22} & \tilde{\Pi}_{23} & E^T M^T E & 0 & 0 & 0 & 0 & 0 \\ \star & \star & \tilde{\Pi}_{33} & \tilde{\Pi}_{34} & -\tilde{L}_d^T S & d_1 \tilde{A}_d^T H_1^T & \bar{d} \tilde{A}_d^T H_2^T & \tilde{L}_d^T Q_-^{\frac{1}{2}} & \tilde{A}_d^T T^T \\ \star & \star & \star & \tilde{\Pi}_{44} & 0 & 0 & 0 & 0 & 0 \\ \star & \star & \star & \star & \tilde{\Pi}_{55} & d_1 \tilde{B}^T H_1^T & \bar{d} \tilde{B}^T H_2^T & \tilde{G}^T Q_-^{\frac{1}{2}} & \tilde{B}^T T^T \\ \star & \star & \star & \star & \star & -H_1 S_1^{-1} H_1^T & 0 & 0 & 0 \\ \star & \star & \star & \star & \star & \star & -H_2 S_2^{-1} H_2^T & 0 & 0 \\ \star & \star & \star & \star & \star & \star & \star & -I & 0 \\ \star & \star & \star & \star & \star & \star & \star & \star & -TP^{-1} T^T \end{bmatrix} < 0. \tag{5.79}$$

In addition, matrices $H_j, j = 1, 2$, and T are nonsingular due to $S_j - H_j^T - H_j < 0$, $j = 1, 2$, and $P - T^T - T < 0$. Then, pre- and postmultiplying (5.79) by $\text{diag}\{I, I, I, I, I, S_1 H_1^{-1}, S_2 H_2^{-1}, PT^{-1}\}$ and its transpose yield (5.65). Therefore the equivalence between (5.78) and (5.65) is proved. □

Then a sufficient condition for robust admissibility and strict (Q, S, R)-dissipativity of uncertain singular system (5.62) can be derived by using a parameter-dependent Lyapunov functional.

Theorem 5.37. *Let matrices Q, S, and R be given with Q and R real symmetric and $Q \leq 0$. Then the system in (5.62) with sensor failure and polytopic uncertainty (5.60) is robustly admissible and strictly (Q, S, R)-dissipative if there exist matrices $P_i > 0$, $\bar{Q}_{li} > 0$, $l = 1, 2, 3$, $S_{ji} > 0$, H_j, $j = 1, 2$, T, M_i, and $\tilde{Z}$ such that the following set of LMIs hold for $i = 1, \ldots, q$:*

$$\begin{bmatrix} S_{2i} & M_i \\ \star & S_{2i} \end{bmatrix} \geq 0, \tag{5.80}$$

$$\tilde{\Omega}_i = \begin{bmatrix} \tilde{\Pi}_{11i} & \tilde{\Pi}_{12i} & \tilde{\Pi}_{13i} & 0 & \tilde{\Pi}_{15i} & \tilde{\Pi}_{16i} H_1^T & \tilde{\Pi}_{17i} H_2^T & \tilde{L}_i^T Q_-^{\frac{1}{2}} & \tilde{A}_i^T T^T \\ \star & \tilde{\Pi}_{22i} & \tilde{\Pi}_{23i} & E^T M_i^T E & 0 & 0 & 0 & 0 & 0 \\ \star & \star & \tilde{\Pi}_{33i} & \tilde{\Pi}_{34i} & -\tilde{L}_{di}^T S & d_1 \tilde{A}_{di}^T H_1^T & \bar{d} \tilde{A}_{di}^T H_2^T & \tilde{L}_{di}^T Q_-^{\frac{1}{2}} & \tilde{A}_{di}^T T^T \\ \star & \star & \star & \tilde{\Pi}_{44i} & 0 & 0 & 0 & 0 & 0 \\ \star & \star & \star & \star & \tilde{\Pi}_{55i} & d_1 \tilde{B}_i^T H_1^T & \bar{d} \tilde{B}_i^T H_2^T & \tilde{G}_i^T Q_-^{\frac{1}{2}} & \tilde{B}_i^T T^T \\ \star & \star & \star & \star & \star & \tilde{\Pi}_{66i} & 0 & 0 & 0 \\ \star & \star & \star & \star & \star & \star & \tilde{\Pi}_{77i} & 0 & 0 \\ \star & \star & \star & \star & \star & \star & \star & -I & 0 \\ \star & \star & \star & \star & \star & \star & \star & \star & \tilde{\Pi}_{99i} \end{bmatrix} < 0, \tag{5.81}$$

where

$$\tilde{\Pi}_{11i} = -\tilde{E}^T P_i \tilde{E} + Q_{1i} + Q_{2i} + (\tilde{d}+1)Q_{3i} + \mathrm{sym}(\tilde{A}_i^T \tilde{U}\tilde{Z}) - \Phi^T E^T S_{1i} E \Phi,$$
$$\tilde{\Pi}_{12i} = \Phi^T E^T S_{1i} E,\ \tilde{\Pi}_{13i} = \tilde{Z}^T \tilde{U}^T \tilde{A}_{di},$$
$$\tilde{\Pi}_{22i} = -E^T S_{2i} E - \bar{Q}_{1i} - E^T S_{1i} E,\ \tilde{\Pi}_{23i} = E^T S_{2i} E - E^T M_i^T E,$$
$$\tilde{\Pi}_{33i} = -\bar{Q}_{3i} - 2E^T S_{2i} E + \mathrm{sym}(E^T M_i E),\ \tilde{\Pi}_{34} = E^T S_{2i} E - E^T M_i^T E,$$
$$\tilde{\Pi}_{44i} = -\bar{Q}_{2i} - E^T S_{2i} E,\ \tilde{\Pi}_{15i} = \tilde{Z}^T \tilde{U}^T \tilde{B}_i - \tilde{L}_i^T S,$$
$$\tilde{\Pi}_{55i} = -\tilde{G}_i^T S - S^T \tilde{G}_i - R,\ \tilde{\Pi}_{16i} = d_1 \Phi^T (A_i - E)^T,\ \tilde{\Pi}_{17i} = \tilde{d}\Phi^T (A_i - E)^T,$$
$$\tilde{\Pi}_{66i} = S_{1i} - H_1^T - H_1,\ \tilde{\Pi}_{77i} = S_{2i} - H_2^T - H_2,\ \tilde{\Pi}_{99i} = P_i - T^T - T,$$
$$Q_{li} = \mathrm{diag}\{\bar{Q}_{li}, \bar{Q}_{li}\},\ l = 1,2,3,\ \tilde{d} = d_2 - d_1,$$

and $\tilde{U} \in \mathbb{R}^{2n\times(2n-r)}$ *is any matrix with full column rank satisfying* $\tilde{E}^T\tilde{U} = 0$.

Proof. Let P, $\bar{Q}_l$, $l = 1,2,3$, $S_j, j = 1,2$, and M be expressed as

$$\begin{pmatrix} P & \bar{Q}_l & S_j & M \end{pmatrix} = \sum_{i=1}^{q} \lambda_i \begin{pmatrix} P_i & \bar{Q}_{li} & S_{ji} & M_i \end{pmatrix},$$

where λ_i is defined in (5.60). By multiplying conditions (5.80), (5.81) with λ_i and summing up from 1 to q, we obtain (5.77) and (5.78), respectively. □

5.3.3 Filter design

Now, our attention will be devoted to design a filter in the form of (5.61) such that filtering error system (5.62) subject to possible sensor failures is admissible and strictly (Q, S, R)-dissipative. Based on the result of Theorem 5.36, the reliable dissipative filter design method for singular system (5.58) is presented in the following theorem:

Theorem 5.38. *Let matrices* Q, S, *and* R *be given with* Q *and* R *real symmetric and* $Q \le 0$. *Then the system in (5.62) with sensor failure and polytopic uncertainty (5.60) is robustly admissible and strictly* (Q, S, R)*-dissipative if there exist matrices* $\begin{bmatrix} P_{1i} & P_{2i} \\ \star & P_{3i} \end{bmatrix} > 0$, $\tilde{Q}_{li} > 0$, $\bar{Q}_{li} > 0$, $l = 1,2,3$, $S_{ji} > 0$, H_j, $F_j, j = 1,2$, *diagonal matrix* $N > 0$, T_1, M_i, *and* Z *such that the following set of LMIs hold for* $i = 1, \ldots, q$:

$$\begin{bmatrix} S_{2i} & M_i \\ \star & S_{2i} \end{bmatrix} \ge 0, \tag{5.82}$$

$$\Omega_i = \begin{bmatrix} \Xi_i & \Gamma_i \\ \star & \Lambda_i \end{bmatrix} < 0, \tag{5.83}$$

where

$$\Xi_i = \begin{bmatrix} \Xi_{11i} & -E^T P_{2i} & E^T S_{1i} E & Z^T U^T A_{di} + C_i^T N C_{di} & 0 & \Xi_{16i} \\ \star & \Xi_{22i} & 0 & 0 & 0 & \bar{C}_f^T S \\ \star & \star & \Xi_{33i} & E^T(S_{2i} - M_i^T)E & E^T M_i^T E & 0 \\ \star & \star & \star & \Xi_{44i} & E^T(S_{2i} - M_i^T)E & \Xi_{46i} \\ \star & \star & \star & \star & \Xi_{55i} & 0 \\ \star & \star & \star & \star & \star & \Xi_{66i} \end{bmatrix},$$

$$\Gamma_i = \begin{bmatrix} \Gamma_{11i} & \Gamma_{12i} & \Gamma_{13i} & A_i^T T_1^T + C_i^T \mathcal{A}_{s0} \bar{B}_f^T & A_i^T F_1^T + C_i^T \mathcal{A}_{s0} \bar{B}_f^T & 0 \\ 0 & 0 & \bar{C}_f^T Q_{-}^{\frac{1}{2}} & \bar{A}_f^T & \bar{A}_f^T & 0 \\ 0 & 0 & 0 & 0 & 0 & 0 \\ d_1 A_{di}^T H_1^T & \tilde{d} A_{di}^T H_2^T & \Gamma_{43i} & A_{di}^T T_1^T + C_{di}^T \mathcal{A}_{s0} \bar{B}_f^T & A_{di}^T F_1^T + C_{di}^T \mathcal{A}_{s0} \bar{B}_f^T & 0 \\ 0 & 0 & 0 & 0 & 0 & 0 \\ d_1 B_i^T H_1^T & \tilde{d} B_i^T H_2^T & \Gamma_{63i} & B_i^T T_1^T + D_i^T \mathcal{A}_{s0} \bar{B}_f^T & B_i^T F_1^T + D_i^T \mathcal{A}_{s0} \bar{B}_f^T & S^T \bar{D}_f \mathcal{B} \end{bmatrix},$$

$$\Lambda_i = \begin{bmatrix} S_{1i} - H_1 - H_1^T & 0 & 0 & 0 & 0 & 0 \\ \star & S_{2i} - H_2 - H_2^T & 0 & 0 & 0 & 0 \\ \star & \star & -I & 0 & 0 & -Q_{-}^{\frac{1}{2}} \bar{D}_f \mathcal{B} \\ \star & \star & \star & P_{1i} - T_1 - T_1^T & P_{2i} - F_2 - F_1^T & \bar{B}_f \mathcal{B} \\ \star & \star & \star & \star & P_{3i} - F_2 - F_2^T & \bar{B}_f \mathcal{B} \\ \star & \star & \star & \star & \star & -N \end{bmatrix},$$

$$\Xi_{11i} = -E^T(P_{1i} + S_{1i})E + \bar{Q}_{1i} + \bar{Q}_{2i} + (\tilde{d} + 1)\bar{Q}_{3i} + \mathrm{sym}(A_i^T U Z) + C_i^T N C_i,$$

$$\Xi_{22i} = -P_{3i} + \tilde{Q}_{1i} + \tilde{Q}_{2i} + (\tilde{d} + 1)\tilde{Q}_{3i},$$

$$\Xi_{33i} = -E^T S_{2i} E - \bar{Q}_{1i} - E^T S_{1i} E,$$

$$\Xi_{44i} = -\bar{Q}_{3i} + \mathrm{sym}(E^T(M_i - S_{2i})E) + C_{di}^T N C_{di},$$

$$\Xi_{55i} = -\bar{Q}_{2i} - E^T S_{2i} E, \quad \Gamma_{63i} = (G_i - \bar{D}_f \mathcal{A}_{s0} D_i)^T Q_{-}^{\frac{1}{2}},$$

$$\Xi_{66i} = \mathrm{sym}(-G_i^T S - R + S^T \bar{D}_f \mathcal{A}_{s0} D_i) + D_i^T N D_i,$$

$$\Xi_{16i} = Z^T U^T B_i - L_i^T S + C_i^T \mathcal{A}_{s0} \bar{D}_f S + C_i^T N D_i,$$

$$\Xi_{46i} = -L_{di}^T S + C_{di}^T \mathcal{A}_{s0} \bar{D}_f^T S + C_{di}^T N D_i,$$

$$\Gamma_{11i} = d_1(A_i - E)^T H_1^T, \quad \Gamma_{12i} = \tilde{d}(A_i - E)^T H_2^T,$$

$$\Gamma_{13i} = (L_i - \bar{D}_f \mathcal{A}_{s0} C_i)^T Q_{-}^{\frac{1}{2}}, \quad \Gamma_{43i} = (L_{di} - \bar{D}_f \mathcal{A}_{s0} C_{di})^T Q_{-}^{\frac{1}{2}},$$

where $U \in \mathbb{R}^{n\times(n-r)}$ *is any matrix with full column rank satisfying* $E^T U = 0$. *Moreover, a suitable robustly reliable dissipative filter is given by*

$$A_f = \bar{A}_f F_2^{-1},\ B_f = \bar{B}_f,\ C_f = \bar{C}_f F_2^{-1},\ D_f = \bar{D}_f. \tag{5.84}$$

Proof. By using Schur complement equivalence, (5.83) is equivalent to

$$\begin{bmatrix} \bar{\Xi}_i & \bar{\Gamma}_i \\ \star & \bar{\Lambda}_i \end{bmatrix} + \Psi_{1i}^T N \Psi_{1i} + \Psi_2^T \mathcal{B} N^{-1} \mathcal{B} \Psi_2 < 0, \tag{5.85}$$

where $\bar{\Xi}_i$ is Ξ_i in (5.83) with the terms, including N disappear; $\bar{\Gamma}_i$ is Γ_i in (5.83) without the last column; $\bar{\Lambda}_i$ is Λ_i in (5.83) without the last row and last column, and

$$\begin{aligned} \Psi_{1i} &= \begin{bmatrix} C_i & 0 & 0 & C_{di} & 0 & D_i & 0 & 0 & 0 & 0 & 0 \end{bmatrix}, \\ \Psi_{2i} &= \begin{bmatrix} 0 & 0 & 0 & 0 & 0 & \bar{D}_f^T S & 0 & 0 & -\bar{D}_f^T Q_-^{\frac{1}{2}} & \bar{B}_f^T & \bar{B}_f^T \end{bmatrix}. \end{aligned}$$

Then, using the elementary inequality $x^T y + y^T x \leq \varepsilon x^T x + \varepsilon^{-1} y^T y$, we have

$$\bar{\Omega}_i = \begin{bmatrix} \bar{\Xi}_i & \bar{\Gamma}_i \\ \star & \bar{\Lambda}_i \end{bmatrix} + \mathrm{sym}\left(\Psi_{1i}^T \Delta_s \Psi_2\right) < 0. \tag{5.86}$$

Introduce four matrices T_1, T_2, T_3, and T_4 with T_4 invertible and define

$$\begin{aligned} J_1 &= \begin{bmatrix} I & 0 \\ 0 & T_2 T_4^{-1} \end{bmatrix},\ F_1 = T_2 T_4^{-1} T_3,\ F_2 = T_2 T_4^{-T} T_2^T, \\ \tilde{Q}_i &= T_2 T_4^{-1} \bar{Q}_i T_4^{-T} T_2,\ J = \mathrm{sym}\{J_1, I, I, I, I, I, I, I, J_1\}, \\ \tilde{U} &= \begin{bmatrix} U \\ 0 \end{bmatrix},\ \tilde{Z} = \begin{bmatrix} Z & 0 \end{bmatrix},\ \begin{bmatrix} P_{1i} & P_{2i} \\ \star & P_{3i} \end{bmatrix} = J_1 P_i J_1^T, \\ T &= \begin{bmatrix} T_1 & T_2 \\ T_3 & T_4 \end{bmatrix}, \\ \begin{bmatrix} \bar{A}_f & \bar{B}_f \\ \bar{C}_f & \bar{D}_f \end{bmatrix} &= \begin{bmatrix} T_2 & 0 \\ 0 & I \end{bmatrix} \begin{bmatrix} A_f & B_f \\ C_f & D_f \end{bmatrix} \begin{bmatrix} T_4^{-T} T_2^T & 0 \\ 0 & I \end{bmatrix}. \end{aligned} \tag{5.87}$$

From (5.83), we have $F_2 + F_2^T = T_2 T_4^{-T} T_2^T + T_2 T_4^{-1} T_2^T > 0$, which implies that T_2 is nonsingular. Hence, J is nonsingular. Noting that,

$$\bar{\Omega}_i = J \tilde{\Omega}_i J^T,$$

we have $\tilde{\Omega}_i < 0$, and the filtering error system in (5.62) is robustly admissible and strictly (Q, S, R)-dissipative from (5.81). Because T_2 and T_4 cannot be obtained from (5.83), we cannot determine the filters from (5.87). From (5.87), we have

$$\begin{aligned} A_f &= T_2^{-1}\bar{A}_f F_2^{-1} T_2, \\ B_f &= T_2^{-1}\bar{B}_f, \\ C_f &= \bar{C}_f F_2^{-1} T_2^{-1}, \\ D_f &= \bar{D}_f. \end{aligned}$$

Hence, the systems (A_f, B_f, C_f, D_f) and $(\bar{A}_f F_2^{-1}, \bar{B}_f, \bar{C}_f F_2^{-1}, \bar{D}_f)$ are algebraically equivalent, and the desired filter can be obtained from (5.84). This completes the proof. □

Remark 5.39. The number of LMI decision variables in Theorem 5.38 is $7(q+1)n^2+(5q+p+m-r)n+(p+1)m$, where n, m, q, p, and r are defined in (5.58). If the delay-partitioning method is employed, the number of decision variables are dependent not only on system dimensions, but also the partitioning number l. The computational burden will be significantly increased with the growth of l. Consequently, a longer central processing unit time will be consumed when testing the criterion. Therefore the reciprocal convex approach reduces the conservatism with less variables, whereas the delay-partitioning method improves the result by costing more computer processing unit time. The reciprocal convex approach can be seen as a method striking a balance between time-consumption and reduction of conservatism.

Based on Theorem 5.36, the following by-product condition can be easily obtained for the dissipative filtering of the delay singular system in (5.63) without sensor failures.

Corollary 5.40. *Let matrices Q, S, and R be given with Q and R real symmetric and $Q \leq 0$. Then the system in (5.62) with $\mathcal{A}_s = I$ and polytopic uncertainty (5.60) is robustly admissible and strictly (Q, S, R)-dissipative if there exist matrices $\begin{bmatrix} P_{1i} & P_{2i} \\ \star & P_{3i} \end{bmatrix} > 0$, $\tilde{Q}_{li} > 0$, $\bar{Q}_{li} > 0$, $l = 1, 2, 3$, $S_{ji} > 0$, H_j, F_j, $j = 1, 2$, T_1, M_i, and Z such that the following set of LMIs hold for $i = 1, \ldots, q$:*

$$\begin{bmatrix} S_{2i} & M_i \\ \star & S_{2i} \end{bmatrix} \geq 0, \tag{5.88}$$

$$\Omega_i = \begin{bmatrix} \bar{\Xi}_i & \bar{\Gamma}_i \\ \star & \bar{\Lambda}_i \end{bmatrix} < 0, \tag{5.89}$$

where $\bar{\Xi}_i$, $\bar{\Gamma}_i$, *and* $\bar{\Lambda}_i$ *are defined in (5.85).*

Remark 5.41. The reliably dissipative filtering result in Theorem 5.38 includes reliable H_∞ filtering and passivity filtering as special cases:

1. When $Q=-I$, $S=0$, $R=\gamma^2 I$, it reduces to the reliable H_∞ filtering case, such as [111], [110].
2. When $Q=0$, $S=I$, $R=0$, and $\mathcal{A}_s=1$, it reduces to the passivity filtering case without considering reliability, such as [1], [35], [65].

5.3.4 Illustrative examples

In this section, three examples are provided to illustrate the effectiveness of the proposed approach. The examples chosen here have been widely used in existing references [78], [150], [241], and the reduced conservatism can be easily illustrated and compared.

Example 5.8. To illustrate the reduced conservatism of the robust H_∞ filtering result and the applicability of the filter design method, consider a delay singular system in (5.58) with parameters borrowed from [78]:

$$\begin{aligned} E &= \begin{bmatrix} 1 & 0 \\ 0 & 0 \end{bmatrix}, \ A=\begin{bmatrix} 0.9 & 0 \\ 0 & 0.7+\Delta_1 \end{bmatrix}, \ A_d=\begin{bmatrix} -0.1 & \Delta_2 \\ -0.1 & -0.1 \end{bmatrix}, \\ B &= \begin{bmatrix} 0 \\ 1 \end{bmatrix}, \ C=\begin{bmatrix} 1 & 1 \end{bmatrix}, \ C_d=\begin{bmatrix} 0.2 & 0.5 \end{bmatrix}, \ L=\begin{bmatrix} 1 & 2 \end{bmatrix}, \\ L_d &= \begin{bmatrix} 0.5 & 0.6 \end{bmatrix}, \ D=1, \ G=-0.5, \ |\Delta_1| \le 0.2, \ |\Delta_2| \le 0.1. \end{aligned}$$

- H_∞ case: $Q=-I$, $S=0$, $R=\gamma^2 I$, $\mathcal{A}_s=I$. To compare with the H_∞ result in [78], the detailed comparisons on the minima of γ for given d_1 and d_2 are listed in Table 5.10. From Table 5.10, we can see that the minimum values of the H_∞ performance index obtained from [78] and Corollary 5.40 are 4.9515 and 4.7221, with $d_1=1$ and $d_2=5$,

Table 5.10 Comparisons of minimum allowed γ for $d_2=5$.

Methods \ d_1	1	2	3	4	5
[78]	4.9515	4.5463	4.2144	3.9073	3.5960
Corollary 5.40	4.7221	4.3631	4.0815	3.8359	3.5960

respectively. When $d_1 = 3$, $d_2 = 5$, and $\gamma = 4.0815$, a feasible solution can be found by applying Corollary 5.40, whereas it fails to do that by using the results in [78]. It is clearly seen that the minimum γ value obtained in this section is less than that in [78] when the time delays d_1 and d_2 are the same.
Moreover, when $d_1 = 1$, $d_2 = 5$, the corresponding robust filter is

$$\begin{aligned} A_f &= \begin{bmatrix} 0.2470 & -0.0006 \\ 6.4106 & -0.0158 \end{bmatrix}, \; B_f = \begin{bmatrix} -0.0218 \\ 0.6068 \end{bmatrix} \times 10^{-3} \\ C_f &= \begin{bmatrix} 0.3723 & 0.0002 \end{bmatrix}, \; D_f = 1.5823. \end{aligned}$$

- General dissipative case: $\mathcal{Q} = -0.25$, $\mathcal{S} = -2$, $\mathcal{R} = 4$, $\mathcal{A}_s = 1$, $d_1 = 2$, $d_2 = 6$. By solving the LMIs in Corollary 5.40, the corresponding robust filter is

$$\begin{aligned} A_f &= \begin{bmatrix} 0.5884 & -0.0001 \\ 0.0861 & -0.0002 \end{bmatrix}, \; B_f = \begin{bmatrix} 0.0156 \\ -0.0000 \end{bmatrix}, \\ C_f &= \begin{bmatrix} 0.1701 & -0.0002 \end{bmatrix}, \; D_f = 0.9548. \end{aligned} \tag{5.90}$$

The estimate error $\tilde{z}$ of all vertices of the polytopic system is given in Fig. 5.3.

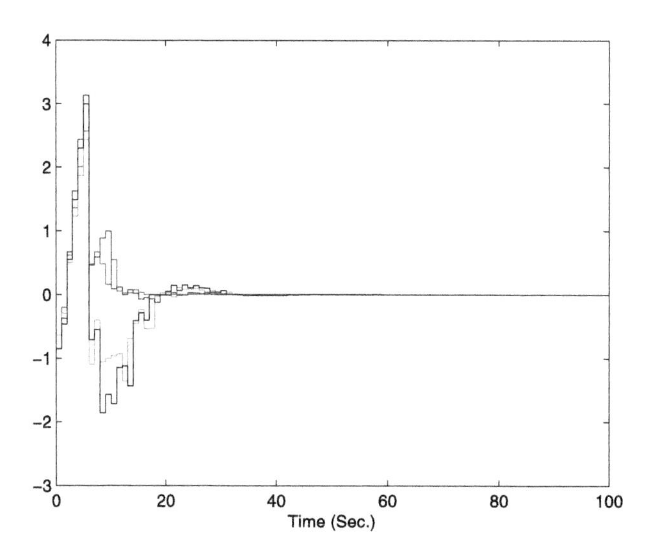

Figure 5.3 The estimate error $\tilde{z}$ of all vertices of the polytopic system.

- Reliably dissipative case: $Q = -0.25$, $S = -2$, $R = 4$, $\underline{\mathcal{A}}_s = 0.4$, $\bar{\mathcal{A}}_s = 0.6$, $d_1 = 2$, $d_2 = 6$. By solving the LMIs in Theorem 5.38, the corresponding robust reliability filter is

$$\begin{aligned} A_f &= \begin{bmatrix} 0.5060 & 0.0009 \\ 0.4604 & 0.0011 \end{bmatrix}, \; B_f = \begin{bmatrix} 0.0066 \\ 0.0386 \end{bmatrix}, \\ C_f &= \begin{bmatrix} 0.3256 & 0.0008 \end{bmatrix}, \; D_f = 1.9694. \end{aligned} \tag{5.91}$$

The estimate error $\tilde{z}$ of all vertices of the polytopic system with the filter in (5.91) for different sensor failure cases is given in Figs. 5.4–5.6, respectively.

Example 5.9. To show the generality of the dissipative filtering, some system matrices in Example 5.9 are changed as follows:

$$D = -1, \; G = 1. \tag{5.92}$$

- H_∞ case: $Q = -\gamma^{-1}I$, $S = 0$, $R = \gamma I$, $\mathcal{A}_s = 1$, $d_1 = 2$, and $d_2 = 6$. By solving the LMIs in Corollary 5.40 with γ being 5 and the correspond-

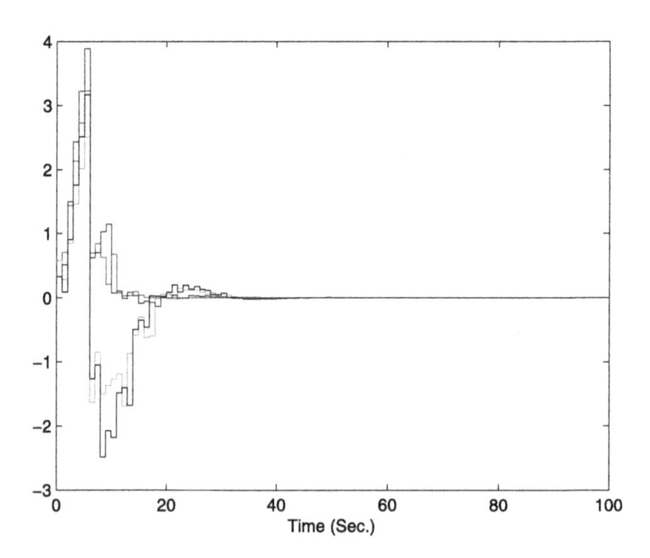

Figure 5.4 $\mathcal{A}_s = 0.4$.

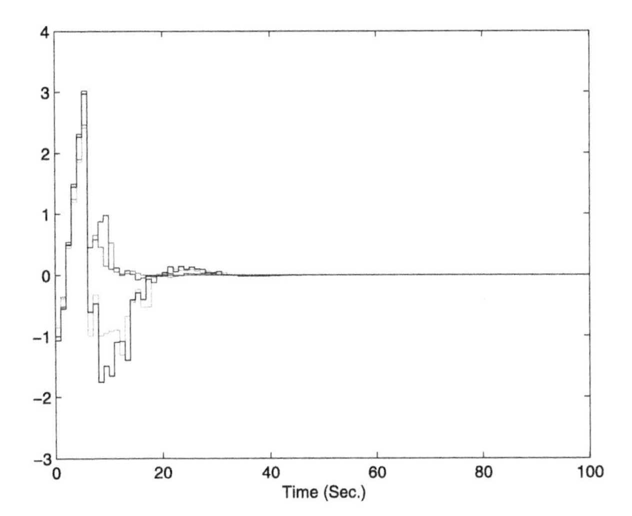

Figure 5.5 $\mathcal{A}_S = 0.5$.

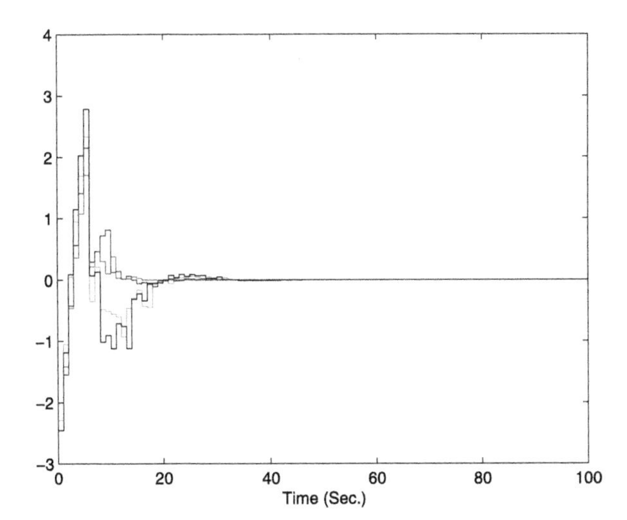

Figure 5.6 $\mathcal{A}_S = 0.6$.

ing robust filter is

$$\begin{aligned} A_f &= \begin{bmatrix} 0.5203 & 0.0011 \\ 0.6648 & 0.0013 \end{bmatrix}, \; B_f = \begin{bmatrix} 0.0053 \\ 0.0132 \end{bmatrix}, \\ C_f &= \begin{bmatrix} 2.4475 & 0.0020 \end{bmatrix}, \; D_f = 1.3850. \end{aligned} \tag{5.93}$$

- Passive case: $Q=0$, $S=I$, $R=0$, $\mathcal{A}_s=1$, $d_1=2$, and $d_2=6$. By solving the LMIs in Corollary 5.40, the corresponding robust filter is

$$\begin{aligned} A_f &= \begin{bmatrix} 0.3823 & -0.0013 \\ -0.1285 & -0.0007 \end{bmatrix}, \ B_f = \begin{bmatrix} -0.1832 \\ -0.2188 \end{bmatrix}, \\ C_f &= \begin{bmatrix} 0.4744 & -0.0017 \end{bmatrix}, \ D_f = 21.8321. \end{aligned} \tag{5.94}$$

- Dissipative case: $Q=-\gamma^{-1}\theta I$, $S=(1-\theta)I$, $R=\gamma\theta I$, $\theta \in (0,1)$, $\mathcal{A}_s=1$, $d_1=2$, and $d_2=6$. It can be seen that it is the trade-off between H_∞ and passivity performance with $\theta \in (0,1)$. By solving the LMIs in Corollary 5.40 with $\theta=0.5$ and $\gamma=3.8565$, the corresponding robust dissipative filter is

$$\begin{aligned} A_f &= \begin{bmatrix} 0.5115 & 0.0007 \\ 0.1282 & 0.0005 \end{bmatrix}, \ B_f = \begin{bmatrix} 0.0051 \\ 0.0074 \end{bmatrix}, \\ C_f &= \begin{bmatrix} 3.3265 & 0.0001 \end{bmatrix}, \ D_f = 1.8731. \end{aligned} \tag{5.95}$$

Example 5.10. To show the reduction of conservatism for the stability criterion, we consider the state-space system borrowed from [53]:

$$x(k+1) = \begin{bmatrix} 0.8 & 0 \\ 0.05 & 0.9 \end{bmatrix} x(k) + \begin{bmatrix} -0.1 & 0 \\ -0.2 & -0.1 \end{bmatrix} x(k-d(k)).$$

To compare with the results in [150], which proposed latest results on stability condition for discrete-time with time-varying delay, Corollary 5.35 is used with $E=I$ and $U=0$. When the lower bounds of $d(k)$ are 7 and 10, the upper bounds such that this system is asymptotically stable are 16 and 18, respectively, by using Proposition 2 in [150]. However, by employing the result in Corollary 5.35, the upper bounds can reach up to 18 and 20, respectively. Moreover, for different value of d_1, the admissible upper bound d_2 is obtained from different methods in Table 5.11. It can be seen that Corollary 5.35 can tolerate larger delay upper bound with the same delay lower bound than Proposition 2 in [150], and it can get the same delay upper bound with Proposition 1 in [150]. However, the decision variables in Proposition 1 is larger than that in Corollary 5.35 due to directly dealing with the inversely weighted convex combination of quadratic terms of summable quantities by utilizing the reciprocally convex approach, which illustrates the reduced conservatism of our method.

Table 5.11 Comparisons of maximum allowed d_2 for given d_1.

Methods \ d_1	**7**	**10**	**15**	**20**	**25**	**Variables**
[150] Proposition 2	16	18	21	25	30	18
[150] Proposition 1	18	20	23	27	31	38
Corollary 5.35	18	20	23	27	31	22

5.3.5 Conclusion

In this subchapter, the problem of robust reliable dissipative control for uncertain discrete-time singular systems with time-varying delay and sensor failures has been studied. The sufficient conditions in terms of LMIs have been proposed for rendering considered filtering error singular systems admissible and strictly (Q, S, R)-dissipative. Based on these results, a desired filter is designed to guarantee the filtering error singular system to be admissible and strictly (Q, S, R)-dissipative. The results presented in this section are in terms of strict LMIs, which make the conditions more tractable. Moreover, the results benefiting from the reciprocally convex approach proposed in [140] have improved the existing results, and H_∞ filtering and passive filtering of singular systems are unified in the proposed results. Finally, numerical examples are given to demonstrate the effectiveness of our methods.

CHAPTER 6

State-feedback control for singular Markovian systems

In this chapter, we consider the problem of state-feedback control for continuous singular Markovian systems. Firstly, for delay-free singular Markovian systems, necessary and sufficient conditions are proposed for the system to be admissible and for the state-feedback control design by applying equivalent sets technique. For time-delay singular Markovian systems, a new bounded real lemma is proposed and the corresponding H_∞ control problem is studied. Secondly, we consider the problem of reliable dissipative control for continuous-time singular Markovian systems with actuator failure. Our attention is focused on the state-feedback controller design method such that the closed-loop system is regular, impulse free, stochastically stable, and strictly (Q, S, R)-dissipative. Numerical examples are given to illustrate the effectiveness of the theoretic results developed.

6.1 Admissibilization and H_∞ control for singular Markovian systems

In this section, the issues of admissibilization and H_∞ control for continuous singular Markovian systems and singular delay Markovian systems are addressed, respectively, by applying equivalent sets technique.

6.1.1 Admissibility of singular Markovian jump systems

Consider the following singular Markovian jump systems:

$$E\dot{x}(t) = A(r_t)x(t) + B(r_t)u(t), \tag{6.1}$$

where $x(t) \in \mathbb{R}^n$ is the state vector; $u(t) \in \mathbb{R}^m$ is the control vector. The matrices $A(\cdot)$ and $B(\cdot)$, which are functions of r_t, are known real matrices with appropriate dimensions. $\{r_t, t \geq 0\}$ is a continuous-time Markov process, which takes values in a finite set $\mathcal{N} = \{1, 2, \ldots, N\}$ and describes the evolution of the mode at time t. For notational simplicity, for each $r_t = i$, $i \in \mathcal{N}$, matrix $A(r_t)$ will be denoted by A_i. The Markov process describes

Analysis and Synthesis of Singular Systems
https://doi.org/10.1016/B978-0-12-823739-7.00013-6

the switching between different modes, and its evolution is governed by the following probability transitions:

$$\Pr\{r_{t+\delta}=j|r_t=i\}=\begin{cases}\pi_{ij}\delta+o(\delta) & i\neq j\\ 1+\pi_{ii}\delta+o(\delta) & i=j,\end{cases} \tag{6.2}$$

where $\delta>0$ and $\lim_{\delta\to 0}\frac{o(\delta)}{\delta}=0$, and $\pi_{ij}\geq 0$, for $i\neq j$, is the transition rate from mode i at time t to mode j at time $t+\delta$, which satisfies $\pi_{ii}=-\sum_{j=1,i\neq j}^{N}\pi_{ij}$.

We recall the following lemma:

Lemma 6.1. [209] *System (6.1) with $u(t)=0$ is stochastically admissible if and only if there exist matrices X_i, $i\in\mathcal{N}$, such that the following coupled LMIs holds for each $i\in\mathcal{N}$:*

$$E^TX_i={X_i}^TE\geq 0, \tag{6.3}$$

$${X_i}^TA_i+A_i^TX_i+\sum_{j=1}^{N}\pi_{ij}E^TX_j<0. \tag{6.4}$$

Remark 6.2. The matrices X_i, $i\in\mathcal{N}$, in Lemma 6.1 are nonsingular. If they are singular, then there exist nonzero vectors ξ_i, $i\in\mathcal{N}$, such that $X_i\xi_i=0$. Then for the nonzero vector ξ_i, we have the following inequality from (6.4):

$$\begin{aligned}&\sum_{j=1,j\neq i}^{N}\pi_{ij}{\xi_i}^TE^TX_j\xi_i+\pi_{ii}{\xi_i}^TE^TX_i\xi_i+{\xi_i}^T{X_i}^TA_i\xi_i+{\xi_i}^TA_i^TX_i\xi_i\\ &=\sum_{j=1,j\neq i}^{N}\pi_{ij}{\xi_i}^TE^TX_j\xi_i<0.\end{aligned} \tag{6.5}$$

As $E^TX_j\geq 0$ and $\pi_{ij}\geq 0$, $i\neq j$, the inequality in (6.5) cannot hold, which implies the inequality in (6.4) does not hold. Based on the discussions, matrices X_i, $i\in\mathcal{N}$, are nonsingular.

Then combining Lemma 2.5 with Remark 6.2, a new necessary and sufficient admissibility condition of singular Markovian jump systems can be obtained from Lemma 6.1, which is given in the following theorem:

Theorem 6.3. *System (6.1) with $u(t)=0$ is stochastically admissible if and only if there exist symmetric matrices P_i, $i\in\mathcal{N}$, and nonsingular matrices Φ_i, $i\in\mathcal{N}$, such that the following coupled LMIs hold for each $i\in\mathcal{N}$:*

$$E_L^T P_i E_L > 0, \tag{6.6}$$

$$\mathrm{sym}(A_i^T(P_iE + U^T\Phi_i\Lambda^T)) + \sum_{j=1}^{N}\pi_{ij}E^TP_jE < 0, \tag{6.7}$$

where E_L, U, and Λ are defined in Lemma 2.5.

Remark 6.4. It should be mentioned that the obtained admissibility condition in Theorem 6.3 does not require positive definite matrices P_i, $i \in \mathcal{N}$. From this point of view, the result in Theorem 6.3 is less conservative.

Now we consider the state feedback control problem for system (6.1) with $u(t) = K(r_t)x(t)$ such that the closed-loop system

$$E\dot{x}(t) = \bar{A}(r_t)x(t) = (A(r_t) + B(r_t)K(r_t))x(t) \tag{6.8}$$

is stochastically admissible.

A necessary and sufficient state feedback controller design condition is proposed in the following result, based on Theorem 6.3:

Theorem 6.5. *There exists a state feedback controller such that the closed-loop system in (6.8) is stochastically admissible if and only if there exist symmetric matrices $\bar{P}_i$, $i \in \mathcal{N}$; nonsingular matrices $\bar{\Phi}_i$, $i \in \mathcal{N}$, and matrices H_i, $i \in \mathcal{N}$, such that the following LMIs hold for each $i \in \mathcal{N}$:*

$$E_R^T\bar{P}_iE_R > 0, \tag{6.9}$$

$$\begin{bmatrix} \mathrm{sym}(A_iY_i + B_iH_i) + \pi_{ii}E\bar{P}_iE^T & Y_i^T\Omega_i \\ \star & -\Psi_i \end{bmatrix} < 0, \tag{6.10}$$

where

$$Y_i = \bar{P}_iE^T + \Lambda\bar{\Phi}_iU,$$
$$\Omega_i = \begin{bmatrix} \sqrt{\pi_{i1}}E_R & \sqrt{\pi_{i2}}E_R & \cdots & \sqrt{\pi_{i(i-1)}}E_R & \sqrt{\pi_{i(i+1)}}E_R & \cdots & \sqrt{\pi_{iN}}E_R \end{bmatrix},$$
$$\Psi_i = \mathrm{diag}\{E_R^T\bar{P}_1E_R,\ E_R^T\bar{P}_2E_R,\ \ldots,\ E_R^T\bar{P}_{i-1}E_R,\ E_R^T\bar{P}_{i+1}E_R,\ \ldots, E_R^T\bar{P}_NE_R\}.$$

E_R, U, and Λ are defined in Lemma 2.5. Then, the desired controller can be obtained by $K_i = H_iY_i^{-1}$.

Proof. Based on Theorem 6.3, the closed-loop system (6.8) is stochastically admissible if and only if there exist symmetric matrices P_i, $i \in \mathcal{N}$ and nonsingular matrices Φ_i, $i \in \mathcal{N}$, such that the following coupled LMIs hold for each $i \in \mathcal{N}$:

$$E_L^T P_i E_L > 0, \tag{6.11}$$

$$\mathrm{sym}((A_i + B_i K_i)^T (P_i E + U^T \Phi_i \Lambda^T)) + \sum_{j=1}^{N} \pi_{ij} E^T P_j E < 0. \tag{6.12}$$

Considering $\pi_{ij} > 0$, $i \neq j$, $E_L^T P_i E_L > 0$, and using Schur complement equivalence, (6.12) is equivalent to

$$\begin{bmatrix} \mathrm{sym}((A_i + B_i K_i)^T (P_i E + U^T \Phi_i \Lambda^T)) + \pi_{ii} E^T P_i E & \Omega_i \\ \star & -\Psi_i \end{bmatrix}$$
$$= \begin{bmatrix} \mathrm{sym}((A_i + B_i K_i)^T (P_i E + U^T \Phi_i \Lambda^T)) + \pi_{ii} E^T (P_i E + U^T \Phi_i \Lambda^T) & \Omega_i \\ \star & -\Psi_i \end{bmatrix}$$
$$<0.$$

Then performing congruence transformation to above inequality with matrix $\begin{bmatrix} Y_i & 0 \\ 0 & I \end{bmatrix}$ with $Y_i = \bar{P}_i E^T + \Lambda \bar{\Phi}_i U = (P_i E + U^T \Phi_i \Lambda^T)^{-1}$, and setting $H_i = K_i Y_i$, the inequality in (6.10) is obtained. Because of $E_R^T \bar{P}_i E_R = (E_L^T P_i E_L)^{-1}$, the inequalities in (6.9) and (6.11) are equivalent. □

Remark 6.6. It should be mentioned that Theorem 6.5 provides a necessary and sufficient admissibility condition of state feedback control of singular Markovian systems, whereas the admissibilization condition proposed in [199] is only a necessary condition. On the other hand, to obtain a feedback control method in terms of linear matrix inequalities, the term $\pi_{ii} Y_i^T E_R (E_R^T \bar{P}_i E_R)^{-1} E_R^T Y_i$ is enlarged as $\pi_{ii}(Y_i^T E^T + E Y_i - E \bar{P}_i E^T)$, and a condition is obtained in [199]. Because $E\Lambda = E_L E_R^T \Lambda = 0$ and E_L is of full column rank, it yields that $E_R^T \Lambda = 0$. Moreover, noting that

$$\begin{aligned} \pi_{ii} Y_i^T E_R (E_R^T \bar{P}_i E_R)^{-1} E_R^T Y_i &= \pi_{ii} E_L E_R^T \bar{P}_i E_R (E_R^T \bar{P}_i E_R)^{-1} E_R^T \bar{P}_i E_R E_L^T \\ &= \pi_{ii} E \bar{P}_i E^T, \end{aligned}$$

we have the necessary and sufficient state feedback control result in Theorem 6.5.

6.1.2 H_∞ control of singular Markovian jump systems with time delay

Consider a class of singular Markovian jump systems with time delay described by

$$\begin{cases} E\dot{x}(t) = A(r_t)x(t) + A_d(r_t)x(t-d) + B(r_t)u(t) + B_w(r_t)w(t) \\ z(t) = C(r_t)x(t) + C_d(r_t)x(t-d) + D(r_t)u(t) + D_w(r_t)w(t) \\ x(t) = \varphi(t),\ t \in [-d, 0], \end{cases} \tag{6.13}$$

where $x(t)$, $u(t)$, E, $A(r_t)$, and $B(r_t)$ are defined in system (6.1); $w(t) \in \mathbb{R}^p$ represents the exogenous input (which includes disturbances to be rejected), and $z(t) \in \mathbb{R}^q$ is the controlled output; $A_d(r_t)$, $C_d(r_t)$, $B_w(r_t)$, $C(r_t)$, $D(r_t)$, and $D_w(r_t)$ denote known real matrices with appropriate dimensions; d is the constant time delay and $\varphi(t)$ represents initial condition.

Following a similar line as that in [184,185], a new bounded real lemma for time-delay singular Markovian jump system (6.13) is proposed in the following lemma:

Lemma 6.7. *Given a scalar $\gamma > 0$, the singular Markovian jump system in (6.13) with $u(t) = 0$ is stochastically admissible with an H_∞ performance γ if there exist nonsingular matrices X_i, $i \in \mathcal{N}$; matrices $Q > 0$, $R > 0$ and W_i, $i \in \mathcal{N}$, such that the following LMIs hold for each $i \in \mathcal{N}$:*

$$E^T X_i = X_i^T E \geq 0, \tag{6.14}$$

$$\begin{bmatrix} \Psi_{11i} & X_i^T A_{di} - W_i E & X_i^T B_{wi} & dW_i E & dA_i^T R & C_i^T \\ \star & -Q & 0 & 0 & dA_{di}^T R & C_{di}^T \\ \star & \star & -\gamma^2 I & 0 & dB_{wi}^T R & D_{wi}^T \\ \star & \star & \star & -dR & 0 & 0 \\ \star & \star & \star & \star & -dR & 0 \\ \star & \star & \star & \star & \star & -I \end{bmatrix} < 0, \tag{6.15}$$

where $\Psi_{11i} = X_i^T A_i + A_i^T X_i + Q + W_i E + E^T W_i^T + \sum_{j=1}^N \pi_{ij} E^T X_j$.

Proof. The result can be obtained by choosing the following Lyapunov function and free-weighting equation:

$$V(x_t) = x^T(t)E^T X_i x(t) + \int_{t-d}^{t} x(\tau)^T Q x(\tau) d\tau + \int_{-d}^{0}\int_{t+\theta}^{t} \dot{x}(\tau)^T E^T R E \dot{x}(\tau) d\tau d\theta,$$

$$0 = 2x(t)^T W_i E \left[x(t) - x(t-d) - \int_{t-d}^{t} \dot{x}(\tau)d\tau \right].$$

To prove the invertibility of matrices X_i, $i \in \mathcal{N}$, the following inequality will be used

$$\begin{bmatrix} \Psi_{11i} & X_i^T A_{di} - W_i E \\ \star & -Q \end{bmatrix} < 0, \tag{6.16}$$

which can be derived from (6.15). Then pre- and postmultiplying (6.16) by $[I, \ I]$ and its transpose, respectively, we obtain

$$\text{sym}(X_i^T(A_i + A_{di})) + \sum_{j=1}^{N} \pi_{ij} E^T X_j < 0,$$

which implies the matrices X_i, $i \in \mathcal{N}$, are nonsingular based on discussions in Remark 6.2. □

Noting that the invertibility of matrices X_i, $i \in \mathcal{N}$, in Lemma 6.7, a new bounded real lemma for singular system (6.13) is presented below in terms of strict LMIs based on Lemma 2.5.

Theorem 6.8. *Given a scalar $\gamma > 0$, the singular Markovian jump system in (6.13) is stochastically admissible with an H_∞ performance γ if there exist symmetric matrices P_i, $i \in \mathcal{N}$, and nonsingular matrices Φ_i, $i \in \mathcal{N}$, such that the following LMIs hold for each $i \in \mathcal{N}$:*

$$E_L^T P_i E_L > 0, \tag{6.17}$$

$$\begin{bmatrix} \bar{\Psi}_{11i} & X_i^T A_{di} - W_i E & X_i^T B_{wi} & dW_i E & dA_i^T R & C_i^T \\ \star & -Q & 0 & 0 & dA_{di}^T R & C_{di}^T \\ \star & \star & -\gamma^2 I & 0 & dB_{wi}^T R & D_{wi}^T \\ \star & \star & \star & -dR & 0 & 0 \\ \star & \star & \star & \star & -dR & 0 \\ \star & \star & \star & \star & \star & -I \end{bmatrix} < 0, \tag{6.18}$$

where

$$\bar{\Psi}_{11i} = X_i^T A_i + A_i^T X_i + Q_i + W_i E + E^T W_i^T + \sum_{j=1}^{N} \pi_{ij} E^T P_j E,$$

$$X_i = P_i E + U^T \Phi_i \Lambda^T.$$

E_L, U, and Λ are defined in Lemma 2.5.

Remark 6.9. A bounded real lemma for singular Markovian systems with time delay is proposed in Lemma 1 of [184]. However, the matrix E is restricted to be $\begin{bmatrix} I_r & 0 \\ 0 & 0 \end{bmatrix}$, and the free-weighting matrix W_i needs to satisfy $W_iE = W_i$ in the process of deriving the bounded real lemma. In Theorem 6.8, these constraints are both removed. From this point of view, the result in Theorem 6.8 is more general and less conservative.

Based on Theorem 6.8, the state feedback controller $u(t) = K_ix(t)$ is designed in the following theorem:

Theorem 6.10. *Given a scalar $\gamma > 0$, the singular system in (6.13) is stochastically admissible with an H_∞ performance γ if there exist symmetric matrices $\bar{P}_i$, $i \in \mathcal{N}$; nonsingular matrices $\bar{\Phi}_i$, $i \in \mathcal{N}$; matrices $\bar{Q} > 0$, $\bar{R} > 0$, $\bar{W}_i$ and H_i, $i \in \mathcal{N}$, such that the following LMIs hold for each $i \in \mathcal{N}$:*

$$E_R^T \bar{P}_i E_R > 0, \tag{6.19}$$

$$\begin{bmatrix} \tilde{\Psi}_{11i} & A_{di}Y_i - \bar{W}_iE^T & B_{wi} & d\bar{W}_iE^T & d(A_iY_i + B_iH_i)^T & (C_iY_i + D_iH_i)^T & Y_i^T & Y_i^T\Omega_i \\ \star & \tilde{\Psi}_{22i} & 0 & 0 & d(A_{di}Y_i)^T & (C_{di}Y_i)^T & 0 & 0 \\ \star & \star & -\gamma^2 I & 0 & dB_{wi}^T & D_{wi}^T & 0 & 0 \\ \star & \star & \star & \tilde{\Psi}_{44i} & 0 & 0 & 0 & 0 \\ \star & \star & \star & \star & -d\bar{R} & 0 & 0 & 0 \\ \star & \star & \star & \star & \star & -I & 0 & 0 \\ \star & \star & \star & \star & \star & \star & -\bar{Q} & 0 \\ \star & \star & \star & \star & \star & \star & \star & -\Psi_i \end{bmatrix}$$
$$< 0, \tag{6.20}$$

where

$$\tilde{\Psi}_{11i} = \text{sym}(A_iY_i + B_iH_i + \bar{W}_iE^T) + \pi_{ii}E\bar{P}_iE^T, \quad \tilde{\Psi}_{22i} = -Y_i^T - Y_i + \bar{Q},$$
$$\tilde{\Psi}_{44i} = -dY_i^T - dY_i + d\bar{R}, \quad Y_i = \bar{P}_iE^T + \Lambda\bar{\Phi}_iU.$$

E_R, U, and Λ are defined in Lemma 2.5. Moreover, if the above LMIs are feasible, then the state feedback controller K_i can be given by $K_i = H_iY_i^{-1}$.

Proof. Based on Theorem 6.8 and using Schur complement equivalence for closed-loop system, it is stochastically admissible with an H_∞ performance if the following inequalities hold:

$$E_L^T P_i E_L > 0, \quad (6.21)$$

$$\begin{bmatrix} \hat{\Psi}_{11i} & X_i^T A_{di} - W_i E & X_i^T B_{wi} & dW_i E & d\bar{A}_i^T R & \bar{C}_i^T & I & \Omega_i \\ \star & -Q & 0 & 0 & dA_{di}^T R & C_{di}^T & 0 & 0 \\ \star & \star & -\gamma^2 I & 0 & dB_{wi}^T R & D_{wi}^T & 0 & 0 \\ \star & \star & \star & -dR & 0 & 0 & 0 & 0 \\ \star & \star & \star & \star & -dR & 0 & 0 & 0 \\ \star & \star & \star & \star & \star & -I & 0 & 0 \\ \star & \star & \star & \star & \star & \star & -Q^{-1} & 0 \\ \star & \star & \star & \star & \star & \star & \star & -\Psi_i \end{bmatrix} < 0, \quad (6.22)$$

where $\hat{\Psi}_{11i} = \text{sym}(X_i^T \bar{A}_i + W_i E) + \pi_{ii} E^T P_i E$, $\bar{A}_i = A_i + B_i K_i$, $\bar{C}_i = C_i + D_i K_i$, Ω_i, and Ψ_i are defined in Theorem 6.5.

Defining $Y_i = X_i^{-1}$, $\bar{Q} = Q^{-1}$, $\bar{R} = R^{-1}$, $H_i = K_i Y_i$, and $\bar{W}_i = Y_i^T W_i Y_i^T$, pre- and postmultiplying the inequality in (6.22) by diag$\{Y_i^T, Y_i^T, I, Y_i^T, R^{-1}, I, Y_i^T, I\}$ and its transpose, and factoring in $EY_i = Y_i^T E^T = E\bar{P}_i E^T$, we have

$$\begin{bmatrix} \tilde{\Psi}_{11i} & A_{di} Y_i - \bar{W}_i E^T & B_{wi} X_i & d\bar{W}_i E^T & d(A_i Y_i + B_i H_i)^T & (C_i Y_i + D_i H_i)^T & Y_i^T & Y_i^T \Omega_i \\ \star & -Y_i^T Q Y_i & 0 & 0 & dY_i^T A_{di}^T & Y_i^T C_{di}^T & 0 & 0 \\ \star & \star & -\gamma^2 I & 0 & dB_{wi}^T & D_{wi}^T & 0 & 0 \\ \star & \star & \star & -dY_i^T R Y_i & 0 & 0 & 0 & 0 \\ \star & \star & \star & \star & -d\bar{R} & 0 & 0 & 0 \\ \star & \star & \star & \star & \star & -I & 0 & 0 \\ \star & \star & \star & \star & \star & \star & -\bar{Q} & 0 \\ \star & \star & \star & \star & \star & \star & \star & -\Psi_i \end{bmatrix} < 0. \quad (6.23)$$

Considering $-Y_i^T Q Y_i \leq -Y_i^T - Y_i + \bar{Q}$, and $-Y_i^T R Y_i \leq -Y_i^T - Y_i + \bar{R}$, the inequality in (6.20) implies that the inequality in (6.23) holds. On the other hand, the inequality in (6.21) is equivalent to the inequality in (6.19). Therefore the proof is completed. □

Remark 6.11. Note that the H_∞ control problem is also investigated in [184], where the invertibility of matrix $\mathcal{P}_j E + \sigma \mathcal{N}$ with $\mathcal{N} = \begin{bmatrix} 0_{r \times r} & 0_{r \times (n-r)} \\ \star & I_{(n-r) \times (n-r)} \end{bmatrix}$ is required, that is, $(\mathcal{P}_j E + \sigma \mathcal{N})^{-1} = \mathcal{P}_j^{-1} E + \sigma \mathcal{N}$. The above constraints have the following two disadvantages: one is that the inverse of matrix $\mathcal{P}_j E + \sigma \mathcal{N}$ may not exist when matrix E is not in the form of $\begin{bmatrix} I_r & 0 \\ 0 & 0 \end{bmatrix}$; on the other hand, applying the parameter σ to enlarge matrix $\mathcal{P}_j E$, that is $\mathcal{P}_j E \leq \mathcal{P}_j E + \sigma \mathcal{N}$, will derive conservatism. In Theorem 6.10, the invertibility of matrix $X_i = P_i E + U^T \Phi_i \Lambda^T$ is needed. From Lemma 2.5, we can see the invertibility of matrix X_i can be guaranteed, regardless of the form of matrix E and without any enlargement.

Remark 6.12. In [174], based on the equivalence of the H_∞ norms of transfer functions of original system

$$\begin{cases} E\dot{x}(t) = (A_i + B_iK_i)x(t) + A_{di}x(t-d) + B_{wi}\omega(t) \\ z(t) = (C_ix(t) + D_iK_i)x(t) \end{cases} \tag{6.24}$$

and its dual system

$$\begin{cases} E^T\dot{x}(t) = (A_i + B_iK_i)^Tx(t) + A_{di}^Tx(t-d) + (C_i + D_iK_i)^T\omega(t) \\ z(t) = B_{wi}^Tx(t), \end{cases} \tag{6.25}$$

an H_∞ control method is given in [174]. However, the method cannot be used when the delay is time-varying, because the transfer function of a system with time-varying delay cannot be expressed explicitly. The two equivalent sets method without using transfer function in this section can be extended to systems with time-varying delay.

6.1.3 Examples

In this subsection, numerical examples are provided to show the advantages on numerical computations and the state feedback control of the equivalent sets approach.

Example 6.1. Consider the singular Markovian jump system in (6.1) with two operating modes, that is, $N = 2$ and the following parameters:

$$E = \begin{bmatrix} 1 & 0 & 0 \\ 0 & 1 & 0 \\ 0 & 0 & 0 \end{bmatrix}, \quad A_1 = \begin{bmatrix} 1.3 & 0.8 & 1.0 \\ 0.7 & 0.8 & 0.9 \\ 0.4 & 0.2 & -0.7 \end{bmatrix}, \quad B_1 = \begin{bmatrix} 1.5 & -1.5 \\ 0.9 & 0.6 \\ 1.1 & 5 \end{bmatrix},$$

$$A_2 = \begin{bmatrix} 0.7 & 0.9 & 0.3 \\ 1.1 & 1.4 & -0.7 \\ 0.5 & 0.3 & 1.6 \end{bmatrix}, \quad B_2 = \begin{bmatrix} 2 & 0.9 \\ -2 & 0.6 \\ 0.7 & 1 \end{bmatrix}, \quad E_L = E_R = \begin{bmatrix} 1 & 0 \\ 0 & 1 \\ 0 & 0 \end{bmatrix},$$

$$U = \begin{bmatrix} 0 & 0 & -0.7 \end{bmatrix}, \quad \Lambda = \begin{bmatrix} 0 \\ 0 \\ 0.8 \end{bmatrix}, \quad \begin{bmatrix} \pi_{11} & \pi_{12} \\ \pi_{21} & \pi_{22} \end{bmatrix} = \begin{bmatrix} -0.9 & 0.9 \\ 0.5 & -0.5 \end{bmatrix}.$$

By solving the characteristic equation $sE - A_1 = 0$, two finite characteristic roots are 2.6825 and 0.24604, respectively. Therefore the open-loop system is not admissible. The controller design method in Theorem 6 and Theorem 7 in [199] cannot be used. The purpose is to design a state feedback controller to guarantee the closed-loop system to be stochastically admissible. By solving the LMIs in Theorem 6.5, we have

$$K_1 = \begin{bmatrix} 34.8763 & 28.5998 & 0.6234 \\ 37.9459 & 28.4043 & 0.0044 \end{bmatrix},$$

$$K_2 = \begin{bmatrix} -102.5126 & -18.2368 & -0.5429 \\ -490.4724 & -90.5122 & -1.2175 \end{bmatrix}.$$

Example 6.2. A direct current (DC) motor has been modeled as a singular Markovian jump system in [146], [170], [171], [238]. The singular Markovian model is given as follows:

$$E\dot{x}(t) = A(r_t)x(t) + B(r_t)u(t). \tag{6.26}$$

The matrix parameters are borrowed from Example 2 in [170]:

$$E = \begin{bmatrix} 1 & 0 & 0 \\ 0 & 1 & 0 \\ 0 & 0 & 0 \end{bmatrix}, \ A_1 = \begin{bmatrix} 0 & 1 & 0 \\ 9.8 & 0 & 1 \\ -20 & -3 & -1 \end{bmatrix},$$

$$A_2 = \begin{bmatrix} 0 & 1 & 0 \\ 9.8 & 0 & 1 \\ -20 & -3 & -0.5 \end{bmatrix}, \ B_1 = \begin{bmatrix} 0.4 \\ -0.2 \\ 0.5 \end{bmatrix},$$

$$B_2 = \begin{bmatrix} -1.2 \\ 0.5 \\ -0.2 \end{bmatrix}, \ \begin{bmatrix} \pi_{11} & \pi_{12} \\ \pi_{21} & \pi_{22} \end{bmatrix} = \begin{bmatrix} -0.6 & 0.6 \\ 0.2 & -0.2 \end{bmatrix},$$

$$E_L = E_R = \begin{bmatrix} 1 & 0 \\ 0 & 1 \\ 0 & 0 \end{bmatrix}, \ U = \begin{bmatrix} 0 & 0 & -0.7 \end{bmatrix}, \ \Lambda = \begin{bmatrix} 0 \\ 0 \\ 0.8 \end{bmatrix}.$$

By solving the condition in Theorem 6.5, the state-feedback controllers are obtained:

$$K_1 = \begin{bmatrix} 186.3514 & 38.5301 & 2.0267 \end{bmatrix},$$

$$K_2 = \begin{bmatrix} 188.1730 & 21.4539 & -2.3464 \end{bmatrix}.$$

Then the closed-loop system is

$$E\dot{x}(t) = \bar{A}(r_t)x(t)$$

with

$$A_1 = \begin{bmatrix} 74.5406 & 16.4120 & 0.8107 \\ -27.4703 & -7.7060 & 0.5947 \\ 73.1757 & 16.2650 & 0.0133 \end{bmatrix},$$

$$A_2 = \begin{bmatrix} -225.8076 & -24.7447 & 2.8157 \\ 103.8865 & 10.7270 & -0.1732 \\ -57.6346 & -7.2908 & -0.0307 \end{bmatrix}.$$

To testify the effectiveness of the state-feedback control method, Theorem 2.2 in [171], Theorem 6.5 in this section, and Theorem 4 in [199] are utilized. By solving the LMIs in these theorems, we note that all the conditions are feasible, which implies the state-feedback controller is applicable.

Example 6.3. Consider the following singular Markovian jump time-delay system with two operating modes ($N = 2$) and the following parameters:

$$A_1 = \begin{bmatrix} 0.5023 & 2.0125 & 0.0150 \\ 0.3025 & 0.4004 & -4.0020 \\ -0.1002 & 0.3002 & -3.5001 \end{bmatrix},$$

$$A_2 = \begin{bmatrix} 0.5005 & 0.5052 & -0.1002 \\ 0.1256 & -0.0552 & 0.3003 \\ 0.1033 & 1.0015 & -2.0045 \end{bmatrix},$$

$$A_{d1} = \begin{bmatrix} -0.1669 & 0.0802 & 1.6820 \\ -0.8162 & -0.9373 & 0.5936 \\ 2.0941 & 0.6357 & 0.7902 \end{bmatrix},$$

$$A_{d2} = \begin{bmatrix} 0.1053 & -0.1948 & -0.6855 \\ 0.1586 & 0.0755 & -0.2684 \\ 0.7709 & -0.5266 & -1.1883 \end{bmatrix},$$

$$E = \begin{bmatrix} 1 & 0 & 0 \\ 0 & 1 & 0 \\ 0 & 0 & 0 \end{bmatrix}, \; B_1 = \begin{bmatrix} 0.9 \\ 1.8 \\ 1.4 \end{bmatrix}, \; B_2 = \begin{bmatrix} 1.5 \\ 0.9 \\ 1.1 \end{bmatrix},$$

$$B_{w1} = \begin{bmatrix} 0.1 \\ 0.2 \\ 0.4 \end{bmatrix}, \; B_{w2} = \begin{bmatrix} -0.6 \\ 0.5 \\ 0.8 \end{bmatrix},$$

$$C_1 = \begin{bmatrix} 0.8 & 0.3 & 0.9 \end{bmatrix},\ C_2 = \begin{bmatrix} -0.5 & 0.2 & 0.3 \end{bmatrix},\ D_1 = D_2 = 0,$$

$$C_{d1} = \begin{bmatrix} 0.2486 & 0.1025 & -0.0410 \end{bmatrix},$$

$$C_{d2} = \begin{bmatrix} -2.2476 & -0.5108 & 0.2492 \end{bmatrix},$$

$$D_{w1} = 0.2,\ D_{w2} = 0.5,\ U = \begin{bmatrix} 0 & 0 & -0.7 \end{bmatrix},$$

$$\Lambda = \begin{bmatrix} 0 & 0 & 0.8 \end{bmatrix},\ E_R = E_L = \begin{bmatrix} 1 & 0 \\ 0 & 1 \\ 0 & 0 \end{bmatrix},$$

$$\begin{bmatrix} \pi_{11} & \pi_{12} \\ \pi_{21} & \pi_{22} \end{bmatrix} = \begin{bmatrix} -0.3 & 0.3 \\ 0.5 & -0.5 \end{bmatrix},\ d = 0.2.$$

The purpose is to design a state feedback controller such that the closed-loop system is stochastically admissible with an H_∞ performance γ. By solving LMIs in Theorem 6.10, the minimal value of γ is obtained as $\gamma_{min} = 2.43$. When $\gamma = 2.43$, the state feedback controllers are obtained as follows:

$$K_1 = \begin{bmatrix} -4.4763 & -2.2181 & -0.1445 \end{bmatrix},$$

$$K_2 = \begin{bmatrix} -5.3581 & -0.4309 & -1.1303 \end{bmatrix}.$$

By using the method in Theorem 1 of [184], the minimal value of γ is solved as $\gamma = 2.68$, which illustrates the reduced conservatism of out result in Theorem 6.10.

On the other hand, when some system parameters change to

$$E = \begin{bmatrix} 2 & \sqrt{2} & 0 \\ \sqrt{2} & 2 & 0 \\ 0 & 0 & 0 \end{bmatrix},\ E_R = E_L = \begin{bmatrix} 1 & 1 \\ 0 & \sqrt{2} \\ 0 & 0 \end{bmatrix},\ D_1 = 0.1,\ D_2 = -0.5,$$

the condition in Theorem 1 of [184] cannot be applied. By solving the conditions with $\gamma = 5.60$ in Theorem 6.10 in this note, the corresponding state feedback controller is given as follows:

$$K_1 = \begin{bmatrix} -4.7783 & -0.8063 & -0.9794 \end{bmatrix},$$

$$K_2 = \begin{bmatrix} -4.4531 & 0.3942 & -4.0966 \end{bmatrix}.$$

6.1.4 Conclusion

In this section, the application of the equivalent sets is extended to admissibility and corresponding state feedback control of singular Markovian jump systems, which gives a novel admissibility condition and an effective state feedback control method. H_∞ control of time-delay singular Markovian jump systems is also addressed. A new bounded real lemma and improved H_∞ control result are given. The effectiveness and improvement of the presented method have been illustrated by numerical examples. Compared with existing results, the main contributions of this section are not only presenting the new two equivalent sets, but also developing some improved results as follows:

- New necessary and sufficient admissibilization condition is obtained for singular Markovian jump systems.
- For singular Markovian jump systems with time-delay, a new bounded real lemma is given in terms of strict LMIs without any constraint on system matrix E.

6.2 Reliable dissipative control for singular Markovian systems

In this section, the problem of reliable dissipative control is investigated for continuous-time singular Markovian system with actuator failure. Firstly, a sufficient condition is established in terms of LMIs, which guarantees a singular Markovian system to be stochastically admissible and strictly (Q, S, R)-dissipative. Based on the criterion, a state feedback controller design method is given in terms of strict LMIs. Moreover, the dissipative control results include the results of H_∞ control and passive control as special cases. The effectiveness of the controller designed in this section is illustrated by a numerical example.

6.2.1 Problem statement

Consider a class of linear continuous-time singular Markovian systems described by

$$\left\{\begin{array}{rcl} E\dot{x}(t) & = & A(r_t)x(t) + B(r_t)u(t) + B_w(r_t)w(t) \\ z(t) & = & C(r_t)x(t) + D(r_t)u(t) + D_w(r_t)w(t) \\ x(t_0) & = & x_0, \end{array}\right. \tag{6.27}$$

where $x(t) \in \mathbb{R}^n$ is the state vector; $u(t) \in \mathbb{R}^m$ is the control input; $w(t) \in \mathbb{R}^l$ represents a set of exogenous inputs (which belongs to $\mathcal{L}_2[0, \infty)$), and $z(t) \in \mathbb{R}^q$ is the controlled output; $\{r_t, t \geq 0\}$ is a continuous-time Markov process taking values in a finite space $\mathbb{S} \triangleq \{1, 2, \ldots, N\}$ and describing the evolution of the mode at time t; $A(r_t)$, $B(r_t)$, $B_w(r_t)$, $C(r_t)$, $D(r_t)$, and $D_w(r_t)$ denote constant matrices with appropriate dimensions. In contrast with standard linear state-space systems with $E = I$, the matrix $E \in \mathbb{R}^{n\times n}$ has $\text{rank}(E) = p \leq n$. The transition probability rate matrix of the Markov process $\{r_t, t \geq 0\}$ is $\Pi \triangleq \{\pi_{ij}\}$, and we have

$$\Pr\{r_{t+\delta} = j | r_t = i\} = \begin{cases} \pi_{ij}\delta + o(\delta) & i \neq j \\ 1 + \pi_{ii}\delta + o(\delta) & i = j, \end{cases} \tag{6.28}$$

where $\delta > 0$ and $\lim_{\delta\to 0} \frac{o(\delta)}{\delta} = 0$, and $\pi_{ij} \geq 0$, for $i \neq j$, is the transition rate from mode i at time t to mode j at time $t + \delta$, which satisfies $\pi_{ii} = -\sum_{j=1, i\neq j}^{N} \pi_{ij}$. For notational simplicity, a matrix $W(r_t)$ will be denoted by W_i for each possible $r_t = i$, $i \in \mathbb{S}$; for example, $A(r_t)$ is denoted by A_i and so on. For the control input u_i, $i = 1, 2, \ldots, m$, let u_i^F denote the signal from the actuator that may be faulty. The following failure model is adopted here:

$$u_i^F = \alpha_{ai} u_i, i = 1, 2, \ldots, m,$$

where

$$0 \leq \underline{\alpha}_{ai} \leq \alpha_{ai} \leq \bar{\alpha}_{ai}, i = 1, 2, \ldots, m$$

with $\underline{\alpha}_{ai} \leq 1 \leq \bar{\alpha}_{ai}$. Then we have

$$u^F = \mathcal{A}_a u, \ \mathcal{A}_a = \text{diag}\{\alpha_{a1}, \alpha_{a2}, \ldots, \alpha_{am}\}.$$

Remark 6.13. In the model mentioned above, when $\underline{\alpha}_{ai} = \bar{\alpha}_{ai} = 1$, it corresponds to the normal fully operating case, that is, $u_i^F = u_i$; when $\underline{\alpha}_{ai} = 0$, then it covers the outage case in [167]; when $\alpha_{ai} \neq 0$ and $\alpha_{ai} \neq 1$, then it corresponds to the case that the intensity of the feedback signal from actuator changes.

Denote

$$\begin{aligned} \bar{\mathcal{A}}_a &= \text{diag}\{\bar{\alpha}_{a1}, \bar{\alpha}_{a2}, \ldots, \bar{\alpha}_{am}\}, \\ \underline{\mathcal{A}}_a &= \text{diag}\{\underline{\alpha}_{a1}, \underline{\alpha}_{a2}, \ldots, \underline{\alpha}_{am}\}, \\ \mathcal{B}_{a0} &= \text{diag}\{\beta_{a01}, \beta_{a02}, \ldots, \beta_{a0m}\}, \end{aligned}$$

$$\mathcal{B} = \text{diag}\{\beta_1, \beta_2, \ldots, \beta_m\},$$
$$\Delta_a = \text{diag}\{\Delta_{a1}, \Delta_{a2}, \ldots, \Delta_{am}\},$$

where $\beta_{a0i} = \frac{\bar{\alpha}_{ai}+\underline{\alpha}_{ai}}{2}$, $\beta_i = \frac{\bar{\alpha}_{ai}-\underline{\alpha}_{ai}}{\bar{\alpha}_{ai}+\underline{\alpha}_{ai}}$ and $\Delta_{ai} = \frac{\alpha_{ai}-\beta_{a0i}}{\beta_{a0i}}$. Then we have

$$\mathcal{A}_a = (I + \Delta_a)\mathcal{B}_{a0},$$
$$\text{diag}\{\mid \Delta_{a1} \mid, \mid \Delta_{a2} \mid, \ldots, \mid \Delta_{am} \mid\} \leq \mathcal{B} \leq I. \tag{6.29}$$

Consider the following memoryless state-feedback controller for system (6.27):

$$u(t) = K(r_t)x(t). \tag{6.30}$$

By applying controller (6.30) to system (6.27), the closed-loop system can be described by

$$\begin{cases} E\dot{x}(t) &= (A(r_t) + B(r_t)\mathcal{A}_a K(r_t))x(t) + B_w(r_t)w(t) \\ z(t) &= (C(r_t) + D(r_t)\mathcal{A}_a K(r_t))x(t) + D_w(r_t)w(t) \\ x(t_0) &= x_0. \end{cases} \tag{6.31}$$

Before moving on, we give some definitions and lemmas concerning the following nominal unforced counterpart of the system in (6.27) with $w(t) = 0$:

$$E\dot{x}(t) = A(r_t)x(t). \tag{6.32}$$

Then, we introduce the following definition for singular Markovian system (6.32):

Definition 6.14. [209]

1. The singular Markovian system in (6.32) is said to be regular and impulse free if $\det(sE - A_i)$ is not identically zero $\forall i \in \mathbb{S}$.
2. The singular Markovian system in (6.32) is said to be impulse free if $\deg\{\det(sE - A_i)\} = \text{rank } E$, $\forall i \in \mathbb{S}$.
3. The singular Markovian system in (6.32) is said to be stochastically stable, if, for any $x_0 \in \mathbb{R}^n$ and $r_0 \in \mathbb{S}$, there exists a scalar $\delta(x_0, r_0) > 0$ such that

$$\lim_{t\to\infty} \mathcal{E}\left\{\int_0^t x^T(s, x_0, r_0)x(s, x_0, r_0)ds \mid x_0, r_0\right\} \leq \delta(x_0, r_0),$$

where $\mathcal{E}$ is the mathematical expectation, and $x(t, x_0, r_0)$ denotes the solution to system (6.32) at time t under the initial conditions x_0 and r_0.

4. The singular Markovian system in (6.32) is said to be stochastically admissible if the system is regular, impulse free, and stochastically stable.

Lemma 6.15. [200] *For matrices $Y > 0$, M, and N with appropriate dimensions, the following inequality holds:*

$$M^T N + N^T M \leq N^T Y N + M^T Y^{-1} M.$$

Lemma 6.16. [7] *For any matrices U and $V > 0$, the following inequality holds:*

$$U V^{-1} U^T \geq U + U^T - V.$$

Definition 6.17. [234] Given matrices Q, R, and S with Q and R real symmetric, systems (6.27) with $u(t) = 0$ is called strictly (Q, S, R)-dissipative if, for any $\tau \geq 0$, under zero initial state, the following condition is satisfied for some scalar $\alpha > 0$:

$$\mathcal{E}\left\{\langle z, Qz\rangle_\tau\right\} + 2\mathcal{E}\left\{\langle z, Sw\rangle_\tau\right\} + \mathcal{E}\left\{\langle w, Rw\rangle_\tau\right\} \geq \alpha\mathcal{E}\left\{\langle w, w\rangle_\tau\right\}. \tag{6.33}$$

As in [202], we assume that $Q \leq 0$. Then we can get

$$-Q = (Q_-^{\frac{1}{2}})^2$$

for some $Q_-^{\frac{1}{2}} \geq 0$.

Our main objective is to design a state-feedback controller in the form of (6.30) such that the closed-loop systems in (6.31) is stochastically admissible and strictly (Q, S, R)-dissipative.

In this section, a sufficient condition is derived to guarantee that the unforced system of (6.27) is stochastically admissible and strictly (Q, S, R)-dissipative. Based on this result, a state-feedback controller is designed to render the closed-loop system stochastically admissible and strictly (Q, S, R)-dissipative.

6.2.2 Reliable dissipativity analysis

The result of reliable dissipativity analysis for system (6.27) is presented in the following theorem:

Theorem 6.18. *Let matrices Q, S, and R be given with Q and R real symmetric and $Q \leq 0$. Then the system in (6.27) is stochastically admissible and strictly*

(Q, S, R)-dissipative if there exist matrices P_i, $i = 1, 2, \ldots, N$, such that the following set of LMIs hold for $i = 1, 2, \ldots, N$:

$$E^T P_i = P_i^T E \geq 0, \tag{6.34}$$

$$\Omega_i < 0, \tag{6.35}$$

where

$$\Omega_i = \begin{bmatrix} P_i^T A_i + A_i^T P_i + \sum_{j=1}^{N} \pi_{ij} E^T P_j & P_i^T B_{wi} - C_i^T S & C_i^T Q_-^{\frac{1}{2}} \\ \star & -D_{wi}^T S - S^T D_{wi} - R & D_{wi}^T Q_-^{\frac{1}{2}} \\ \star & \star & -I \end{bmatrix}.$$

Proof. From the LMI in (6.35), we have

$$A_i^T P_i + P_i^T A_i + \sum_{j=1}^{N} \pi_{ij} E^T P_j < 0.$$

Combining with the condition in (6.34), the stochastic admissibility of the system in (6.32) is guaranteed, based on Lemma 6.1. Next, we show the dissipativity of system (6.27). To this end, we choose the stochastic Lyapunov function as

$$V(x(t), r_t = i) = x^T(t) E^T P_i x(t).$$

Let $\mathcal{L}$ be the weak infinitesimal generator of the stochastic process $\{x(t), r_t\}$. Then, for each $r_t = i$, $i \in \mathbb{S}$, we have

$$\begin{aligned} \mathcal{L}V(x(t), r_t = i) = {} & x^T(t)(A_i^T P_i + P_i^T A_i + \sum_{j=1}^{N} \pi_{ij} E^T P_j) x(t) \\ & + x^T(t) P_i^T B_{wi} w(t) + w^T(t) B_{wi}^T P_i x(t). \end{aligned} \tag{6.36}$$

Denote

$$J(\tau) = \mathcal{E}\left\{ \int_0^{\tau} \left[z^T(t) Q z(t) + 2 z^T(t) S w(t) + w^T(t) R w(t) \right] dt \right\}.$$

Then under the zero initial condition, that is, $x_0 = 0$, it can be shown that for any nonzero $w \in \mathcal{L}_2[0, \infty)$,

$$\begin{aligned}
&\mathcal{E}\{V(x(\tau), r_\tau)\} - J(\tau) \\
&= \mathcal{E}\left\{\int_0^\tau \left[\mathcal{L}V(x(t), r_t) - z^T(t)Qz(t) - 2z^T(t)Sw(t) - w^T(t)Rw(t)\right] dt\right\} \\
&= \mathcal{E}\left\{\int_0^\tau \left[\xi^T(t)\Omega_{1i}\xi(t)\right] dt\right\},
\end{aligned}$$

where

$$\begin{aligned}
\Omega_{1i} &= \begin{bmatrix} P_i^T A_i + A_i^T P_i + \sum_{j=1}^N \pi_{ij} E^T P_j & P_i^T B_{wi} - C_i^T S \\ \star & -D_{wi}^T S - S^T D_{wi} - R \end{bmatrix} \\
&\quad - \begin{bmatrix} C_i^T \\ D_{wi}^T \end{bmatrix} Q \begin{bmatrix} C_i & D_{wi} \end{bmatrix}, \\
\xi(t) &= \begin{bmatrix} x(t) \\ w(t) \end{bmatrix}.
\end{aligned}$$

By using Schur complement equivalence, the inequality in (6.35) is equivalent to $\Omega_{1i} < 0$. Then we have $J(\tau) > 0$, and there exists a sufficiently small $\alpha > 0$ such that

$$J(\tau) \geq \alpha \mathcal{E}\left\{\int_0^\tau w^T(t)Rw(t)dt\right\}.$$

By the definition, the system in (6.27) is strictly (Q, S, R)-dissipative. This completes the proof. □

Remark 6.19. When $\bar{A}_a = \underline{A}_a = \text{diag}\{1, 1, \ldots, 1\}$, $Q = -I$, $S = 0$ and $R = \gamma^2 I$, the result in Theorem 6.18 reduces to the H_∞ result Lemma 1 in [235], which implies the more generality of our result.

6.2.3 Controller design

Now, our attention will be devoted to design a state-feedback controller in the form of (6.30), such that closed-loop system (6.31) subject to possible actuator failures is stochastically admissible and strictly (Q, S, R)-dissipative.

There are equality constraints $EP_i = P_i^T E^T$ in Theorem 6.18, which will lead to numerical problems when solving such nonstrict LMIs. However, the method to deal with the constraints $EP_i = P_i^T E^T$ in [7] is difficult to make the equality exactly hold. Moreover, by employing the state feedback controller design approach in [235], a strict LMI condition guaranteeing the stochastic admissibility and strict dissipativity of the closed-loop system in (6.31) is proposed in the next theorem.

Theorem 6.20. *Let matrices Q, S, and R be given with Q and R real symmetric and $Q \le 0$. Then the system in (6.27) is stochastically admissible and strictly (Q, S, R)-dissipative if there exist matrices N_i, nonsingular matrix $\bar{\Phi}_i$, $\bar{Z}_i > 0$, $i = 1, 2, \ldots, N$, and diagonal matrix $G > 0$ such that the following set of LMIs hold for $i = 1, 2, \ldots, N$:*

$$\Xi_i = \begin{bmatrix} \Xi_{11i} & \Xi_{12i} & \Xi_{13i} & N_i^T \mathcal{B}_{a0}\mathcal{B} & \Xi_{15i} \\ \star & \Xi_{22i} & \Xi_{23i} & 0 & 0 \\ \star & \star & \Xi_{33i} & 0 & 0 \\ \star & \star & \star & -G & 0 \\ \star & \star & \star & \star & \Xi_{55i} \end{bmatrix} < 0, \tag{6.37}$$

where

$$\begin{aligned}
\Xi_{11i} &= \mathrm{sym}(A_i M_i + B_i \mathcal{B}_{a0} N_i) + \pi_{ii} M_i^T E^T + B_i G B_i^T, \\
\Xi_{12i} &= B_{wi} - (M_i^T C_i^T + N_i^T \mathcal{B}_{a0} D_i^T) S - B_i G D_i^T S, \\
\Xi_{13i} &= (M_i^T C_i^T + N_i^T \mathcal{B}_{a0} D_i^T) Q_-^{\frac{1}{2}} + B_i G D_i^T Q_-^{\frac{1}{2}}, \\
\Xi_{15i} &= \left[\sqrt{\pi_{i1}} M_i^T E_R \quad \cdots \quad \sqrt{\pi_{ii-1}} M_i^T E_R \quad \sqrt{\pi_{ii+1}} M_i^T E_R \quad \cdots \quad \sqrt{\pi_{iN}} M_i^T E_R\right], \\
\Xi_{22i} &= -D_{wi}^T S - S^T D_{wi} - R + S^T D_i G D_i^T S, \\
\Xi_{23i} &= D_{wi}^T Q_-^{\frac{1}{2}} - S^T D_i G D_i^T Q_-^{\frac{1}{2}}, \\
\Xi_{33i} &= -I + Q_-^{\frac{1}{2}} D_i G D_i^T Q_-^{\frac{1}{2}}, \\
\Xi_{55i} &= -\mathrm{diag}(E_R^T \bar{Z}_1 E_R, \ldots, E_R^T \bar{Z}_{i-1} E_R, E_R^T \bar{Z}_{i+1} E_R, \ldots, E_R^T \bar{Z}_N E_R), \\
M_i &= \bar{Z}_i E^T + V \bar{\Phi}_i U^T,
\end{aligned}$$

and U, V are defined in Lemma 2.4. When the above conditions are satisfied, an admissible and (Q, S, R)-dissipative controller is given by $K_i = N_i M_i^{-1}$.

Proof. Employing Schur complement equivalence, (6.37) is equivalent to the following inequality:

$$\begin{aligned}
&\tilde{\Psi}_i + \begin{bmatrix} B_i \\ -S^T D_i \\ Q_-^{\frac{1}{2}} D_i \end{bmatrix} G \begin{bmatrix} B_i^T & -D_i^T S & D_i^T Q_-^{\frac{1}{2}} \end{bmatrix} \\
&+ \begin{bmatrix} N_i^T \mathcal{B}_{a0}\mathcal{B} \\ 0 \\ 0 \end{bmatrix} G^{-1} \begin{bmatrix} \mathcal{B}\mathcal{B}_{a0} N_i & 0 & 0 \end{bmatrix} < 0,
\end{aligned} \tag{6.38}$$

where

$$\tilde{\Psi}_i = \begin{bmatrix} \tilde{\Psi}_{11i} & B_{wi} - (C_iM_i + D_i\mathcal{B}_{a0}N_i)^T S & (C_iM_i + D_i\mathcal{B}_{a0}N_i)^T Q_-^{\frac{1}{2}} \\ \star & -D_{wi}^T S - S^T D_{wi} - R & D_{wi}^T Q_-^{\frac{1}{2}} \\ \star & \star & -I \end{bmatrix},$$

$$\begin{aligned} \tilde{\Psi}_{11i} &= \mathrm{sym}(A_iM_i + B_i\mathcal{B}_{a0}N_i + \pi_{ii}EM_i) - \pi_{ii}E\bar{Z}_iE^T \\ &+ \sum_{j=1, j\neq i}^{N} \pi_{ij}M_i^T E_R(E_R^T\bar{Z}_jE_R)^{-1}E_R^T M_i. \end{aligned}$$

By using the inequality $\pi_{ii}M_i^T E_R(E_R^T\bar{Z}_iE_R)^{-1}E_R^TM_i \leq \pi_{ii}(M_i^TE^T + EM_i - E\bar{Z}_iE^T)$, based on Lemma 6.15, (6.38) implies the following inequality holds:

$$\begin{aligned} &\breve{\Psi}_i + \begin{bmatrix} B_i \\ -S^TD_i \\ Q_-^{\frac{1}{2}}D_i \end{bmatrix} G \begin{bmatrix} B_i^T & -D_i^TS & D_i^TQ_-^{\frac{1}{2}} \end{bmatrix} \\ &+ \begin{bmatrix} N_i^T\mathcal{B}_{a0}\mathcal{B} \\ 0 \\ 0 \end{bmatrix} G^{-1} \begin{bmatrix} \mathcal{B}\mathcal{B}_{a0}N_i & 0 & 0 \end{bmatrix} < 0, \end{aligned} \tag{6.39}$$

where

$$\breve{\Psi}_i = \begin{bmatrix} \breve{\Psi}_{11i} & B_{wi} - (C_iM_i + D_i\mathcal{B}_{a0}N_i)^T S & (C_iM_i + D_i\mathcal{B}_{a0}N_i)^T Q_-^{\frac{1}{2}} \\ \star & -D_{wi}^T S - S^T D_{wi} - R & D_{wi}^T Q_-^{\frac{1}{2}} \\ \star & \star & -I \end{bmatrix},$$

$$\begin{aligned} \breve{\Psi}_{11i} &= \mathrm{sym}(A_iM_i + B_i\mathcal{B}_{a0}N_i) + \pi_{ii}M_i^TE_R(E_R^T\bar{Z}_iE_R)^{-1}E_R^TM_i \\ &+ \sum_{j=1, j\neq i}^{N} \pi_{ij}M_i^T E_R(E_R^T\bar{Z}_jE_R)^{-1}E_R^T M_i \\ &= \mathrm{sym}(A_iM_i + B_i\mathcal{B}_{a0}N_i) + \sum_{j=1}^{N} \pi_{ij}M_i^T E_R(E_R^T\bar{Z}_jE_R)^{-1}E_R^T M_i. \end{aligned}$$

Then, according to Lemma 2.4, we have $M_i^{-1} = Z_iE + U\Phi_iV^T$, and let

$$\begin{aligned} T_{2i} &= \mathrm{diag}(M_i^{-1}, I_l, I_q) \in \mathbb{R}^{(n+l+q)\times(n+l+q)}, \\ N_i &= K_iM_i. \end{aligned}$$

Pre- and postmultiplying to (6.39) by T_{2i}^T and T_{2i}, and using Lemma 6.1 yields

$$\bar{\Psi}_i + \begin{bmatrix} (Z_iE + U\Phi_iV^T)^T B_i \\ -S^T D_i \\ Q_-^{\frac{1}{2}} D_i \end{bmatrix} G \begin{bmatrix} B_i^T(Z_iE + U\Phi_iV^T) & -D_i^T S & D_i^T Q_-^{\frac{1}{2}} \end{bmatrix}$$
$$+ \begin{bmatrix} K_i^T \mathcal{B}_{a0} \\ 0 \\ 0 \end{bmatrix} \mathcal{B} G^{-1} \mathcal{B} \begin{bmatrix} \mathcal{B}_{a0} K_i & 0 & 0 \end{bmatrix} < 0, \tag{6.40}$$

where

$$\bar{\Psi}_i = \begin{bmatrix} \bar{\Psi}_{11i} & (Z_iE + U\Phi_iV^T)^T B_{wi} - (C_i + D_i\mathcal{B}_{a0}K_i)^T S & (C_i + D_i\mathcal{B}_{a0}K_i)^T Q_-^{\frac{1}{2}} \\ \star & -D_{wi}^T S - S^T D_{wi} - R & D_{wi}^T Q_-^{\frac{1}{2}} \\ \star & \star & -I \end{bmatrix},$$

$$\bar{\Psi}_{11i} = \mathrm{sym}((Z_iE + U\Phi_iV^T)^T(A_i + B_i\mathcal{B}_{a0}K_i)) + \sum_{j=1}^{N} \pi_{ij} E^T Z_j E.$$

From (6.29) and the elementary inequality $x^T y + y^T x \leq \varepsilon x^T x + \varepsilon^{-1} y^T y$, (6.40) implies

$$\bar{\Psi}_i + \mathrm{sym}\left(\begin{bmatrix} (Z_iE + U\Phi_iV^T)^T B_i \\ -S^T D_i \\ Q_-^{\frac{1}{2}} D_i \end{bmatrix} \Delta_a \begin{bmatrix} \mathcal{B}_{a0} K_i & 0 & 0 \end{bmatrix} \right)$$
$$= \begin{bmatrix} \Theta_i & (Z_iE + U\Phi_iV^T)^T B_{wi} - \bar{C}_i^T S & \bar{C}_i^T Q_-^{\frac{1}{2}} \\ \star & -D_{wi}^T S - S^T D_{wi} - R & D_{wi}^T Q_-^{\frac{1}{2}} \\ \star & \star & -I \end{bmatrix} < 0, \tag{6.41}$$

where

$$\Theta_i = \mathrm{sym}((Z_iE + U\Phi_iV^T)^T \bar{A}_i) + \sum_{j=1}^{N} \pi_{ij} E^T Z_j E,$$
$$\bar{A}_i = A_i + B_i\mathcal{A}_a K_i, \;\; \bar{C}_i = C_i + D_i\mathcal{A}_a K_i.$$

Replacing A_i, C_i, and P_i in (6.34) and (6.35) with $\bar{A}_i$, $\bar{C}_i$, and $Z_iE + U\Phi_iV^T$, respectively, the conditions in (6.34) and (6.35) hold from (6.41), which guarantees the stochastic admissibility and the strict dissipativity of the closed-loop system in (6.31). □

Remark 6.21. When $\bar{A}_a = \underline{A}_a = \text{diag}\{1, 1, \ldots, 1\}$, $Q = -I$, $S = 0$, and $R = \gamma^2 I$, the result in Theorem 6.5 reduces to Theorem 2 in [235]. Therefore our result is more general.

6.2.4 Illustrative example

Example 6.4. Consider a two-modes singular system in (6.27) with following parameters:

- mode 1

$$A_1 = \begin{bmatrix} -0.7 & -0.3 & 0 \\ 0.7 & -0.7 & -0.5 \\ 0.1 & 0 & 1.2 \end{bmatrix}, \quad B_1 = \begin{bmatrix} 1 & -0.2 \\ 1 & -3.5 \\ 0 & -1 \end{bmatrix},$$

$$B_{w1} = \begin{bmatrix} 1 & 0.3 \\ 0.2 & 2 \\ 0 & -0.5 \end{bmatrix}, \quad C_1 = \begin{bmatrix} -1 & 0 & 1 \\ 0 & 1 & 1 \end{bmatrix},$$

$$D_1 = \begin{bmatrix} 2.7 & 0.1 \\ 0 & 0 \end{bmatrix}, \quad D_{w1} = \begin{bmatrix} 0.7 & 0 \\ 0 & -1 \end{bmatrix};$$

- mode 2

$$A_2 = \begin{bmatrix} -0.5 & 1.8 & 1.3 \\ -0.2 & -2.1 & -0.1 \\ 1.2 & 2.5 & -1 \end{bmatrix}, \quad B_2 = \begin{bmatrix} 1 & 2 \\ -1.6 & 1 \\ -1 & -0.7 \end{bmatrix},$$

$$B_{w2} = \begin{bmatrix} 0.1 & -1 \\ 0.5 & 1 \\ 0.1 & 0 \end{bmatrix}, \quad C_2 = \begin{bmatrix} -2 & 1 & 1 \\ -1 & 0 & 0.3 \end{bmatrix},$$

$$D_2 = \begin{bmatrix} 0.1 & 0 \\ 0 & 2 \end{bmatrix}, \quad D_{w2} = \begin{bmatrix} -3 & 0 \\ -0.8 & 0.2 \end{bmatrix}.$$

The matrix E is

$$E = \begin{bmatrix} 1 & 0.5 & 0 \\ 0 & 1 & 0 \\ 0 & 0 & 0 \end{bmatrix},$$

and the switching between the two modes is described by the probability rate matrix

$$\Pi = \begin{bmatrix} -1.4 & 1.4 \\ 1.1 & -1.1 \end{bmatrix}.$$

It can be verified by using Lemma 6.1 that the system considered in this example is not stochastically admissible. The purpose is to design state feedback controller such that the singular system in (6.31) is stochastically admissible and strictly (Q, S, R)-dissipative. To this end, we choose

$$S = \begin{bmatrix} 0.5 & 0.2 \\ 0.1 & 0.6 \end{bmatrix}, \quad Q = \begin{bmatrix} -1 & 0 \\ 0 & -1 \end{bmatrix}, \quad R = \begin{bmatrix} 16 & 0 \\ 0 & 16 \end{bmatrix}.$$

To use Theorem 6.20, we choose

$$U = V = \begin{bmatrix} 0 & 0 & 1 \end{bmatrix}^T, \quad E_R = \begin{bmatrix} 1 & 0 \\ 0 & 1 \\ 0 & 0 \end{bmatrix}.$$

Then, the state feedback controller in the form of (6.30) can be calculated by using standard software. Table 6.1 presents the obtained controllers by using Theorem 6.20 for different actuator failure cases, respectively. To illustrate the effectiveness of our obtained controllers, we give the simulation results for the first case, that is, $\mathcal{A}_a = \text{diag}\{1, 1\}$ and the fourth case, that is, $\bar{\mathcal{A}}_a = \text{diag}\{1.5, 1.5\}$, $\underline{\mathcal{A}}_a = \text{diag}\{0.5, 0.5\}$. One of the possible realizations of the jumping mode $r(t)$ is presented in Fig. 6.1. Figs. 6.2 and 6.3 illustrate the state trajectories of closed-loop system without actuator failure and with $\mathcal{A}_a = \text{diag}\{1.1, 0.8\}$, respectively, which show the stochastic stability of the closed-loop systems are guaranteed.

6.2.5 Conclusion

In this section, the problem of reliable dissipative control for continuous-time singular Markovian systems with actuator failure has been studied. A sufficient condition in terms of LMIs has been proposed for guaranteeing singular Markovian systems stochastically admissible and strictly (Q, S, R)-dissipative. Based on the result, a state-feedback controller characterization has been given to guarantee the stochastically admissibility and strictly (Q, S, R)-dissipativity of the closed-loop system. The results presented in this section are in terms of strict LMIs, which make the conditions

Table 6.1 State feedback controllers by using Theorem 6.20 with different actuator failure cases.

Actuator cases	Theorem 6.20
$\bar{\mathcal{A}}_a = \text{diag}\{1, 1\}$ $\underline{\mathcal{A}}_a = \text{diag}\{1, 1\}$	$K_1 = \begin{bmatrix} -0.4860 & -0.7418 & -0.2308 \\ 15.0454 & 13.2740 & -4.2807 \end{bmatrix}$ $K_2 = \begin{bmatrix} 33.5587 & -2.9683 & -11.6581 \\ -2.0174 & -2.1034 & -0.1496 \end{bmatrix}$
$\bar{\mathcal{A}}_a = \text{diag}\{1, 1\}$ $\underline{\mathcal{A}}_a = \text{diag}\{0.5, 0.3\}$	$K_1 = \begin{bmatrix} -0.5375 & -0.7746 & -0.4725 \\ 3.6674 & 4.1089 & -0.9051 \end{bmatrix}$ $K_2 = \begin{bmatrix} -31.4550 & 9.5591 & 12.7751 \\ -6.8894 & -3.8126 & 0.8952 \end{bmatrix}$
$\bar{\mathcal{A}}_a = \text{diag}\{1, 1\}$ $\underline{\mathcal{A}}_a = \text{diag}\{0.05, 0.1\}$	$K_1 = \begin{bmatrix} -4.0080 & -4.2692 & -0.5972 \\ 9.7568 & 13.9286 & -0.8025 \end{bmatrix}$ $K_2 = \begin{bmatrix} -0.5840 & 5.0246 & 1.2211 \\ -15.7071 & -18.3959 & 0.3777 \end{bmatrix}$
$\bar{\mathcal{A}}_a = \text{diag}\{1.5, 1.5\}$ $\underline{\mathcal{A}}_a = \text{diag}\{0.5, 0.5\}$	$K_1 = \begin{bmatrix} -0.3771 & -0.5744 & -0.3424 \\ 1.9874 & 3.1286 & -0.7061 \end{bmatrix}$ $K_2 = \begin{bmatrix} -7.3258 & 2.1828 & 2.8738 \\ -3.9245 & -3.1026 & 0.3638 \end{bmatrix}$

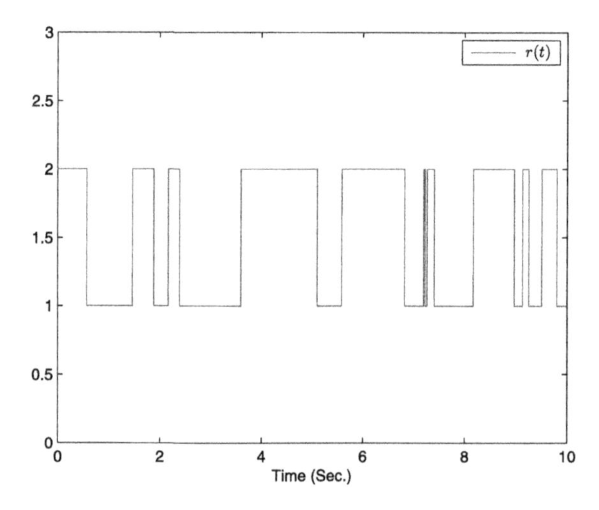

Figure 6.1 Random jumping mode $r(t)$.

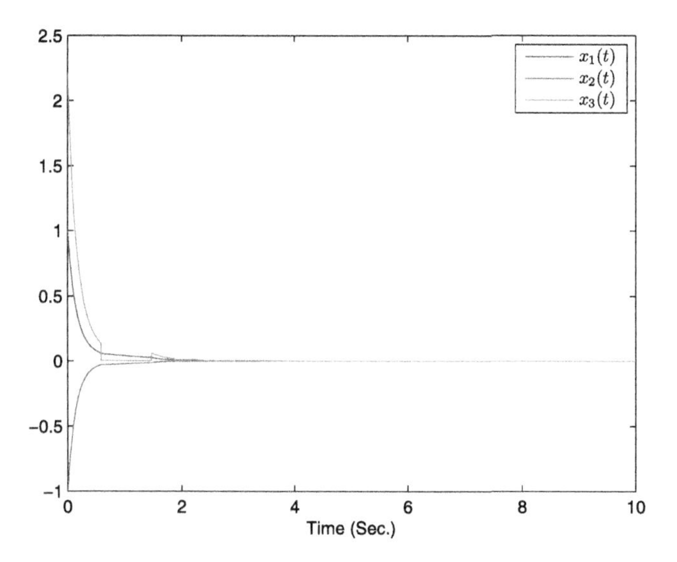

Figure 6.2 State trajectories of closed-loop system without actuator failure.

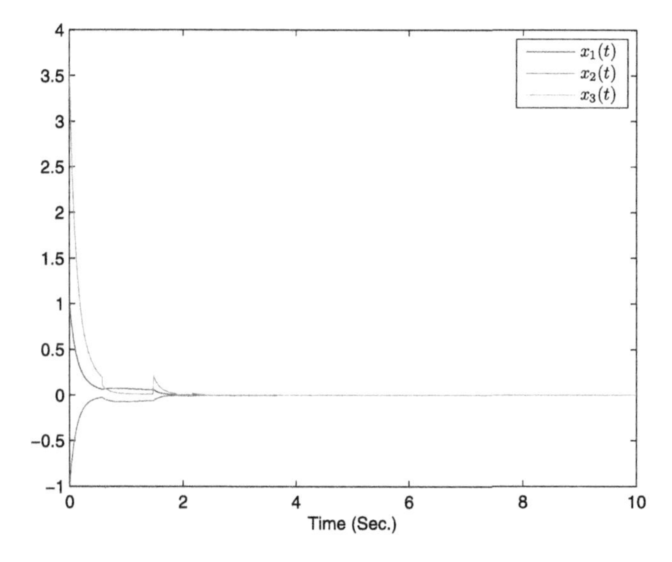

Figure 6.3 State trajectories of closed-loop system with $\mathcal{A}_a = \text{diag}\{1.1, 0.8\}$.

more tractable. Moreover, the results on H_∞ control and passive control of singular Markovian systems are unified in the proposed results. Finally, a numerical example is given to demonstrate the effectiveness of our methods.

CHAPTER 7

Sliding mode control of singular stochastic Markov jump systems

An investigation has been made into the SMC problem for SSMSs in this chapter. Firstly, by using replacement of matrix variables, a new mean square admissibility criterion of SSMSs is proposed. Based on this admissibility criterion, the desired state-feedback controller is designed to guarantee the closed-loop system to be mean square admissible. Then, the obtained results, given in the form of strict LMIs, are applied to solve SMC problem of SSMSs. To illustrate the workability and applicability of the theoretic results developed, numerical examples are provided.

7.1 Problem formulation

Let SSMSs be described as follows:

$$\begin{cases} Edx(t) = (A(r_t)x(t) + B(r_t)u(t))dt + C(r_t)x(t)dw(t) \\ y(t) = D(r_t)x(t), \end{cases} \tag{7.1}$$

where $x(t) \in \mathbb{R}^n$ and $u(t) \in \mathbb{R}^m$ are the system state vector and the control input, respectively; $w(t)$ is one-dimensional Brownian motion defined on the filtered probability space $(\Omega, \mathcal{F}, \mathcal{F}_t, \mathcal{P})$ with a filtering $\{\mathcal{F}_t\}_{\{t\geq 0\}}$. Matrix $E \in \mathbb{R}^{n\times n}$; we assume that $\text{rank}(E) = r \leq n$. The matrices $A(\cdot)$, $B(\cdot)$, $C(\cdot)$, and $D(\cdot)$, which are functions of r_t are real matrices with appropriate dimensions. $\{r_t, t \geq 0\}$ be a right-continuous Markov chain that takes values in a finite state set $\mathcal{N} = \{1, 2, \ldots, N\}$ and illustrates the jumping mode at time t. For notational simplicity, we write $A(r_t = i) \triangleq A_i$. The jumping feature between different modes is illustrated by Markov process with generator $\pi = [\pi_{ij}]$, $(i, j \in \mathcal{N})$ given by

$$\Pr\{r_{t+\sigma} = j | r_t = i\} = \begin{cases} \pi_{ij}\sigma + o(\sigma) & i \neq j \\ 1 + \pi_{ii}\sigma + o(\sigma) & i = j, \end{cases} \tag{7.2}$$

where $\sigma > 0$, $\lim_{\sigma\to 0} \frac{o(\sigma)}{\sigma} = 0$, and $\pi_{ij} \geq 0$ for $i \neq j$ stands for the transition rate from mode i to mode j, which satisfies $\pi_{ii} = -\sum_{j=1, i\neq j}^{N} \pi_{ij}$.

Analysis and Synthesis of Singular Systems
https://doi.org/10.1016/B978-0-12-823739-7.00014-8

As that in [245], the assumption $\text{rank}(E, C_i) = \text{rank}(E)$ for every $i \in \mathcal{N}$ is needed. Considering this assumption with other conditions, the SSMS in (7.1) is guaranteed to have a solution, which is impulse free. The conservatism of this assumption is less than in previous work, such as [69], and the details can be referred to in Remark 7 in [245].

7.2 Admissibilization of SSMSs

The matrix variables replacement method is employed to deal with the admissibility of SSMSs in this section. Based on the admissibility criterion, the admissibilization problem is solved.

Firstly, an existing admissibility criterion of system (7.1) is provided.

Lemma 7.1. [245] *For $u(t) = 0$, system (7.1) is mean square admissible if there exists a matrix X_i for each $i \in \mathcal{N}$ such that the following coupled matrix inequalities are satisfied $\forall i \in \mathcal{N}$:*

$$E^T X_i = X_i^T E \geq 0, \tag{7.3}$$

$$\text{sym}(X_i^T A_i) + \sum_{j=1}^{N} \pi_{ij} E^T X_j + C_i^T E^{+T} E^T X_i E^+ C_i < 0. \tag{7.4}$$

Remark 7.2. Notice that the nonstrict LMI and equality constraint exist in Lemma 7.1, which will cause numerical difficulties in checking the nonstrict inequality condition, because of round-off errors in numerical calculation and fragility and nonperfect satisfaction of equality constraints.

Now a new admissibility condition of system (7.1) is given in the following theorem:

Theorem 7.3. *For each $i \in \mathcal{N}$, there exist a matrix $P_i = P_i^T$, an invertible matrix Φ_i, such that the following LMIs hold:*

$$E_L^T P_i E_L > 0, \tag{7.5}$$

$$\begin{aligned} &\text{sym}((P_i E + U^T \Phi_i \Lambda^T)^T A_i) + \sum_{j=1}^{N} \pi_{ij} E^T P_j E \\ &+ C_i^T E^{+T} E^T P_i E E^+ C_i < 0, \end{aligned} \tag{7.6}$$

where E_L, U, and Λ are defined in Lemma 2.4, then system (7.1) is mean square admissible.

Proof. We first prove that the matrix X_i, for each $i \in \mathcal{N}$ in Lemma 7.1, is nonsingular. Due to $E^T X_i \geq 0$, we have

$$\text{sym}(X_i^T A_i) + \sum_{j=1}^{N} \pi_{ij} E^T X_j < 0. \tag{7.7}$$

If matrix X_i is noninvertible, there is a nonzero vector ξ_i for each $i \in \mathcal{N}$ leading to $X_i \xi_i = 0$. Then for the nonzero vector ξ_i, we have the following inequality from (7.7):

$$\begin{aligned} &\sum_{j=1, j\neq i}^{N} \pi_{ij} \xi_i^T E^T X_j \xi_i + \pi_{ii} \xi_i^T E^T X_i \xi_i + \xi_i^T X_i^T A_i \xi_i + \xi_i^T A_i^T X_i \xi_i \\ &= \sum_{j=1, j\neq i}^{N} \pi_{ij} \xi_i^T E^T X_j \xi_i < 0. \end{aligned} \tag{7.8}$$

Due to $E^T X_j \geq 0$ and $\pi_{ij} \geq 0$, $i \neq j$, the inequality in (7.8) cannot hold, which implies the inequality in (7.7) does not hold. Based on the discussions, matrix X_i, for each $i \in \mathcal{N}$, is nonsingular.

Then combining Lemma 2.5, equivalent conditions of inequalities (7.3) and (7.4) in Lemma 7.1 are obtained in (7.5) and (7.6), which guarantee the mean square admissibility of system (7.1). □

Remark 7.4. The sufficient condition in Theorem 7.3 is given in the form of strict LMIs, which are reliable and tractable in numerical computation. Although the strict LMIs are also proposed in Proposition 2 of [181] and in Theorem 3 of [245] for admissibility of SSMSs, the condition in Theorem 7.3 does not require positive definite matrix P_i for each $i \in \mathcal{N}$, which is needed in Proposition 2 of [181] and in Theorem 3 of [245]. From this point of view, the result in Theorem 7.3 is less conservative. Additionally, the condition in Theorem 7.3 is easily applied to the problem of admissibilization, which can be discussed in the theorem that follow.

Now we are ready to tackle with the admissibilization problem for system (7.1) with $u(t) = K(r_t)x(t)$ such that the resultant closed-loop system

$$Edx(t) = (A(r_t) + B(r_t)K(r_t))x(t)dt + C(r_t)x(t)dw(t) \tag{7.9}$$

is mean square admissible.

Theorem 7.5. *If there is a matrix $\bar{P}_i = \bar{P}_i^T$, an invertible matrix $\bar{\Phi}_i$, and a matrix H_i such that the following LMIs hold for each $i \in \mathcal{N}$:*

$$E_R^T \bar{P}_i E_R > 0, \tag{7.10}$$

$$\begin{bmatrix} \Upsilon_i & Y_i^T \Omega_i & Y_i^T C_i^T E^{+T} E_R \\ \star & -\Psi_i & 0 \\ \star & \star & -E_R^T \bar{P}_i E_R \end{bmatrix} < 0, \tag{7.11}$$

where

$$\Upsilon_i = \text{sym}(A_i Y_i + B_i H_i) + \pi_{ii} E \bar{P}_i E^T, \ Y_i = \bar{P}_i E^T + \Lambda \bar{\Phi}_i U,$$

$$\Omega_i = \Big[\sqrt{\pi_{i1}} E_R \quad \sqrt{\pi_{i2}} E_R \quad \cdots \quad \sqrt{\pi_{i(i-1)}} E_R \quad \sqrt{\pi_{i(i+1)}} E_R \quad \cdots \quad \sqrt{\pi_{iN}} E_R \Big],$$

$$\Psi_i = \text{diag}\left\{ E_R^T \bar{P}_1 E_R, \ E_R^T \bar{P}_2 E_R, \ \ldots, \ E_R^T \bar{P}_{i-1} E_R, \ E_R^T \bar{P}_{i+1} E_R, \ \ldots, E_R^T \bar{P}_N E_R \right\}.$$

E_R, U, and Λ are given in Lemma 2.5, then the resultant closed-loop system (7.9) is mean square admissible, and the desired state-feedback controller is obtained by $K_i = H_i Y_i^{-1}$.

Proof. Based on Theorem 7.3, system (7.9) is mean square admissible if there is a matrix $P_i = P_i^T$, a nonsingular matrix Φ_i, guaranteeing that the following coupled LMIs hold for each $i \in \mathcal{N}$:

$$E_L^T P_i E_L > 0, \tag{7.12}$$

$$\text{sym}((A_i + B_i K_i)^T (P_i E + U^T \Phi_i V^T)) + \sum_{j=1}^{N} \pi_{ij} E^T P_j E + C_i^T E^{+T} E^T P_i E E^+ C_i < 0. \tag{7.13}$$

Considering $\pi_{ij} > 0$, $i \neq j$, $E_L^T P_i E_L > 0$, and applying Schur complement equivalence, inequality (7.13) is equivalent to the following inequality:

$$\begin{bmatrix} \Gamma_{1i} & \Omega_i & C_i^T E^{+T} E_R \\ \star & -\Psi_i & 0 \\ \star & \star & -E_R^T \bar{P}_i E_R \end{bmatrix} = \begin{bmatrix} \Gamma_{2i} & \Omega_i & C_i^T E^{+T} E_R \\ \star & -\Psi_i & 0 \\ \star & \star & -E_R^T \bar{P}_i E_R \end{bmatrix} < 0,$$

where

$$\Gamma_{1i} = \text{sym}((A_i + B_i K_i)^T (P_i E + U^T \Phi_i V^T)) + \pi_{ii} E^T P_i E,$$

$$\Gamma_{2i} = \text{sym}((A_i + B_iK_i)^T(P_iE + U^T\Phi_iV^T)) + \pi_{ii}E^T(P_iE + U^T\Phi_iV^T).$$

Then performing congruence transformation to above inequality with matrix

$$\begin{bmatrix} Y_i & 0 & 0 \\ 0 & I & 0 \\ 0 & 0 & I \end{bmatrix},$$

in which

$$Y_i = \bar{P}_iE^T + \Lambda\bar{\Phi}_iU = (P_iE + U^T\Phi_iV^T)^{-1},$$

and setting $H_i = K_iY_i$, the inequality in (7.11) is obtained. Because of $E_R^T\bar{P}_iE_R = (E_L^TP_iE_L)^{-1}$, the inequalities in (7.10) and (7.12) are equivalent. □

Remark 7.6. For SSMSs, a sufficient admissibilization condition by state-feedback control is proposed in Theorem 7.5, whereas only the analysis result is provided in Theorem 3 in [245]. Although strict LMIs are given in Theorem 3 in [245], it is difficult to apply on state-feedback control problem, because matrices $K_i(P_iE + FQ_i)^{-1}$, P_i^{-1}, and $(P_iE + FQ_i)^{-1}$ will appear simultaneously when the state-feedback controller is designed.

When there is no the Brownian motion $w(t)$, $t > 0$, then the resultant system in (7.9) will become the following SMS:

$$E\dot{x}(t) = (A_i + B_iK_i)x(t). \tag{7.14}$$

Combining the admissibility condition in Theorem 10.1 in [209], and by employing similar procedure in Theorem 7.5, a necessary and sufficient admissibility condition of the system in (7.14) will be obtained:

Corollary 7.7. [47] *The system in (7.14) is stochastically admissible if and only if there exist a matrix $\bar{P}_i = \bar{P}_i^T$, an invertible matrix $\bar{\Phi}_i$, and a matrix H_i, resulting in the following LMIs holding for each $i \in \mathcal{N}$:*

$$E_R^T\bar{P}_iE_R > 0, \tag{7.15}$$

$$\begin{bmatrix} \text{sym}(A_iY_i + B_iH_i) + \pi_{ii}E\bar{P}_iE^T & Y_i^T\Omega_i \\ \star & -\Psi_i \end{bmatrix} < 0, \tag{7.16}$$

where Y_i, Ω_i, Ψ_i are defined in Theorem 7.5. Moreover, E_R, U, and Λ are the same as that in Lemma 2.5. Then, we can design the state-feedback controller by $K_i = H_i Y_i^{-1}$.

7.3 Application to SMC

The observer-based SMC problem is solved firstly when the system involves unmeasured states in this section. When the states are available, the SMC law is also developed to assure the admissibility of resultant closed-loop system.

The obtained results in Theorem 7.3 and Corollary 7.7 will be used to solve the observer-based SMC problem for SSMSs. For this aim, the following observer is designed to estimate state of system (7.1):

$$\begin{cases} E\dot{\breve{x}}(t) = A_i\breve{x}(t) + B_iu(t) + L_i(y(t) - \breve{y}(t)) \\ \breve{y}(t) = D_i\breve{x}(t), \end{cases} \tag{7.17}$$

where $\breve{x}(t)$ is the estimation of state $x(t)$, and we will design the observer L_i later.

By denoting the error $e(t) = x(t) - \breve{x}(t)$, we write the estimation error dynamics from systems (7.1) and (7.17) in the following formula:

$$Ede(t) = (A_i - L_iD_i)e(t)dt + C_ix(t)dw(t). \tag{7.18}$$

For every Markov mode $i \in \mathcal{N}$, the integral sliding surface function will be chosen as the following form:

$$s(t) = G_iE\breve{x}(t) - \int_0^t G_i(A_i + B_iK_i)\breve{x}(\delta)d\delta, \tag{7.19}$$

where controller gain K_i is designed to guarantee

$$E\dot{\breve{x}}(t) = (A_i + B_iK_i)\breve{x}(t) \tag{7.20}$$

to be stochastically admissible. Moreover, the matrix G_i is chosen to satisfy that G_iB_i is nonsingular.

Remark 7.8. It can be seen from systems (7.14) and (7.20) that the controller gain K_i can be designed by using Corollary 7.7. Although the work in [43], [182] provides a controller design approach, the conditions given there are sufficient only, which may have some conservatism. Corollary 7.7 provides a necessary and sufficient admissibilization result, which shows the novelty of the result.

Following SMC theory, when the system states reach onto the sliding surface, we have $s(t)=0$ and $\dot{s}(t)=0$, which implies the equivalent control input can be chosen as

$$u_{eq}(t)=K_i\breve{x}(t)-(G_iB_i)^{-1}G_iL_iD_ie(t). \tag{7.21}$$

Replacing $u(t)$ in (7.17) with (7.21), the resultant SMDs in the state space of $\breve{x}(t)$ are obtained:

$$E\dot{\breve{x}}(t)=(A_i+B_iK_i)\breve{x}(t)+(I-B_i(G_iB_i)^{-1}G_i)L_iD_ie(t). \tag{7.22}$$

Considering the system in (7.18), the SMDs both in $\breve{x}(t)$ and $e(t)$ are

$$\begin{cases} Ed\breve{x}(t)=\big((A_i+B_iK_i)\breve{x}(t)+(I-B_i(G_iB_i)^{-1}G_i)L_iD_ie(t)\big)\,dt \\ Ede(t)=(A_i-L_iD_i)e(t)dt+C_ix(t)dw(t). \end{cases} \tag{7.23}$$

Then by denoting augmented state as $\eta(t)=\begin{bmatrix}\breve{x}(t)\\ e(t)\end{bmatrix}$, the SMDs are rewritten as

$$\mathcal{E}d\eta(t)=\bar{A}_i\eta(t)dt+\bar{C}_i\eta(t)dw(t), \tag{7.24}$$

where

$$\mathcal{E}=\begin{bmatrix}E & 0\\ 0 & E\end{bmatrix},\ \bar{A}_i=\begin{bmatrix}A_i+B_iK_i & (I-B_i(G_iB_i)^{-1}G_i)L_iD_i\\ 0 & A_i-L_iD_i\end{bmatrix}$$

$$\bar{C}_i=\begin{bmatrix}0 & 0\\ C_i & C_i\end{bmatrix}.$$

Because $\text{rank}(\mathcal{E},\bar{C}_i)=\text{rank}(\mathcal{E})=2r$, where $r=\text{rank}(E)$, the rank constraint above Section 7.2 is still satisfied. Based on Theorem 7.3, the observer gain L_i design method is given in the following theorem to guarantee the admissibility of SMDs in (7.24):

Theorem 7.9. *Given scalars $\mu_1>0$, $\mu_2>0$, $\mu_3>0$, if there exist matrices $P_i>0$, $Z_i=Z_i^T$, $\bar{L}_i$ and an invertible matrix $\bar{Y}_i$ such that following LMIs are satisfied $\forall i\in\mathcal{N}$:*

$$\mathcal{E}_L^T Z_i \mathcal{E}_L > 0, \tag{7.25}$$

$$\begin{bmatrix} \Pi_{11} & \Pi_{12} & \Pi_{13} & \Pi_{14} & \Pi_{15} & 0 \\ \star & \Pi_{22} & 0 & 0 & 0 & \Pi_{26} \\ \star & \star & -B_i^T P_i B_i & 0 & 0 & 0 \\ \star & \star & \star & -P_i & 0 & 0 \\ \star & \star & \star & \star & -P_i & 0 \\ \star & \star & \star & \star & \star & -B_i^T P_i B_i \end{bmatrix} < 0, \tag{7.26}$$

where

$$\Pi_{11} = \begin{bmatrix} \mu_1 \mathrm{sym}\,(P_i(A_i + B_i K_i)) & \mu_1 \bar{L}_i D_i \\ \star & \mu_2 \mathrm{sym}(P_i A_i - \bar{L}_i D_i) \end{bmatrix}$$
$$+ \sum_{j=1}^{N} \pi_{ij} \mathcal{E}^T Z_j \mathcal{E} + \bar{C}_i^T (\mathcal{E}^+)^T \mathcal{E}^T Z_i \mathcal{E} \mathcal{E}^+ \bar{C}_i,$$

$$\Pi_{12} = Q_i^T + \begin{bmatrix} (A_i + B_i K_i)^T P_i - \mu_1 P_i & 0 \\ D_i^T \bar{L}_i^T & \mu_3 (A_i^T P_i^T - D_i^T \bar{L}_i^T) - \mu_2 P_i \end{bmatrix},$$

$$\Pi_{13} = \sqrt{\mu_1} \begin{bmatrix} P_i B_i \\ 0 \end{bmatrix}, \ \Pi_{14} = \sqrt{\mu_1} \begin{bmatrix} 0 \\ D_i^T \bar{L}_i^T \end{bmatrix}, \ \Pi_{15} = \begin{bmatrix} 0 \\ D_i^T \bar{L}_i^T \end{bmatrix},$$

$$\Pi_{22} = \begin{bmatrix} -2P_i & 0 \\ 0 & -2\mu_3 P_i \end{bmatrix}, \ \Pi_{26} = \begin{bmatrix} P_i B_i \\ 0 \end{bmatrix}, \ Q_i = Z_i \mathcal{E} + \bar{U}^T \bar{Y}_i \bar{\Lambda}^T.$$

Similar as in Lemma 2.4, $\mathcal{E} = \mathcal{E}_L \mathcal{E}_R^T$ meets $\mathrm{rank}(\mathcal{E}_L) = \mathrm{rank}(\mathcal{E}_R) = 2r$ and $\mathcal{E}_L, \mathcal{E}_R \in \mathbb{R}^{2n \times 2r}$. $\bar{U} \in \mathbb{R}^{(2n-2r) \times 2n}$ with $\mathrm{rank}(\bar{U}) = 2n - 2r$ and $\bar{\Lambda} \in \mathbb{R}^{2n \times (2n-2r)}$ with $\mathrm{rank}(\bar{\Lambda}) = 2n - 2r$ satisfy $\bar{U}\mathcal{E} = 0$ and $\mathcal{E}\bar{\Lambda} = 0$, respectively. Then system (7.24) with $G_i = B_i^T P_i$ is mean square admissible. Then the observer gain L_i can be obtained by $L_i = P_i^{-1} \bar{L}_i$.

Proof. Let $G_i = B_i^T P_i$. Due to matrix $P_i > 0$ and matrix B_i with full column rank, the matrix $G_i B_i$ is nonsingular. By using the Schur complement equivalence and noting that $\bar{L}_i = P_i L_i$, the inequality in (7.26) is equivalent to

$$\begin{bmatrix} \bar{\Pi}_{11} & \Pi_{12} \\ \star & \bar{\Pi}_{22} \end{bmatrix} < 0, \tag{7.27}$$

where

$$\bar{\Pi}_{11} = \Pi_{11} + \begin{bmatrix} \mu_1 P_i B_i (B_i^T P_i B_i)^{-1} B_i^T P_i & 0 \\ 0 & (\mu_1 + 1) D_i^T L_i^T P_i L_i D_i \end{bmatrix}$$

$$\bar{\Pi}_{22} = \Pi_{22} + \begin{bmatrix} P_i B_i (B_i^T P_i B_i)^{-1} B_i^T P_i & 0 \\ 0 & 0 \end{bmatrix}.$$

On the other hand, due to

$$\begin{bmatrix} P_i & I \\ I & P_i^{-1} \end{bmatrix} \geq 0$$

by premultiplying and postmultiplying the above inequality by

$$\begin{bmatrix} P_i B_i (B_i^T P_i B_i)^{-1} B_i^T & 0 \\ 0 & D_i^T L_i^T P_i \end{bmatrix}$$

and its transpose, the following inequality is obtained:

$$\begin{bmatrix} P_i B_i (B_i^T P_i B_i)^{-1} B_i^T P_i & P_i B_i (B_i^T P_i B_i)^{-1} B_i^T P_i L_i D_i \\ \star & D_i^T L_i^T P_i L_i D_i \end{bmatrix} \geq 0. \tag{7.28}$$

Similarly, the following inequality holds:

$$\begin{bmatrix} D_i^T L_i^T P_i L_i D_i & D_i^T L_i^T P_i B_i (B_i^T P_i B_i)^{-1} B_i^T P_i \\ \star & P_i B_i (B_i^T P_i B_i)^{-1} B_i^T P_i \end{bmatrix} \geq 0. \tag{7.29}$$

Combining inequalities (7.27) to (7.29), we have

$$\tilde{\Pi} = \begin{bmatrix} \tilde{\Pi}_{11} & \tilde{\Pi}_{12} \\ \star & \Pi_{22} \end{bmatrix} < 0, \tag{7.30}$$

where

$$\tilde{\Pi}_{11} = \Pi_{11} + \begin{bmatrix} 0 & -\mu_1 P_i B_i (B_i^T P_i B_i)^{-1} B_i^T P_i L_i D_i \\ \star & 0 \end{bmatrix}$$

$$\tilde{\Pi}_{12} = \Pi_{12} + \begin{bmatrix} 0 & 0 \\ -D_i^T L_i^T P_i B_i (B_i^T P_i B_i)^{-T} B_i^T P_i & 0 \end{bmatrix}.$$

Noting that $G_i = B_i^T P_i$, the inequality in (7.30) is rewritten as

$$\tilde{\Pi} = \begin{bmatrix} \hat{\Pi}_i & Q_i + \bar{A}_i^T \bar{F}_i^T - \bar{G}_i \\ \star & -\bar{F}_i - \bar{F}_i^T \end{bmatrix} < 0, \tag{7.31}$$

where

$$\hat{\Pi}_i = \text{sym}(\bar{G}_i\bar{A}_i) + \sum_{j=1}^{N} \pi_{ij}\mathcal{E}^T Q_j + \bar{C}_i^T(\mathcal{E}^+)^T\mathcal{E}^T Q_i\mathcal{E}^+\bar{C}_i,$$

$$\bar{G}_i = \begin{bmatrix} \mu_1 P_i & 0 \\ 0 & \mu_2 P_i \end{bmatrix}, \ \bar{F}_i = \begin{bmatrix} P_i & 0 \\ 0 & \mu_3 P_i \end{bmatrix}, \ Q_i = Z_i\mathcal{E} + \bar{U}^T\bar{Y}_i\bar{\Lambda}^T.$$

Then carrying out the pre- and postmultiplications on inequality (7.31) with matrix $\begin{bmatrix} I & \bar{A}_i^T \end{bmatrix}$ and its transpose, we have

$$\text{sym}(\bar{A}_i^T Q_i) + \sum_{j=1}^{N} \pi_{ij}\mathcal{E}^T Q_j + \bar{C}_i^T(\mathcal{E}^+)^T\mathcal{E}^T Q_i\mathcal{E}^+\bar{C}_i < 0.$$

Combining with inequality (7.25), we can see that there exists a matrix $Z_i = Z_i^T$, an invertible matrix $\bar{Y}_i$, assuring the following LMIs to be satisfied $\forall i \in \mathcal{N}$:

$$\mathcal{E}_L^T Z_i\mathcal{E}_L > 0,$$

$$\text{sym}((Z_i\mathcal{E} + \bar{U}^T\bar{Y}_i\bar{\Lambda}^T)^T\bar{A}_i) + \sum_{j=1}^{N} \pi_{ij}\mathcal{E}^T Z_j\mathcal{E} + \bar{C}_i^T(\mathcal{E}^+)^T\mathcal{E}^T Z_i\mathcal{E}\mathcal{E}^+\bar{C}_i < 0.$$

Based on Theorem 7.3, the mean square admissibility of system (7.24) is guaranteed. □

Remark 7.10. An invertible matrix $\bar{Y}_i$ can be obtained by solving the LMI in (7.26). If the obtained matrix $\bar{Y}_i$ is singular, we can tune one of the elements of matrix $\bar{Y}_i$ slightly such that the new $\bar{Y}_i$ is invertible, which does not affect the holding of inequality (7.26).

Remark 7.11. Although some observer-based SMC conditions have been published for different systems in literature references, such as Markov neutral-type stochastic systems [75], stochastic time-delay systems [106], no results about observer-based SMC for SSMSs have been reported. The result in Theorem 7.9 provides an effective method, and the condition is given in the form of strict LMIs, which is one of contributions in this case.

Remark 7.12. The system considered in this section is without uncertainty. Consider an uncertain system:

$$Edx(t) = (A_i + \Delta A_i(t))x(t) + B_i u(t))dt + C_i x(t)dw(t),$$

where $\Delta A_i(t)$ is parameter uncertainty satisfying $\Delta A_i(t) = \bar{M}_i F_i(t) \bar{N}_i$. Here $\bar{M}_i$ and $\bar{N}_i$ are known constant matrices, and $F_i(t)$ is an unknown matrix function satisfying $F_i^T(t) F_i(t) \leq I$. The main technique in this manuscript is to use the equivalent sets in Lemma 2.5 to solve the admissibilization and SMC problems. The appearing of uncertainty does not affect the utilization of this technique. For this uncertain system, the basic admissibility criterion will be obtained by replacing matrix A_i in (7.4) with $A_i + \Delta A_i(t)$, that is,

$$\mathrm{sym}(X_i^T \Delta A_i(t)) + \Theta_i < 0, \tag{7.32}$$

where

$$\Theta_i = \mathrm{sym}(X_i^T A_i) + \sum_{j=1}^{N} \pi_{ij} E^T X_j + C_i^T E^{+T} E^T X_i E^+ C_i.$$

By using Lemma 1 in [105], the inequality in (7.32) can be guaranteed by the following inequality:

$$\begin{bmatrix} \Theta_i + \varepsilon \bar{N}_i^T \bar{N}_i & X_i^T \bar{M}_i \\ \star & -\varepsilon I \end{bmatrix} < 0, \tag{7.33}$$

where ε is a positive scalar. Combining inequality (7.33) and the inequality in (7.3), the admissibility condition of SSMSs with uncertainty is obtained. Then using the two equivalent sets and following similar line as that in the section, the admissibilization and SMC problems of uncertain SSMSs will be solved.

Remark 7.13. The reason of choosing the special structures is to make the inequality in (7.26) easier to be solved, because it is LMI when scalars μ_i, $i = 1, 2, 3$ are given. The obtaining LMI condition introduce some conservatism, since the matrices $\bar{G}_i$ and $\bar{F}_i$ are not completely free. Different structures of matrices $\bar{G}_i$ and $\bar{F}_i$ will create different conservatism, and thus no inclusion property can be found.

We now design the SMC law:

$$u(t) = K_i \breve{x}(t) + \sum_{j=1}^{N} \pi_{ij} (G_j B_j)^{-1} s(t) - (\lambda + \varepsilon(t)) \mathrm{sign}(s(t)), \tag{7.34}$$

where

$$\varepsilon(t) = \max_{i \in \mathcal{N}} (\|(G_i B_i)^{-1} G_i L_i y(t)\| + \|(G_i B_i)^{-1} G_i L_i D_i \breve{x}(t)\|),$$

$$G_i = B_i^T P_i, \ \lambda > 0.$$

In the following result, the trajectory of the observer in (7.17) being driven on to the designed sliding surface $s(t) = 0$ in a finite time will be proved.

Theorem 7.14. *By utilizing SMC law (7.34) and integral sliding surface (7.19), the state trajectories of (7.17) will be driven onto the sliding surface $s(t) = 0$ in a finite time.*

Proof. Constructing Lyapunov function:

$$V(t) = \frac{1}{2} s^T(t)(G_i B_i)^{-1} s(t).$$

From (7.19), we have

$$\begin{aligned} \dot{s}(t) = & - G_i B_i (\lambda + \varepsilon(t)) \text{sign}(s(t)) + G_i L_i (y(t) - \check{y}(t)) \\ & + G_i B_i \sum_{j=1}^{N} \pi_{ij} (G_j B_j)^{-1} s(t). \end{aligned}$$

Let $\mathcal{L}$ be the weak infinitesimal generator. Then considering $|s(t)| \geq \|s(t)\|$, we have

$$\begin{aligned} & \mathcal{L}V(t) \\ = & s^T(t)(G_i B_i)^{-1} \dot{s}(t) + s^T(t) \sum_{j=1}^{N} \pi_{ij} (G_j B_j)^{-1} s(t) \\ = & - s^T(t)(\lambda + \varepsilon(t)) \text{sign}(s(t)) + s^T(t)(G_i B_i)^{-1} G_i L_i (y(t) - \check{y}(t)) \\ \leq & - (\lambda + \varepsilon(t)) \|s(t)\| + \|(G_i B_i)^{-1} G_i L_i y(t)\| \|s(t)\| \\ & + \|(G_i B_i)^{-1} G_i L_i D_i \check{x}(t)\| \|s(t)\| \\ \leq & - \lambda \|s(t)\|, \end{aligned}$$

which implies that system trajectories of (7.17) converge to the sliding surface in a finite time due to $\lambda > 0$. □

When the states are measurable, the observer is not needed, and the SMC can be solved following the similar lines as that in [181]. Therefore the process will not be given in detail here. When all the states of system (7.1) are available, the sliding surface function can be obtained directly without using the observer to estimate its state. Following the similar lines as that in [181] to solve the SMC problem, SMDs can be obtained as in (7.9) there. To design the controller gain, Theorem 7.5 in this section can be utilized directly. For details, according to Subsection 3.1 of [181], the integral sliding surface function will be chosen as

$$s(t) = G_i Ex(t) - \int_0^t G_i(A_i + B_i K_i)x(\delta)d\delta,$$

where $G_i B_i$ is nonsingular, and $G_i C_i = 0$. The SMC law is given as

$$u(t) = K_i x(t) + \sum_{j=1}^{N} \pi_{ij}(G_j B_j)^{-1} s(t) - \lambda \text{sign}(s(t)),$$

where $\lambda > 0$.

7.4 Examples

Two examples are provided to show the workability of admissibilization results by state feedback control and the observer-based SMC of SSMSs in this section, respectively.

Example 7.1. Let us consider the modeling of a direct current (DC) motor controlling an inverted pendulum (depicted in Fig. 7.1), which has been modeled by the following SMS in [170]:

$$E\dot{x}(t) = A_i x(t) + B_i u(t), \tag{7.35}$$

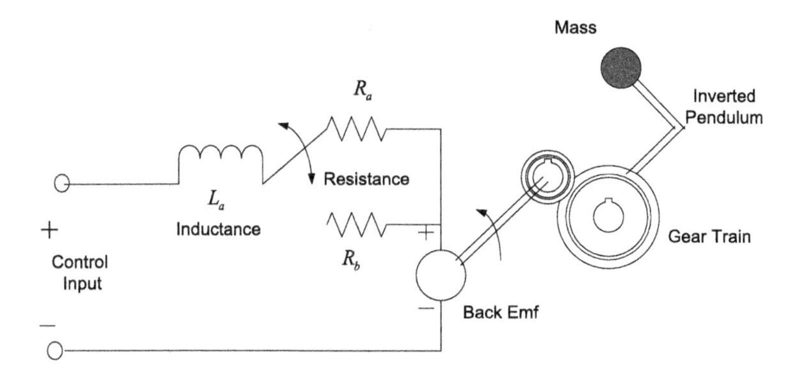

Figure 7.1 Block diagram of a DC motor controller inverted pendulum.

where

$$E=\begin{bmatrix}1&0&0\\0&1&0\\0&0&0\end{bmatrix},\ A_1=\begin{bmatrix}0&1&0\\9.8&0&1\\-20&-3&-1\end{bmatrix},\ B_1=\begin{bmatrix}0.4\\-0.2\\-0.5\end{bmatrix},$$

$$A_2=\begin{bmatrix}0&1&0\\9.8&0&1\\-20&-3&-0.5\end{bmatrix},\ B_2=\begin{bmatrix}-1.2\\0.5\\-0.2\end{bmatrix},$$

$$E_L=\begin{bmatrix}1&0\\0&1\\0&0\end{bmatrix},\ E_R=\begin{bmatrix}1&0\\0&1\\0&0\end{bmatrix},\ E^{+}=\begin{bmatrix}1&0&0\\0&1&0\\0&0&0\end{bmatrix},$$

$$\pi=\begin{bmatrix}-0.6&0.6\\0.2&-0.2\end{bmatrix},\ U=\begin{bmatrix}0\\0\\-0.7\end{bmatrix},\ \Lambda=\begin{bmatrix}0\\0\\0.8\end{bmatrix}.$$

Similarly, as examples in [245], if the system matrix A_i is affected by some random environmental effects, such as A_i becoming $A_i + C_i$"*noise*", the system in (7.35) will be

$$Edx(t) = (A_ix(t) + B_iu(t))dt + C_ix(t)dw(t). \tag{7.36}$$

Set

$$C_1=\begin{bmatrix}0.2&0&1\\0.1&0.2&0\\0&0&0\end{bmatrix},\ C_2=\begin{bmatrix}1&0&0.1\\0.2&0.1&0\\0&0&0\end{bmatrix}.$$

For system (7.36) with $u(t) = 0$, by solving the LMIs in Theorem 7.3, no feasible solutions can be found. By using admissibilization criterion in Theorem 7.5, the LMIs in (7.10) and (7.11) are feasible and the state-feedback controllers are

$$K_1=\begin{bmatrix}-38.9127&-7.1955&-18.8725\end{bmatrix},$$

$$K_2=\begin{bmatrix}-18.4065&-1.1594&-1.6126\end{bmatrix},$$

which implies the closed-loop system is mean square admissible. To check the effectiveness of the feedback control method, on the other hand, for the closed-loop system

$$Edx(t) = (A_i + B_iK_i)x(t)dt + C_ix(t)dw(t)$$

with the given parameters and the obtained controllers K_1 and K_2, the result in Theorem 7.3 will be applied. For this system, by solving the LMIs in (7.5) and (7.6), the following feasible solutions can be obtained:

$$P_1 = \begin{bmatrix} 0.7597 & 0.6504 & 0.2792 \\ 0.6504 & 0.6007 & 0.2291 \\ 0.2792 & 0.2291 & 0 \end{bmatrix}, \ \Phi_1 = 0.1909,$$

$$P_2 = \begin{bmatrix} 0.7020 & 0.5730 & 0.2904 \\ 0.5730 & 0.5254 & 0.2131 \\ 0.2904 & 0.2131 & 0 \end{bmatrix}, \ \Phi_2 = 0.1112,$$

which still shows the closed-loop system is mean square admissible and that the admissibilization method is effective.

Remark 7.15. By now, only a few results about SSMSs have been reported, such as [245] and [181]. In [245], only the stability condition is proposed and the state-feedback controller design is not considered. One method of designing controller is presented in Theorem 1 in [181]. However, the controller design condition in [181] is based on the stability criterion Lemma 1 there, which is cited from [5]. The authors in [240] have pointed out that the stability condition in [5] is improper.

Example 7.2. In this example, the effectiveness of SMC method is demonstrated. The considered SSMS is given with the following parameters:

$$E = \begin{bmatrix} 1 & 1 \\ 0 & 0 \end{bmatrix}, \ A_1 = \begin{bmatrix} 0.2 & 0.7 \\ 0.4 & -0.5 \end{bmatrix}, \ A_2 = \begin{bmatrix} 0.1 & 0.1 \\ 0.3 & -2 \end{bmatrix},$$

$$C_1 = \begin{bmatrix} 0.4 & 0.2 \\ 0 & 0 \end{bmatrix}, \ C_2 = \begin{bmatrix} 0.2 & 0.4 \\ 0 & 0 \end{bmatrix}, \ \pi = \begin{bmatrix} -1 & 1 \\ 2 & -2 \end{bmatrix},$$

$$B_1 = \begin{bmatrix} 1 \\ -9 \end{bmatrix}, \ B_2 = \begin{bmatrix} -2.5 \\ 1 \end{bmatrix}, \ D_1 = \begin{bmatrix} -0.6 \\ -0.9 \end{bmatrix}^T, \ D_2 = \begin{bmatrix} -0.3 \\ 0.7 \end{bmatrix}^T,$$

$$E_L = \begin{bmatrix} 1 \\ 0 \end{bmatrix}, \ E_R = \begin{bmatrix} 1 \\ 1 \end{bmatrix}, \ U = \begin{bmatrix} 0 & 1 \end{bmatrix}, \ \Lambda = \begin{bmatrix} 1 \\ -1 \end{bmatrix}.$$

An observer in (7.17) will be designed to estimate state of system (7.1) and apply the estimation state to design an SMC law $u(t)$ in (7.34), such that overall SMDs are mean square admissible. Firstly, according Remark 7.8, by using Corollary 7.7, the controller K_i can be designed as

$$K_1 = \begin{bmatrix} 1.3847 & 2.6201 \end{bmatrix}, \ K_2 = \begin{bmatrix} 5.5530 & 0.8508 \end{bmatrix}.$$

Then Theorem 7.9 will be sued to obtain the observer gain L_i. By solving the LMIs in (7.25) and (7.26) of Theorem 7.9 with the following parameters:

$$\bar{U} = \begin{bmatrix} 0 & 1 & 0 & 0 \\ 0 & 0 & 0 & 1 \end{bmatrix}, \ \bar{A}^T = \begin{bmatrix} -1 & 1 & 0 & 0 \\ 0 & 0 & -1 & 1 \end{bmatrix},$$

$$\mathcal{E}_L = \begin{bmatrix} 1 & 0 \\ 0 & 0 \\ 0 & 1 \\ 0 & 0 \end{bmatrix}, \ \mathcal{E}_R = \begin{bmatrix} 1 & 0 \\ 1 & 0 \\ 0 & 1 \\ 0 & 1 \end{bmatrix}, \ \mu_1 = 3, \ \mu_2 = 5, \ \mu_3 = 1.5,$$

the feasible variables are given

$$P_1 = \begin{bmatrix} 0.1003 & 0.0308 \\ 0.0308 & 0.0430 \end{bmatrix}, \ P_2 = \begin{bmatrix} 0.0746 & 0.1143 \\ 0.1143 & 0.2532 \end{bmatrix},$$

$$\bar{L}_1 = \begin{bmatrix} -0.1476 \\ -0.1863 \end{bmatrix}, \ \bar{L}_2 = \begin{bmatrix} -0.4231 \\ -0.4424 \end{bmatrix},$$

$$L_1 = P_1^{-1}\bar{L}_1 = \begin{bmatrix} -0.1793 \\ -4.2050 \end{bmatrix}, \ L_2 = P_2^{-1}\bar{L}_2 = \begin{bmatrix} -9.7293 \\ 2.6459 \end{bmatrix},$$

$$G_1 = B_1^T P_1 = \begin{bmatrix} -0.1770 & -0.3561 \end{bmatrix},$$

$$G_2 = B_2^T P_2 = \begin{bmatrix} -0.0721 & -0.0326 \end{bmatrix}.$$

Based on obtained controller and observer, the system parameters of SMDs in (7.24) are given as

$$\mathcal{E} = \begin{bmatrix} 1 & 1 & 0 & 0 \\ 0 & 0 & 0 & 0 \\ 0 & 0 & 1 & 1 \\ 0 & 0 & 0 & 0 \end{bmatrix},$$

$$\bar{A}_1 = \begin{bmatrix} 1.5847 & 3.3201 & 0.4106 & 0.6159 \\ -12.0623 & -24.0809 & -0.2041 & -0.3062 \\ 0 & 0 & 0.0924 & 0.5386 \\ 0 & 0 & -2.1230 & -4.2845 \end{bmatrix},$$

$$\bar{A}_2 = \begin{bmatrix} -13.7825 & -2.0270 & -0.2064 & 0.4817 \\ 5.8530 & -1.1492 & 0.4563 & -1.0647 \\ 0 & 0 & -2.8188 & 6.9105 \\ 0 & 0 & 1.0938 & -3.8522 \end{bmatrix},$$

$$\bar{C}_1 = \begin{bmatrix} 0 & 0 & 0 & 0 \\ 0 & 0 & 0 & 0 \\ 0.4 & 0.2 & 0.4 & 0.2 \\ 0 & 0 & 0 & 0 \end{bmatrix}, \quad \bar{C}_2 = \begin{bmatrix} 0 & 0 & 0 & 0 \\ 0 & 0 & 0 & 0 \\ 0.2 & 0.4 & 0.2 & 0.4 \\ 0 & 0 & 0 & 0 \end{bmatrix}.$$

By solving the conditions in Theorem 7.3 for system (7.24) with above parameters, the feasible variables can be found, which implies that overall SMDs (7.24) are mean square admissible.

7.5 Conclusion

The chapter is to study the SMC with or without the observer for SSMSs in this section. Firstly, sufficient admissibility and admissibilization criteria in terms of strict LMIs for SSMSs are proposed by using matrix replacement approach. Based on these criteria, the observer-based SMC problem and SMC problem without using the observer are solved, respectively. The workability of the proposed methods about admissibilization and SMC has been illustrated by numerical examples. On the other hand, the Markovian jump systems with incomplete transition rate have attracted considerable attention in recent years. By designing an SMC law with different robust terms for known and unknown modes, respectively, a good SMC result proposed in [17] for stochastic Markovian jump systems has been extended to stochastic Markovian jump systems with incomplete transition rate in [16]. Combining the technique provided in [16] and the two equivalent sets in this section, the observer-based SMC problem will be studied in our future work for SSMSs with incomplete transition rate. Moreover, similar to some existing results such as [95], [109], [182], the controller gain K_i and observer gain L_i are designed separately. For how to design the controller gain K_i and observer gain L_i simultaneously, which is also an interesting topic, is addressed in our future work.

CHAPTER 8

Admissibility and admissibilization for fuzzy singular systems

The issues of admissibility analysis for T-S fuzzy singular system with time delay and admissibilization for IT2 fuzzy singular system are investigated in this chapter, respectively. By adopting a novel tighter integral inequality, a sufficient delay-dependent criterion is built on the basis of strict LMIs, which guarantees the admissibility of the T-S fuzzy singular system. And by designing the state feedback controller and static output feedback controller, the problems of admissibility analysis and admissibilization for continuous singular IT2 fuzzy systems have been studied in the second subchapter. Several numerical examples show that the proposed methods are efficient and less conservative.

8.1 Admissibility analysis for Takagi–Sugeno fuzzy singular systems with time delay

This section aims to study the problem of admissibility of T-S fuzzy singular system with time delay by adopting a novel integral inequality, which uses a double integral of the system state and includes the Wirtinger-based inequality. Based on this inequality, a delay-dependent sufficient criterion is put forward to guarantee the admissibility of the considered system. A numerical example shows that the proposed method is efficient and less conservative.

8.1.1 Problem formulation

Consider a class of nonlinear singular system with time delay that is denoted by T-S fuzzy singular model as follows:

Plant Rule i: IF $\theta_1(t)$ is μ_{i1}; $\theta_2(t)$ is μ_{i2}; ... and $\theta_p(t)$ is μ_{ip}, THEN

$$\begin{aligned} E\dot{x}(t) &= A_i x(t) + A_{di} x(t-d), \\ x(t) &= \phi(t),\ t \in [-d, 0],\ i \in \mathbb{S}, \end{aligned} \tag{8.1}$$

Analysis and Synthesis of Singular Systems
https://doi.org/10.1016/B978-0-12-823739-7.00015-X

where $x(t) \in \mathbb{R}^n$ is the state vector; $\theta(t) = [\theta_1(t), \theta_2(t), \ldots, \theta_p(t)]$ is the presumed variable; μ_{ij} is the fuzzy set; $\mathbb{S} = \{1, 2, 3, \ldots, r\}$, and r denotes the number of IF-THEN rules; E probably is singular, and $\text{rank}(E) = g \leq n$; d is the known constant time delay satisfying $0 < d_{\min} \leq d \leq d_{\max}$; $\phi(t)$ means a compatible vector-value initial function, which satisfies certain strict consistent condition and guarantees a unique solution for any sufficiently differentiable input function [34]; A_i, A_{di} represent known real constant matrices.

Afterwards, the whole model of the above systems can be described by the following model:

$$E\dot{x}(t) = \sum_{i=1}^{r} h_i(\theta(t))[A_i x(t) + A_{di} x(t-d)],$$
$$x(t) = \phi(t), \ t \in [-d, 0], \ i \in \mathbb{S}, \tag{8.2}$$

where

$$h_i(t) = \frac{\omega_i(\theta(t))}{\sum_{i=1}^{r} \omega_i(\theta(t))}, \quad \omega_i(\theta(t)) = \prod_{j=1}^{p} \mu_{ij}(\theta(t)),$$

and $\mu_{ij}(\theta(t))$ denotes the grade of membership of $\theta_j(t)$ in μ_{ij}. Clearly, for all t, one can see

$$h_i(\theta(t)) \geq 0, \ \sum_{i=1}^{r} h_i(\theta(t)) = 1.$$

Definition 8.1. [215]

(i) If $\det(sE - \sum_{i=1}^{r} h_i(\theta(t))A_i)$ is not identically zero, the singular system (8.2) is regarded as regular.

(ii) If $\deg(\det(sE - \sum_{i=1}^{r} h_i(\theta(t))A_i)) = \text{rank}(E)$, the singular system (8.2) is referred to as impulse free.

(iii) If, for any $\varepsilon > 0$, there exists a scalar $\delta(\varepsilon) > 0$ such that, for any compatible initial conditions $\phi(t)$ satisfying $\sup_{-d_{max} \leq t \leq 0} \| \phi(t) \| < \delta(\varepsilon)$, the solution $x(t)$ of system (8.2) satisfies $\| x(t) \| < \varepsilon$ for $t \geq 0$, moreover, $\lim_{t \to +\infty} x(t) = 0$, the singular system (8.2) is said to be asymptotically stable.

(iv) If the singular system (8.2) is regular, impulse-free, and asymptotically stable, it is referred to as admissible.

The following nomenclature is adopted to simplify vector and matrix symbolizations:

$$
\begin{aligned}
v_1(t) &= \int_{t-d}^{t} x(s)ds,\ v_2(t) = \int_{t-d}^{t}\int_{t-d}^{s} x(u)duds, \\
\eta_1 &= \begin{bmatrix} x^T(t)E^T & v_1^T(t)E^T & v_2^T(t)E^T \end{bmatrix}^T, \\
\xi(t) &= \begin{bmatrix} x^T(t) & x^T(t-d) & \frac{1}{d}v_1^T(t)E^T & \frac{2}{d^2}v_2^T(t)E^T \end{bmatrix}^T, \\
e_i &= \begin{bmatrix} 0_{n\times(i-1)n} & I_n & 0_{n\times(4-i)n} \end{bmatrix},\ i=1,2,3,4, \\
\Gamma_i &= \begin{bmatrix} A_i & A_{di} & 0_{n\times n} & 0_{n\times n} \end{bmatrix},\ i\in\mathbb{S}.
\end{aligned}
$$

Lemma 8.2. *Presume x as a differentiable function: $[\alpha,\beta]\longrightarrow\mathbb{R}^n$. For symmetric matrices $S\in\mathbb{R}^{n\times n}>0$, N_1, N_2, $N_3\in\mathbb{R}^{4n\times n}$, and $E\in\mathbb{R}^{n\times n}$* (rank$(E)=$ *$g\le n$), the following inequality holds:*

$$
-\int_{\alpha}^{\beta}\dot{x}^T(s)E^TSE\dot{x}(s)ds \le \vartheta^T\Omega\vartheta, \tag{8.3}
$$

where

$$
\begin{aligned}
\Omega =& \tau(\widetilde{E}^TN_1S^{-1}N_1^T\widetilde{E}+\frac{1}{3}\widetilde{E}^TN_2S^{-1}N_2^T\widetilde{E}+\frac{1}{5}\widetilde{E}^TN_3S^{-1}N_3^T\widetilde{E}) \\
&+\mathrm{sym}(\widetilde{E}^TN_1\Pi_1+\widetilde{E}^TN_2\Pi_2+\widetilde{E}^TN_3\Pi_3), \\
\Pi_1 =& E(e_1-e_2),\ \Pi_2=E(e_1+e_2)-2e_3,\ \Pi_3=E(e_1-e_2)-6e_3+6e_4, \\
\widetilde{E} =& \mathrm{diag}(E,E,I,I),\ \widehat{E}=\mathrm{diag}(\widetilde{E},\widetilde{E},\widetilde{E}), \\
\vartheta =& \begin{bmatrix} x^T(\beta) & x^T(\alpha) & \frac{1}{\tau}\int_{\alpha}^{\beta}(Ex(s))^Tds & \frac{2}{\tau^2}\int_{\alpha}^{\beta}\int_{\alpha}^{s}(Ex(u))^Tduds \end{bmatrix}^T,\ \tau=\beta-\alpha.
\end{aligned}
$$

Proof. Define

$$
\begin{aligned}
f_1(s) &= \frac{2s-\beta-\alpha}{\beta-\alpha}, \\
f_2(s) &= \frac{6s^2-6(\beta+\alpha)s+\beta^2+4\beta\alpha+\alpha^2}{(\beta-\alpha)^2}, \\
N &= \begin{bmatrix} N_1^T & N_2^T & N_3^T \end{bmatrix}^T, \\
\zeta(s) &= \begin{bmatrix} \vartheta^T & f_1(s)\vartheta^T & f_2(s)\vartheta^T \end{bmatrix}^T.
\end{aligned}
$$

It is clear that

$$-2\zeta^T(s)\widehat{E}^T NE\dot{x}(s) \leq \zeta^T(s)\widehat{E}^T NS^{-1}N\widehat{E}\zeta(s) + \dot{x}^T(s)E^T SE\dot{x}(s). \qquad (8.4)$$

Integrating (8.4) from α to β derives

$$\begin{aligned}
&-2\vartheta^T\widetilde{E}^T N_1 E(e_1 - e_2)\vartheta - 2\vartheta^T\widetilde{E}^T N_2(Ee_1 + Ee_2 - 2e_3)\vartheta \\
&-2\vartheta^T\widetilde{E}^T N_3(Ee_1 - Ee_2 - 6e_3 + 6e_4)\vartheta \\
\leq&(\beta-\alpha)\vartheta^T\widetilde{E}^T N_1 S^{-1}N_1^T\widetilde{E}\vartheta + \frac{(\beta-\alpha)}{3}\vartheta^T\widetilde{E}^T N_2 S^{-1}N_2^T\widetilde{E}\vartheta \\
&+\frac{(\beta-\alpha)}{5}\vartheta^T\widetilde{E}^T N_3 S^{-1}N_3^T\widetilde{E}\vartheta + \int_\alpha^\beta \dot{x}^T(s)E^T SE\dot{x}(s)d(s). \qquad (8.5)
\end{aligned}$$

Rearranging (8.5) derives (8.3), so the proof is completed. □

Remark 8.3. When $E = I$, Lemma 8.2 becomes into Lemma 1 in [228]. If $\text{rank}(E) = g \leq n$, Lemma 8.2 can be used in singular systems, which shows that Lemma 8.2 has a wider application.

The prime object of this subchapter is to put forward a new admissibility criterion of the considered systems and reduce the conservatism of existing results by utilizing this new inequality technique in Lemma 8.2.

8.1.2 Main results

Theorem 8.4. *Given a constant time delay $d \in [d_{\min}, d_{\max}]$, the T-S fuzzy singular systems with time delay in (8.2) is admissible, if, for all $i \in \mathbb{S}$, there exist $P \in \mathbb{R}^{3n\times 3n} > 0$, $Q \in \mathbb{R}^{n\times n} > 0$, $S \in \mathbb{R}^{n\times n} > 0$, $W \in \mathbb{R}^{n\times g}$, and matrices N_1, N_2, $N_3 \in \mathbb{R}^{4n\times n}$, making the following LMIs hold:*

$$\begin{bmatrix} \Omega_i & \sqrt{d}\widetilde{E}^T N_1 & \sqrt{d}\widetilde{E}^T N_2 & \sqrt{d}\widetilde{E}^T N_3 \\ * & -S & 0 & 0 \\ * & * & -3S & 0 \\ * & * & * & -5S \end{bmatrix} < 0, \qquad (8.6)$$

where $R \in \mathbb{R}^{n\times(n-g)}$ is any matrix that is full column rank and satisfies $E^T R = 0$, and

$$\begin{aligned}
\Omega_i =& \Omega_{1i} + \Omega_2 + \Omega_{3i}, \\
\Omega_{1i} =& \text{sym}(\Pi_4^T P\Pi_{5i}) + e_1^T Qe_1 - e_2^T Qe_2 + d\Gamma_i^T S\Gamma_i, \\
\Omega_2 =& \text{sym}(\widetilde{E}^T N_1\Pi_1 + \widetilde{E}^T N_2\Pi_2 + \widetilde{E}^T N_3\Pi_3),
\end{aligned}$$

$$\Omega_{3i} = \text{sym}(e_1^T W R^T \Gamma_i),$$

$$\Pi_4 = \begin{bmatrix} e_1^T E^T & d e_3^T & \frac{d^2}{2} e_4^T \end{bmatrix}^T,$$

$$\Pi_{5i} = \begin{bmatrix} \Gamma_i^T & (e_1^T - e_2^T) E^T & d(e_3^T - e_2^T E^T) \end{bmatrix}^T,$$

$$P = \begin{bmatrix} P_{11} & P_{12} & P_{13} \\ * & P_{22} & P_{23} \\ * & * & P_{33} \end{bmatrix}, \; N_j = \begin{bmatrix} N_{ja} \\ N_{jb} \\ N_{jc} \end{bmatrix}, \; j = 1, 2, 3,$$

and Π_i, $i = 1, 2, 3$, $\tilde{E}$ *are defined in Lemma 8.2.*

Proof. First of all, system (8.2) will be proved to be regular and impulse-free. Due to $\text{rank}(E) = g$, it is certain that there are two nonsingular matrices F and H making the following equality stand:

$$FEH = \begin{bmatrix} I_g & 0 \\ 0 & 0 \end{bmatrix}. \tag{8.7}$$

Set

$$F\left(\sum_{i=1}^{r} h_i(\theta(t)) A_i\right) H = \begin{bmatrix} A_{11} & A_{12} \\ A_{21} & A_{22} \end{bmatrix}, \; H^T W = \begin{bmatrix} W_1 \\ W_2 \end{bmatrix},$$

$$F^{-T} R = \begin{bmatrix} 0 \\ I_{(n-g)} \end{bmatrix} M, \tag{8.8}$$

where $M \in \mathbb{R}^{(n-g)\times(n-g)}$ is nonsingular. According to (8.6), it can be found $\Omega_i < 0$ such that

$$\begin{bmatrix} \Theta_{1i} & \bullet \\ \bullet & \bullet \end{bmatrix} < 0,$$

where "•" represents the elements in matrix that are not related to next discussions, and

$$\begin{aligned} \Theta_{1i} = {} & E^T P_{11} A_i + A_i^T P_{11} E + E^T P_{12} E + E^T P_{12}^T E + Q + d A_i^T S A_i \\ & + E^T N_{1a} E + E^T N_{1a}^T E + E^T N_{2a} E + E^T N_{2a}^T E \\ & + E^T N_{3a} E + E^T N_{3a}^T E + W R^T A_i + A_i^T R W^T. \end{aligned}$$

Because of $Q > 0$, $S > 0$, we have

$$\tilde{\Theta}_{1i} = E^T P_{11} A_i + A_i^T P_{11} E + E^T P_{12} E + E^T P_{12}^T E + E^T N_{1a} E + E^T N_{1a}^T E$$

$$+ E^T N_{2a} E + E^T N_{2a}^T E + E^T N_{3a} E + E^T N_{3a}^T E$$
$$+ WR^T A_i + A_i^T R W^T < 0.$$

Due to $h_i(\theta(t)) \geq 0$ and $\sum_{i=1}^{r} h_i(\theta(t)) = 1$, we have

$$\begin{aligned}
\Theta_1 =& E^T P_{11} \left(\sum_{i=1}^{r} h_i(\theta(t)) A_i \right) + \left(\sum_{i=1}^{r} h_i(\theta(t)) A_i^T \right) P_{11} E + E^T P_{12} E + E^T P_{12}^T E \\
&+ E^T N_{1a} E + E^T N_{1a}^T E + E^T N_{2a} E + E^T N_{2a}^T E + E^T N_{3a} E + E^T N_{3a}^T E \\
&+ WR^T \left(\sum_{i=1}^{r} h_i(\theta(t)) A_i \right) + \left(\sum_{i=1}^{r} h_i(\theta(t)) A_i^T \right) R W^T < 0.
\end{aligned}$$

Premultiply and postmultiply $\Theta_1 < 0$ by H^T and H, respectively. Substituting (8.7) and (8.8) in the above inequality gets

$$\begin{bmatrix} \bullet & \bullet \\ \bullet & W_2 M^T A_{22} + A_{22}^T M W_2^T \end{bmatrix} < 0.$$

From the above inequality, it is easy to see that $W_2 M^T A_{22} + A_{22}^T M W_2^T < 0$, which means A_{22} is nonsingular. Hence, the pair $\left(E, \left(\sum_{i=1}^{r} h_i(\theta(t)) A_i\right)\right)$ is regular and impulse-free. Based on Definition 8.1, system (8.2) is regular and impulse-free.

To prove the stability of system (8.2), choose Lyapunov–Krasovskii functional as follows:

$$V(x_t) = V_1(x_t) + V_2(x_t) + V_3(x_t),$$

where

$$\begin{aligned}
V_1(x_t) =& \eta_1^T(t) P \eta_1(t), \\
V_2(x_t) =& \int_{t-d}^{t} x^T(s) Q x(s) ds, \\
V_3(x_t) =& \int_{-d}^{0} \int_{t+u}^{t} \dot{x}^T(s) E^T S E \dot{x}(s) ds du.
\end{aligned}$$

According to the solutions of system (8.2), take the derivative of $V(x_t)$, and one can get

$$\dot{V}_1(x_t) = \xi^T(t) \mathrm{sym}(\Pi_4^T P \Pi_{5i}) \xi(t), \tag{8.9}$$
$$\dot{V}_2(x_t) = \xi^T(t)(e_1^T Q e_1 - e_2^T Q e_2) \xi(t), \tag{8.10}$$

$$\dot{V}_3(x_t) = d\xi^T(t)\Gamma_i^T S\Gamma_i\xi(t) - \int_{t-d}^{t} \dot{x}^T(s)E^T SE\dot{x}(s)ds. \tag{8.11}$$

Adopting Lemma 8.2, we have

$$\begin{aligned}
&-\int_{t-d}^{t} \dot{x}^T(s)E^T SE\dot{x}(s)ds \\
\leq &\xi^T(t)\Big[d\widetilde{E}^T N_1 S^{-1} N_1^T \widetilde{E} + \frac{d}{3}\widetilde{E}^T N_2 S^{-1} N_2^T \widetilde{E} + \frac{d}{5}\widetilde{E}^T N_3 S^{-1} N_3^T \widetilde{E} \\
&+\mathrm{sym}(\widetilde{E}^T N_1 \Pi_1 + \widetilde{E}^T N_2 \Pi_2 + \widetilde{E}^T N_3 \Pi_3)\Big]\xi(t) \\
= &\xi^T(t)(\Omega_2 + \Omega_4)\xi(t),
\end{aligned} \tag{8.12}$$

where $\Omega_4 = d\widetilde{E}^T N_1 S^{-1} N_1^T \widetilde{E} + \frac{d}{3}\widetilde{E}^T N_2 S^{-1} N_2^T \widetilde{E} + \frac{d}{5}\widetilde{E}^T N_3 S^{-1} N_3^T \widetilde{E}$. In addition, it is clear to see that

$$2x^T(t)WR^T E\dot{x}(s) \equiv 0. \tag{8.13}$$

Considering the inequalities from (8.9) to (8.13), we have

$$\dot{V}(x_t) \leq \xi^T(t)(\Omega_{1i} + \Omega_2 + \Omega_{3i} + \Omega_4)\xi(t).$$

According to Schur complement, $\Omega_{1i} + \Omega_2 + \Omega_{3i} + \Omega_4 < 0$ is equivalent to (8.6), so we have $\dot{V}(x_t) < 0$. Therefore system (8.2) is asymptotically stable. Then, we complete the proof of Theorem 8.4. □

The following corollary can simplify the computation by eliminating the three free matrices N_1, N_2, N_3 in Theorem 8.4:

Corollary 8.5. *Given a constant time-delay $d \in [d_{\min}, d_{\max}]$, the T-S fuzzy singular systems with time delay in (8.2) is admissible, if there exist $P \in \mathbb{R}^{3n\times 3n} > 0$, $Q \in \mathbb{R}^{n\times n} > 0$, $S \in \mathbb{R}^{n\times n} > 0$, $W \in \mathbb{R}^{n\times g}$ satisfying (8.6) with*

$$\begin{aligned}
N_1 =& \frac{1}{d}\begin{bmatrix} -S & S & 0 & 0 \end{bmatrix}^T, \\
N_2 =& \frac{3}{d}\begin{bmatrix} -S & -S & 2S & 0 \end{bmatrix}^T, \\
N_3 =& \frac{5}{d}\begin{bmatrix} -S & S & 6S & -6S \end{bmatrix}^T.
\end{aligned}$$

Remark 8.6. The prime advantage of this section is to adopt a new inequality in Lemma 8.2, which is an improved version of Lemma 1 in [228] by adding the singular matrix E. Through employing three free matrices

N_1, N_2, and N_3 to cope with the relationships among x_t, x_{t-d}, $\int_{t-d}^{t} x(s)ds$, and $\int_{t-d}^{t}\int_{t-d}^{s} x(u)duds$, we obtain a less conservative result, which is much tighter than Jensen inequality. Corollary 8.5 gives a simple version of Theorem 8.4, which greatly reduces the number of decision variables. Although it is not more conservative than Theorem 8.4, it is a good trade-off between computation and conservatism.

To compare some existing results, according to the approach in [64], we obtain a proposition relevant to the admissibility of the considered system.

Proposition 8.7. *Given an integer $m \geq 1$, the T-S fuzzy singular system with time delay in (8.2) is admissible if, for all $i \in \mathbb{S}$, there exist matrices $Q > 0$, $R > 0$, $P = \begin{bmatrix} P_{11} & 0 \\ P_{21} & P_{22} \end{bmatrix}$, $U = \begin{bmatrix} U_1 & 0_{n\times(n-g)} \end{bmatrix}$ (with $0 < P_{11} \in \mathbb{R}^{g\times g}$, and $U_1 \in \mathbb{R}^{n\times g}$), making the following LMIs hold:*

$$\begin{bmatrix} \Theta_i & \frac{d}{m} W_{Ri}^T R & \frac{d}{m} W_X^T U \\ \star & -\frac{d}{m} R & 0 \\ \star & \star & -\frac{d}{m} R \end{bmatrix} < 0, \tag{8.14}$$

where

$$\begin{aligned} \Theta_i =& \mathrm{sym}(W_X^T P^T A_i W_X + W_X^T U W_X + W_X^T P^T A_{di} W_D \\ & - W_X^T U W_D) + W_Q^T \overline{Q} W_Q, \end{aligned}$$

$$W_X = \begin{bmatrix} I_n & 0_{n,mn} \end{bmatrix},\ W_D = \begin{bmatrix} 0_{n,mn} & I_n \end{bmatrix},\ W_{Ri} = \begin{bmatrix} A_i & 0_{n,(m-1)n} & A_{di} \end{bmatrix},$$

$$\overline{Q} = \begin{bmatrix} Q & 0_{mn} \\ 0_{mn} & -Q \end{bmatrix},\ W_Q = \begin{bmatrix} I_{mn} & 0_{mn,n} \\ 0_{mn,n} & I_{mn} \end{bmatrix}.$$

8.1.3 Numerical example

A numerical example is given to show the effectiveness of our approach in this section.

Example 8.1. Consider the following T-S fuzzy singular system (8.2):

Plant Rule i:

IF $\theta_1(t)$ is μ_{i1}, THEN

$$E\dot{x}(t) = A_i x(t) + A_{di} x(t-d),\ i = 1, 2,$$

where

$$E = \begin{bmatrix} 1 & 0 \\ 0 & 0 \end{bmatrix}, \ A_1 = \begin{bmatrix} 1.5 & 0.2 \\ 0 & -1.1 \end{bmatrix}, \ A_{d1} = \begin{bmatrix} -2 & 0.2 \\ 0.6 & 0.5 \end{bmatrix},$$
$$A_2 = \begin{bmatrix} 1.6 & 0.3 \\ 0 & -1.2 \end{bmatrix}, \ A_{d2} = \begin{bmatrix} -2.2 & 0.1 \\ 0.5 & 0.6 \end{bmatrix}.$$

Applying the approaches of Theorem 8.4, Corollary 8.5, and Proposition 8.7, the admissible maximum values of time delay d_{max} of different methods are calculated by utilizing LMI Tool-box. The results of d_{max} and the corresponding number of decision variables (NDVs) are shown in Table 8.1. Proposition 8.7 adopts the delay partitioning approach. It is clear to see that the results become less conservative, when the delay-partitioning number m gets larger. Nevertheless, the number of variables also increases, which means the computation becomes more complicated. The time-delay result of Theorem 8.4 is a little better than the result of Proposition 8.7 (when $m = 6$), and there are less variables in Theorem 8.4. Particularly, the time-delay result of Corollary 8.5 is close to the above two methods, but the number of variables decreases sharply. Considering both conservatism and computational burden, Corollary 8.5 is a good trade-off.

Table 8.1 Comparison of the number of decision variables and the maximum admissible values d_{max} of different methods.

Methods	d_{max}	NDVs
Proposition 8.7 ($m = 2$)	0.1885	$3.5n^2 + 1.5n + 0.5g(g+1)$
Proposition 8.7 ($m = 4$)	0.3771	$9.5n^2 + 2.5n + 0.5g(g+1)$
Proposition 8.7 ($m = 6$)	0.5657	$19.5n^2 + 3.5n + 0.5g(g+1)$
Theorem 8.4	0.5660	$17.5n^2 + 2.5n$
Corollary 8.5	0.5049	$5.5n^2 + 2.5n$

8.1.4 Conclusion

This subchapter concerns the admissibility of Takagi–Sugeno fuzzy time-delay singular system. According to a new integral inequality and Lyapunov method, a sufficient admissible criterion is developed to guarantee the regularity, absence of impulses and stability of the considered system. A numerical example demonstrates the advantages of this method in reducing the conservatism and computational burden. The proposed method is also easy to apply to uncertain systems, so the issue of the robust dissipative control of the considered system will be considered in our future work.

8.2 Admissibilization of singular IT2 fuzzy systems

In this section, we consider the state feedback and static output feedback control problems of continuous singular IT2 T-S fuzzy systems with mismatched membership functions. Firstly, based on Lyapunov stability theory, a state feedback control criterion is proposed to guarantee the closed-loop system to be admissible. Secondly, the result is extend to static output feedback control problem, and a sufficient condition for synthesis is derived in terms of strict LMIs, which eliminates the disadvantages in some existing results, such as same output matrices and bilinear matrix inequality problem. To obtain less conservative results, the information of mismatched membership functions is employed. Finally, numerical examples are given to illustrate the effectiveness of the proposed techniques.

8.2.1 Preliminaries

Consider the following singular IT2 fuzzy system described by:

Plant Rule i ($i = 1, 2, \ldots, r$): If $f_1(x(t))$ is W_{i1} and $\cdots$ and $f_p(x(t))$ is W_{ip}, THEN

$$\begin{cases} E\dot{x}(t) = A_i x(t) + B_i u(t) \\ y(t) = C_i x(t), \end{cases} \tag{8.15}$$

where W_{is} is an IT2 fuzzy set of rule i corresponding to the function $f_s(x(t))$, $i = 1, 2, \ldots, r$, $s = 1, 2, \ldots, p$, p is the number of premise variables; $x(t) \in \mathbb{R}^n$ is the state vector, $u(t) \in \mathbb{R}^m$ is the input vector, and $y(t) \in \mathbb{R}^l$ is the measured output; $A_i \in \mathbb{R}^{n\times n}$, $B_i \in \mathbb{R}^{n\times m}$, and $C_i \in \mathbb{R}^{l\times n}$ denote constant matrices. In contrast with standard linear systems with $E = I$, the matrix $E \in \mathbb{R}^{n\times n}$ has $0 < \text{rank}(E) = q < n$. The firing strength of the ith rule is of the following interval sets:

$$\bar{W}_i(x(t)) = [\underline{w}_i(x(t)),\ \bar{w}_i(x(t))],\ i = 1, 2, \ldots, r, \tag{8.16}$$

where

$$\underline{w}_i(x(t)) = \prod_{s=1}^{p} \underline{\mu}_{W_{is}}(f_s(x(t))) \geq 0,$$

$$\bar{w}_i(x(t)) = \prod_{s=1}^{p} \bar{\mu}_{W_{is}}(f_s(x(t))) \geq 0,$$

$$\bar{\mu}_{W_{is}}(f_s(x(t))) \geq \underline{\mu}_{W_{is}}(f_s(x(t))) \geq 0,$$

$$\bar{w}_i(x(t)) \geq \underline{w}_i(x(t)) \geq 0$$

with $\underline{\mu}_{W_{is}}(f_s(x(t))) \in [0, 1]$ and $\bar{\mu}_{W_{is}}(f_s(x(t))) \in [0, 1]$ denote the lower grade of membership and upper grade of membership governed by lower and upper membership functions, respectively. The inferred singular IT2 T-S fuzzy model is defined as follows:

$$\begin{cases} E\dot{x}(t) = \sum_{i=1}^{r} w_i(x(t))(A_i x(t) + B_i u(t)), \\ y(t) = \sum_{i=1}^{r} w_i(x(t)) C_i x(t), \end{cases} \tag{8.17}$$

where

$$\begin{aligned} w_i(x(t)) &= \underline{\alpha}_i(x(t))\underline{w}_i(x(t)) + \bar{\alpha}_i(x(t))\bar{w}_i(x(t)) \geq 0 \ \forall i, \\ &\sum_{i=1}^{r} w_i(x(t)) = 1, \\ &0 \leq \underline{\alpha}_i(x(t)) \leq 1 \ \forall i, \\ &0 \leq \bar{\alpha}_i(x(t)) \leq 1 \ \forall i, \\ &\underline{\alpha}_i(x(t)) + \bar{\alpha}_i(x(t)) = 1 \ \forall i, \end{aligned}$$

in which $\underline{\alpha}_i(x(t))$ and $\bar{\alpha}_i(x(t))$ are nonlinear functions not necessarily known but exist, and $w_i(x(t))$ denotes the grade of membership of the embedded membership functions $\underline{\alpha}_i(x(t))$ and $\bar{\alpha}_i(x(t))$. Because the membership degrees are no longer crisp, but characterized by lower and upper membership degrees, the system in (8.17) belongs to type-2 fuzzy system.

Remark 8.8. From the expression in (8.17), we can see that the linear combination of $\underline{w}_i(x(t))$ and $\bar{w}_i(x(t))$ is used to describe the actual grades of membership $w_i(x(t))$ scaled by the nonlinear functions. The uncertainty of nonlinear plant may lead to the uncertain $w_i(x(t))$, $\underline{\alpha}_i(x(t))$, and $\bar{\alpha}_i(x(t))$. The system model presented in (8.17) facilitates the stability and control problems, and is not necessarily implemented.

The first purpose of this section is to design a state feedback IT2 fuzzy controller with r rules of the following format to admissibilize the IT2 T-S fuzzy system in (8.17):

Controller Rule j: IF $g_1(x(t))$ is M_{j1} and $\cdots$ and $g_p(x(t))$ is M_{jp}, THEN

$$u(t) = K_j x(t), \tag{8.18}$$

where M_{js} stands for the jth fuzzy set of the function $g_s(x(t))$, $j = 1, 2, \ldots, r$, $s = 1, 2, \ldots, p$; p is the number of premise variable; $K_j \in R^{m \times n}$ is the state feedback gain matrix of rule j. The firing interval of the jth rule is as follows:

$$\bar{M}_j(x(t)) = [\underline{m}_j(x(t)),\ \bar{m}_j(x(t))], j = 1, 2, \ldots, r, \tag{8.19}$$

where

$$\begin{aligned}
\underline{m}_j(x(t)) &= \prod_{s=1}^{p} \underline{\mu}_{M_{js}}(g_s(x(t))) \geq 0, \\
\bar{m}_j(x(t)) &= \prod_{s=1}^{p} \bar{\mu}_{M_{js}}(g_s(x(t))) \geq 0, \\
\bar{\mu}_{M_{js}}(g_s(x(t))) &\geq \underline{\mu}_{M_{js}}(g_s(x(t))) \geq 0, \\
\bar{m}_j(x(t)) &\geq \underline{m}_j(x(t)) \geq 0
\end{aligned}$$

with $\underline{\mu}_{M_{js}}(g_s(x(t)))$ and $\bar{\mu}_{M_{js}}(g_s(x(t)))$ denote the lower grade of membership, upper grade of membership, respectively. The overall IT2 state feedback control law is given by

$$u(t) = \sum_{j=1}^{r} m_j(x(t))K_j x(t), \tag{8.20}$$

where

$$\begin{aligned}
m_j(x(t)) &= \frac{\underline{\beta}_j(x(t))\underline{m}_j(x(t)) + \bar{\beta}_j(x(t))\bar{m}_j(x(t))}{\sum_{k=1}^{r}\left(\underline{\beta}_k(x(t))\underline{m}_k(x(t)) + \bar{\beta}_k(x(t))\bar{m}_k(x(t))\right)} \geq 0, \\
&\sum_{j=1}^{r} m_j(x(t)) = 1, \\
&0 \leq \underline{\beta}_j(x(t)) \leq 1\ \forall j, \\
&0 \leq \bar{\beta}_j(x(t)) \leq 1\ \forall j, \\
&\underline{\beta}_j(x(t)) + \bar{\beta}_j(x(t)) = 1\ \forall j,
\end{aligned}$$

in which $\underline{\beta}_j(x(t))$ and $\bar{\beta}_j(x(t))$ are predefined functions, and $m_j(x(t))$ denotes the grades of membership of the embedded membership functions. For simple notation, $w_i(x(t))$ and $m_j(x(t))$ are denoted as w_i and m_j, respectively, in the analysis that follows. Then the closed-loop singular IT2 fuzzy system formed by the singular IT2 T-S fuzzy model of (8.17), and the IT2 fuzzy controller of (8.20) can be expressed as

$$E\dot{x}(t) = A_s(w, m)x(t), \tag{8.21}$$

where $A_s(w, m) = A(w) + B(w)K(m) = \sum_{i=1}^{r}\sum_{j=1}^{r} w_i m_j (A_i + B_i K_j)$, $A(w) = \sum_{i=1}^{r} w_i A_i$, $B(w) = \sum_{i=1}^{r} w_i B_i$, $K(m) = \sum_{i=1}^{r} m_i K_i$, $\sum_{i=1}^{r} w_i = 1$, $\sum_{j=1}^{r} m_j = 1$, and $\sum_{i=1}^{r}\sum_{j=1}^{r} w_i m_j = 1$.

The second aim is to design the following form of static output feedback controller:

$$u(t) = \sum_{i=1}^{r} m_i(x(t)) S_i y(t), \tag{8.22}$$

where $m_i(x(t))$ defined in (8.20) and $S_i \in R^{m\times l}$, such that the closed-loop system given by

$$E\dot{x}(t) = A_o(w, m)x(t) \tag{8.23}$$

is admissible, where

$$A_o(w, m) = A(w) + B(w)S(m)C(w) = \sum_{i=1}^{r}\sum_{j=1}^{r}\sum_{k=1}^{r} w_i m_j w_k (A_i + B_i S_j C_k),$$

$$C(w) = \sum_{i=1}^{r} w_i C_i, \quad S(m) = \sum_{i=1}^{r} m_i S_i,$$

$A(w)$, and $B(w)$ are defined in (8.21).

8.2.2 State feedback control of singular systems

In this section, the admissibilization by state feedback control of singular IT2 fuzzy systems is addressed by employing the two equivalent sets.

Theorem 8.9. *The singular IT2 fuzzy system in (8.21) is admissible if the membership functions of the fuzzy model and fuzzy controller satisfy $m_j - \rho_j w_j \geq 0$ for all j with $0 < \rho_j < 1$, and there exist matrices $Q < 0$, P, Δ_i, K_i, $i = 1, 2, \ldots, r$ such that the following matrix inequalities hold:*

$$E^T P = P^T E \geq 0, \tag{8.24}$$

$$\text{sym}[(A_i + B_i K_j)^T P] - \Delta_i < 0, \tag{8.25}$$

$$\text{sym}[\rho_i (A_i + B_i K_i)^T P] + (1 - \rho_i)\Delta_i) < Q_{ii}, \tag{8.26}$$

$$\begin{aligned}&\rho_j \text{sym}[(A_i + B_i K_j)^T P] + (1 - \rho_j)\Delta_i + \rho_i \text{sym}[(A_j + B_j K_i)^T P]\\ &+ (1 - \rho_i)\Delta_j < Q_{ij} + Q_{ji}, \quad i < j,\end{aligned} \tag{8.27}$$

where

$$Q=\begin{bmatrix} Q_{11} & Q_{12} & \cdots & Q_{1r} \\ Q_{21} & Q_{22} & \cdots & Q_{2r} \\ \vdots & \vdots & \ddots & \vdots \\ Q_{r1} & Q_{r2} & \cdots & Q_{rr} \end{bmatrix}.$$

Proof. Firstly, we will prove the stability of the system and choose the following Lyapunov function:

$$V(x(t)) = x(t)^T E^T P x(t), \tag{8.28}$$

where $E^T P = P^T E \geq 0$. Considering the condition in (8.24), the first-order time-derivative of the Lyapunov function along the trajectories of the system in (8.21) is given below:

$$\begin{aligned} \dot{V}(x(t)) &= 2x(t)^T P^T E\dot{x}(t) \\ &= \sum_{i=1}^{r}\sum_{j=1}^{r} w_i m_j x(t)^T \text{sym}[(A_i + B_i K_j)^T P]x(t). \end{aligned} \tag{8.29}$$

To further alleviate the conservatism, the following equalities and some free weighting matrices are introduced:

$$\sum_{i=1}^{r}\sum_{j=1}^{r} w_i(w_j - m_j)\Delta_i = 0, \tag{8.30}$$

where $\Delta_i = \Delta_i^T$ is arbitrary matrix. From (8.29) and (8.30), we have

$$\begin{aligned} &\dot{V}(x(t)) \\ &= \sum_{i=1}^{r}\sum_{j=1}^{r} w_i(m_j + \rho_j w_j - \rho_j w_j)x(t)^T \text{sym}[(A_i + B_i K_j)^T P]x(t) \\ &\quad + \sum_{i=1}^{r}\sum_{j=1}^{r} w_i(w_j - \rho_j w_j)x(t)^T \Delta_i x(t) \\ &\quad - \sum_{i=1}^{r}\sum_{j=1}^{r} w_i(m_j - \rho_j w_j)x(t)^T \Delta_i x(t) \\ &= \sum_{i=1}^{r}\sum_{j=1}^{r} w_i w_j x^T(t)(\text{sym}[\rho_j(A_i + B_i K_j)^T P] + (1 - \rho_j)\Delta_i)x(t) \\ &\quad + \sum_{i=1}^{r}\sum_{j=1}^{r} w_i(m_j - \rho_j w_j)x^T(t)(\text{sym}[(A_i + B_i K_j)^T P] - \Delta_i)x(t). \end{aligned} \tag{8.31}$$

Considering $m_j - \rho_j w_j > 0$ and $\text{sym}[(A_i + B_i K_j)P] - \Delta_i < 0$ for $i, j = 1, 2, \ldots, r$, we have

$$\begin{aligned}\dot{V}(x(t)) \leq & \sum_{i=1}^{r} w_i^2 x(t)^T (\text{sym}[\rho_i (A_i + B_i K_i)^T P] + (1 - \rho_j)\Delta_i) x(t) \\ & + \sum_{i=1}^{r} \sum_{i<j} w_i w_j x(t)^T \left(\rho_j \text{sym}[(A_i + B_i^T K_j)P]\right. \\ & \left. + (1 - \rho_j)\Delta_i + \rho_i \text{sym}[(A_j + B_j K_i)^T P] + (1 - \rho_i)\Delta_j\right) x(t). \end{aligned} \tag{8.32}$$

Due to $(\text{sym}[\rho_i (A_i + B_i K_i)^T P] + (1 - \rho_j)\Delta_i) < Q_{ii}$ and $\rho_j \text{sym}[(A_i + B_i K_j)^T P] + (1 - \rho_j)\Delta_i + \rho_i \text{sym}[(A_j + B_j K_i)^T P] + (1 - \rho_i)\Delta_j < Q_{ij} + Q_{ji}$, it yields from (8.31) that

$$\dot{V}(x(t)) \leq x^T(t) Q x(t), \tag{8.33}$$

where $Q = \begin{bmatrix} Q_{11} & Q_{12} & \cdots & Q_{1r} \\ Q_{21} & Q_{22} & \cdots & Q_{2r} \\ \vdots & \vdots & \ddots & \vdots \\ Q_{r1} & Q_{r2} & \cdots & Q_{rr} \end{bmatrix}$. Because $Q < 0$, we have $\dot{V}(x(t)) < 0$ when $x(t) \neq 0$, which implies the stability of the system in (8.21).

Next, the regularity and nonimpulsiveness of the system in (8.21) are proved. From the equality (8.29) and inequality (8.33), we can see that

$$\text{sym}(A_s^T(w, m)P) < 0. \tag{8.34}$$

For matrix E, there exist two nonsingular matrices G and H such that

$$GEH = \begin{bmatrix} 1 & 0 \\ 0 & 0 \end{bmatrix}. \tag{8.35}$$

Then denote

$$GA_s(w, m)H = \begin{bmatrix} A_{11}(w, m) & A_{12}(w, m) \\ A_{21}(w, m) & A_{22}(w, m) \end{bmatrix}, \quad G^{-1}PH = \begin{bmatrix} P_{11} & P_{12} \\ P_{21} & P_{22} \end{bmatrix}. \tag{8.36}$$

Due to equality (8.24), we have $P_{12} = 0$. Premultiplying and postmultiplying inequality (8.34) by H^T and H, respectively, the following inequality is obtained:

$$\begin{bmatrix} \bullet & \bullet \\ \bullet & \text{sym}(P_{22}^T A_{22}(w, m)) \end{bmatrix} < 0, \tag{8.37}$$

which implies that $A_{22}(w, m)$ is nonsingular, and thus the singular IT2 fuzzy system is regular and impulse-free. □

It can be seen that conditions in Theorem 8.9 are nonlinear matrix inequalities and contain equality constrain, which will lead to difficult computation. Base on Lemma 2.5, a necessary and sufficient condition of Theorem 8.10 is given in the following theorem in terms of strict LMIs:

Theorem 8.10. *The singular IT2 fuzzy system in (8.21) is admissible if the membership functions of the fuzzy model and fuzzy controller satisfy $m_j - \rho_j w_j \geq 0$ for all j with $0 < \rho_j < 1$ and there exist matrices $\bar{Q} < 0$, $\bar{Y} = \bar{Y}^T$, nonsingular matrix $\bar{\Phi}$, $\bar{\Delta}_i$, H_i, $i = 1, 2, \ldots, r$ such that the following matrix inequalities hold:*

$$E_R^T \bar{Y} E_R > 0, \tag{8.38}$$

$$\text{sym}[(A_i \bar{P} + B_i H_j)] - \bar{\Delta}_i < 0, \tag{8.39}$$

$$\text{sym}[\rho_i (A_i \bar{P} + B_i H_j)] + (1 - \rho_i)\bar{\Delta}_i < \bar{Q}_{ii}, \tag{8.40}$$

$$\begin{aligned} &\rho_j \text{sym}[(A_i \bar{P} + B_i H_j)] + (1 - \rho_j)\bar{\Delta}_i + \rho_i \text{sym}[(A_i \bar{P} + B_i H_j)] \\ &+ (1 - \rho_i)\bar{\Delta}_j < \bar{Q}_{ij} + \bar{Q}_{ji}, \ i < j, \end{aligned} \tag{8.41}$$

where

$$\bar{Q} = \begin{bmatrix} \bar{Q}_{11} & \bar{Q}_{12} & \cdots & \bar{Q}_{1r} \\ \bar{Q}_{21} & \bar{Q}_{22} & \cdots & \bar{Q}_{2r} \\ \vdots & \vdots & \ddots & \vdots \\ \bar{Q}_{r1} & \bar{Q}_{r2} & \cdots & \bar{Q}_{rr} \end{bmatrix}, \bar{P} = \bar{Y} E^T + \Lambda \bar{\Phi} U.$$

E_R, U, and Λ are defined in Lemma 2.5. Then the state feedback controllers are given as $K_i = H_i \bar{P}^{-1}$.

Proof. From inequality (8.34), matrix P in Theorem 8.9 is nonsingular. By employing Lemma 2.5 and condition in (8.24), the matrix P in Theorem 8.9 can be replaced by matrix $YE + U\Phi\Lambda$ with $E_L^T Y E_L > 0$, where E_L, U, and Λ are defined in Lemma 2.5. Then, the conditions in (8.24) to (8.27) are equivalent to

$$E_L^T Y E_L > 0, \tag{8.42}$$

$$\text{sym}[(A_i + B_i K_j)^T (YE + U^T \Phi \Lambda^T)] - \Delta_i < 0, \tag{8.43}$$

$$\text{sym}[\rho_i (A_i + B_i K_i)^T (YE + U^T \Phi \Lambda^T)] + (1 - \rho_i)\Delta_i < Q_{ii}, \tag{8.44}$$

$$\begin{aligned} &\rho_j \text{sym}[(A_i + B_i K_j)^T (YE + U^T \Phi \Lambda^T)] + (1 - \rho_j)\Delta_i \\ &+ \rho_i \text{sym}[(A_j + B_j K_i)^T P] + (1 - \rho_i)\Delta_j < Q_{ij} + Q_{ji}, \ i < j. \end{aligned} \tag{8.45}$$

Let $\bar{P} = (YE + U^T \Phi \Lambda^T)^{-1} = \bar{Y}E^T + \Lambda \bar{\Phi} U$. Pre- and postmultiplying inequality (8.43) with $\bar{P}^T$ and $\bar{P}$, we obtain

$$\text{sym}[\bar{P}^T (A_i + B_i K_j)^T] - \bar{P}^T \Delta_i \bar{P} < 0.$$

Denoting $H_j = K_j \bar{P}$, $\bar{\Delta}_i = \bar{P}^T \Delta_i \bar{P}$, the condition in (8.39) is obtained. Similarly, performing the same congruent transformation to inequalities (8.44) and (8.45), respectively, and denoting $\bar{Q}_{ij} = \bar{P}^T Q_{ij} \bar{P}$, the conditions in (8.40) and (8.41) can be derived. Due to $(E_L^T Y E_L)^{-1} = E_R^T \bar{Y} E_R$, the inequality in (8.38) is equivalent to (8.42). □

Remark 8.11. It can be seen that the equality constraint $E^T P = P^T E$ and nonstrict LMI $E^T P \geq 0$ in Theorem 8.9 are removed in the novel admissibility criterion in Theorem 8.10, which makes the condition easy to check by using standard Matlab® toolbox such as LMI or Yalmip.

8.2.3 Static output feedback of singular systems

In this section, the equivalent sets are applied to deal with the SOF control problem of singular IT2 fuzzy systems. To design the SOF controller in terms of LMIs, the following equivalent system of (8.23) is given:

$$\bar{E}\dot{\bar{x}}(t) = \bar{A}(w, m)\bar{x}(t), \tag{8.46}$$

where

$$\bar{A}(w, m) = \begin{bmatrix} A(w) & B(w) \\ S(m)C(w) & -I \end{bmatrix}, \ \bar{x}(t) = \begin{bmatrix} x(t) \\ u(t) \end{bmatrix}, \ \bar{E} = \begin{bmatrix} E & 0 \\ 0 & 0 \end{bmatrix}.$$

Note that the admissibility of system (8.23) is equivalent to that of system (8.46) which is because

$$s\bar{E} - \bar{A}(w, m) = \begin{bmatrix} sE - A(w) & -B(w) \\ -S(m)C(w) & I \end{bmatrix}$$

$$= \begin{bmatrix} I & -B(w) \\ 0 & I \end{bmatrix} \begin{bmatrix} sE - A(w)S(m)C(w) & 0 \\ 0 & I \end{bmatrix} \begin{bmatrix} I & 0 \\ S(m)C(w) & I \end{bmatrix},$$

and $\det(s\bar{E} - \bar{A}(w, m)) = \det(sE - A(w)S(m)C(w))$. Firstly, a new admissibility condition of system (8.46) is proposed in the following theorem:

Theorem 8.12. *The system in (8.46) is admissibility if there exist matrices Z, F, G, and nonsingular matrix Θ satisfying the following matrix inequalities:*

$$\bar{E}_L^T Z \bar{E}_L > 0, \tag{8.47}$$

$$\Gamma(w,m) = \begin{bmatrix} \mathrm{sym}(\bar{A}(w,m)^T G^T) & \Gamma_{12}(w,m) \\ \star & -F - F^T \end{bmatrix} < 0, \tag{8.48}$$

where $\Gamma_{12}(w,m) = (Z\bar{E} + \bar{U}^T \Theta \bar{\Lambda}^T)^T + \bar{A}(w,m)^T F^T - G$, $\bar{E}_L$, *and* $\bar{E}_R$ *are full column rank matrices with* $\bar{E} = \bar{E}_L \bar{E}_R^T$. $\bar{U}^T \in \mathbb{R}^{(n+m)\times(n+m-q)}$ *and* $\bar{\Lambda} \in \mathbb{R}^{(n+m)\times(n+m-q)}$ *with full column rank are the right null matrix of* $\bar{E}^T$ *and* $\bar{E}$, *that is,* $\bar{E}^T \bar{U}^T = 0$ *and* $\bar{E}\bar{\Lambda} = 0$, *respectively.*

Proof. Based on Lemma 2.5, the conditions in (8.47) and (8.48) are equivalent to that when there exists a matrix $X = Z\bar{E} + \bar{U}\Theta\bar{\Lambda}$ such that the following inequalities hold:

$$\bar{E}^T X = X^T \bar{E} \geq 0, \tag{8.49}$$

$$\begin{bmatrix} \mathrm{sym}(\bar{A}(w,m)^T G^T) & X^T + \bar{A}(w,m)^T F^T - G \\ \star & -F - F^T \end{bmatrix} < 0. \tag{8.50}$$

Then premultiplying (8.50) by $\begin{bmatrix} I & \bar{A}(w,m)^T \end{bmatrix}$ and postmultiplying it by $\begin{bmatrix} I & \bar{A}(w,m)^T \end{bmatrix}^T$, we have

$$\bar{A}(w,m)^T X + X^T \bar{A}(w,m) < 0. \tag{8.51}$$

Choosing Lyapunov function $V(\bar{x}(t)) = \bar{x}(t)^T E^T X \bar{x}(t)$, considering (8.47) and (8.51), we can obtain $\dot{V}(\bar{x}(t)) < 0$, which implies that the singular IT2 fuzzy system in (8.46) is asymptotically stable. Following similar lines as that from (8.34) to (8.37), the regularity and nonimpulsiveness of the system can be proved. Therefore the condition in Theorem 8.12 can guarantee the admissibility of the singular IT2 fuzzy system in (8.46). □

The following theorem gives an LMI admissibility condition for closed-loop singular IT2 fuzzy system (8.23):

Theorem 8.13. *The system in (8.23) is admissibility if the membership functions of the fuzzy model and fuzzy controller satisfy* $m_j - \rho_j w_j \geq 0$ *for all j with* $0 < \rho_j < 1$ *and there exist matrices* $Z = Z^T$, F_1, F_2, G_1, G_2, T_i, $i = 1, \ldots, r$, $R < 0$, $\Psi_i = \Psi_i^T$ *and nonsingular matrices* Θ, G_3 *satisfying the following LMIs:*

$$\bar{E}_L^T Z \bar{E}_L > 0, \tag{8.52}$$

$$\Gamma_{ij} - \Psi_i < 0, \tag{8.53}$$

$$\rho_i \Gamma_{ii} + (1 - \rho_i)\Psi_i < R_{ii}, \tag{8.54}$$

$$\rho_j \Gamma_{ij} + (1 - \rho_j)\Psi_i + \rho_i \Gamma_{ji} + (1 - \rho_i)\Psi_j < R_{ij} + R_{ji}, \ i < j, \tag{8.55}$$

where

$$\Gamma_{ij} = \left[\begin{array}{c|c} \Gamma_{ij}^{11} & (Z\bar{E} + \bar{U}^T \Theta \bar{A}^T)^T + \Upsilon_{ij} \\ \hline \star & -\begin{bmatrix} F_1 + F_1^T & LG_3 + F_2^T \\ \star & G_3 + G_3^T \end{bmatrix} \end{array}\right],$$

$$\Gamma_{ij}^{11} = \begin{bmatrix} \mathrm{sym}(G_1 A_i + LT_j C_i) & G_1 B_i - LG_3 + A_i^T G_2^T + C_i^T T_j^T \\ \star & \mathrm{sym}(G_2 B_i - G_3) \end{bmatrix},$$

$$R = \begin{bmatrix} R_{11} & R_{12} & \cdots & R_{1r} \\ R_{21} & R_{22} & \cdots & R_{2r} \\ \vdots & \vdots & \ddots & \vdots \\ R_{r1} & R_{r2} & \ldots & R_{rr} \end{bmatrix},$$

$$\Upsilon_{ij} = \begin{bmatrix} A_i^T F_1^T + C_i^T T_j^T L^T - G_1 & A_i^T F_2^T + C_i^T T_j^T - LG_3 \\ B_i^T F_1^T - G_3^T L^T - G_2 & B(w)^T F_2^T - G_3^T - G_3 \end{bmatrix}.$$

$\bar{E}_R$, $\bar{U}^T \in \mathbb{R}^{(n+m)\times(n+m-r)}$, *and* $\bar{A} \in \mathbb{R}^{(n+m)\times(n+m-r)}$ *are defined in Theorem 8.12. Then the SOF controller is given as* $S_i = T_i G_3^{-1}$.

Proof. Set

$$G = \begin{bmatrix} G_1 & LG_3 \\ G_2 & G_3 \end{bmatrix}, \ F = \begin{bmatrix} F_1 & LG_3 \\ F_2 & G_3 \end{bmatrix},$$

and $T(m) = G_3 S(m)$. We have

$$G\bar{A}(w, m) = \begin{bmatrix} G_1 A(w) + LT(m)C(w) & G_1 B(w) - LG_3 \\ G_2 A(w) + T(m)C(w) & G_2 B(w) - G_3 \end{bmatrix}, \tag{8.56}$$

$$F\bar{A}(w, m) = \begin{bmatrix} F_1 A(w) + LT(m)C(w) & F_1 B(w) - LG_3 \\ F_2 A(w) + T(m)C(w) & F_2 B(w) - G_3 \end{bmatrix}. \tag{8.57}$$

It yields from (8.48) and (8.56)–(8.57) that

$$\Gamma(w, m) = \sum_{i=1}^{r} \sum_{j=1}^{r} w_i m_j \Gamma_{ij}.$$

For any matrices Ψ_i, noting that $\sum_{i=1}^{r}\sum_{j=1}^{r} w_i(w_j - m_j)\Psi_i = 0$, we have

$$
\begin{aligned}
&\Gamma(w,m)\\
&=\sum_{i=1}^{r}\sum_{j=1}^{r} w_i m_j \Gamma_{ij} + \sum_{i=1}^{r}\sum_{j=1}^{r} w_i(w_j - m_j)\Psi_i\\
&=\sum_{i=1}^{r}\sum_{j=1}^{r} w_i(m_j + \rho_j w_j - \rho_j w_j)\Gamma_{ij}\\
&\quad+\sum_{i=1}^{r}\sum_{j=1}^{r} w_i(w_j - \rho_j w_j)\Psi_i - \sum_{i=1}^{r}\sum_{j=1}^{r} w_i(m_j - \rho_j w_j)\Psi_i\\
&=\sum_{i=1}^{r}\sum_{j=1}^{r} w_i w_j(\rho_j \Gamma_{ij} + (1-\rho_j)\Psi_i)\\
&\quad+\sum_{i=1}^{r}\sum_{j=1}^{r} w_j(m_j - \rho_j w_j)(\Gamma_{ij} - \Psi_i).
\end{aligned}
$$

Considering $m_j - \rho_j w_j > 0$, and $\Gamma_{ij} - \Psi_i < 0$ for $i, j = 1, 2, \ldots, r$, we have

$$
\begin{aligned}
\Gamma(w,m) \le &\sum_{i=1}^{r} w_i^2(\rho_i \Gamma_{ii} + (1-\rho_i)\Psi_i) + \sum_{i=1}^{r}\sum_{i<j} w_i w_j(\rho_j \Gamma_{ij}\\
&+ (1-\rho_j)\Psi_i + \rho_i \Gamma_{ji} + (1-\rho_i)\Psi_j).
\end{aligned}
$$

Because $\rho_i \Gamma_{ii} + (1-\rho_i)\Psi_i < R_{ii}$ and $\rho_j \Gamma_{ij} + (1-\rho_j)\Psi_i + \rho_i \Gamma_{ji} + (1-\rho_i)\Psi_j < R_{ij} + R_{ji}$, the following inequality holds:

$$
\Gamma(w,m) \le \begin{bmatrix} w_1 I \\ w_2 I \\ \vdots \\ w_r I \end{bmatrix}^T R \begin{bmatrix} w_1 I & w_2 I & \cdots & w_r I \end{bmatrix},
$$

which implies that $\Gamma(w,m) < 0$, due to $R < 0$. Therefore based on Theorem 8.12, the admissibility of system (8.23) can be guaranteed. □

Remark 8.14. The SOF admissibilization condition is given in terms of strict LMIs in Theorem 8.13. One advantage of the condition is that it is established in terms of strict LMIs, which avoid the computation complexity of involving nonstrict LMI, such as in [221]. The other one is that different output matrices C_i are considered distinct from some existing results with common output matrices $C_i = C$ [18].

Remark 8.15. Based on the results obtained in this section, the main contributions are listed in the following points:

- The admissibility analysis result and the corresponding state feedback control method for singular IT2 fuzzy systems by using two equivalent sets are proposed for the first time.
- The static output feedback controller is designed by using system augmentation approach to guarantee the closed-loop singular IT2 fuzzy systems to be admissible for the first time.
- All the main results are given in terms of strict LMIs, which benefits from the two equivalent sets proposed in Lemma 2.5. This makes the conditions easy to solve by using standard software.

8.2.4 Examples

In this section, numerical examples are provided to show the effectiveness of the proposed results. The first example is about admissibilization of singular IT2 fuzzy systems by state feedback control, and the second is about SOF control. Example 8.4 is given to illustrate how to deal with the uncertain membership functions.

Example 8.2. Consider the following three-rule singular IT2 fuzzy system:

$$E=\begin{bmatrix}1 & 1\\ -1 & -1\end{bmatrix},\ A_1=\begin{bmatrix}-3 & 4\\ 1 & -2\end{bmatrix},\ A_2=\begin{bmatrix}-1 & 1\\ -0.5 & -2\end{bmatrix},$$

$$A_3=\begin{bmatrix}-1 & 0\\ -0.3 & -1\end{bmatrix},\ B_1=\begin{bmatrix}2\\ 1\end{bmatrix},\ B_2=\begin{bmatrix}-3\\ 5\end{bmatrix},\ B_3=\begin{bmatrix}1\\ 1\end{bmatrix},$$

$$E_R=\begin{bmatrix}1\\ 1\end{bmatrix},\ U=\begin{bmatrix}1 & 1\end{bmatrix},\ \Lambda=\begin{bmatrix}1\\ -1\end{bmatrix}.$$

We choose the lower and upper membership functions listed in Table 8.2. The membership functions of the system are depicted in Fig. 8.1. With

Table 8.2 Membership functions of the plant.

Lower membership functions	Upper membership functions
$\underline{w}_1(x_2)=0.95-\frac{0.95}{1+e^{\frac{x_2+4.5}{8}}}$	$\bar{w}_1(x_2)=0.95-\frac{0.925}{1+e^{\frac{x_2+3.5}{8}}}$
$\underline{w}_2(x_2)=0.025+\frac{0.925}{1+e^{\frac{x_2-4.5}{8}}}$	$\bar{w}_2(x_2)=0.025+\frac{0.925}{1+e^{\frac{x_2-3.5}{8}}}$
$\underline{w}_3(x_2)=1-\bar{w}_1(x_2)-\bar{w}_2(x_2)$	$\bar{w}_3(x_2)=1-\underline{w}_1(x_2)-\underline{w}_2(x_2)$

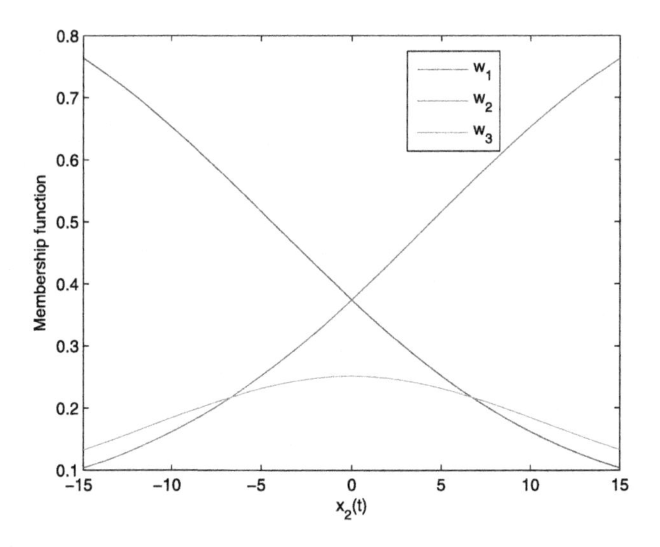

Figure 8.1 Membership functions of the singular IT2 fuzzy system.

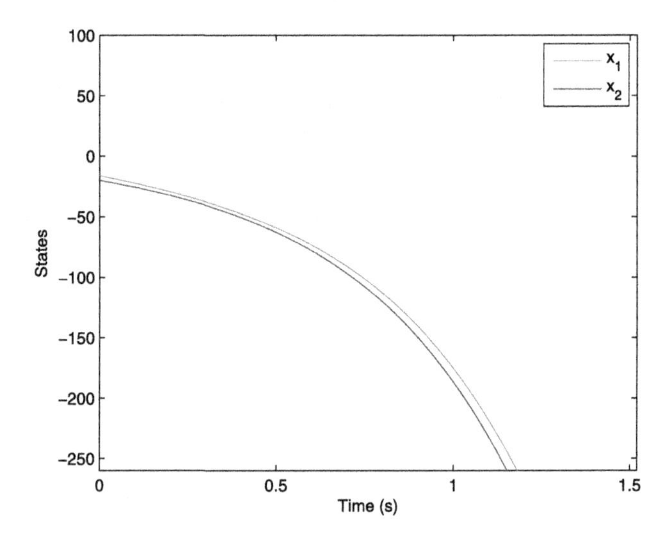

Figure 8.2 State trajectories of open-loop singular IT2 fuzzy system.

these membership functions, the state trajectories of open-loop system with initial condition $x(0) = \begin{bmatrix} -15.9745 \\ -20.0000 \end{bmatrix}$ are depicted in Fig. 8.2, which implies that the open-loop system is not stable. To design the state feedback controller, the lower and upper membership functions of IT2 fuzzy controller

Table 8.3 Membership functions of the IT2 fuzzy controller.

Lower membership functions	Upper membership functions
$\underline{m}_1(x_2) = 1 - \frac{1}{1+e^{\frac{x_2+5}{2}}}$	$\bar{m}_1(x_2) = 1 - \frac{1}{1+e^{\frac{x_2+4}{2}}}$
$\underline{m}_2(x_2) = \frac{1}{1+e^{\frac{x_2-5}{2}}}$	$\bar{m}_2(x_2) = \frac{1}{1+e^{\frac{x_2-4}{2}}}$
$\underline{m}_3(x_2) = 1 - \bar{m}_1(x_2) - \bar{m}_2(x_2)$	$\bar{m}_3(x_2) = 1 - \underline{m}_1(x_2) - \underline{m}_2(x_2)$

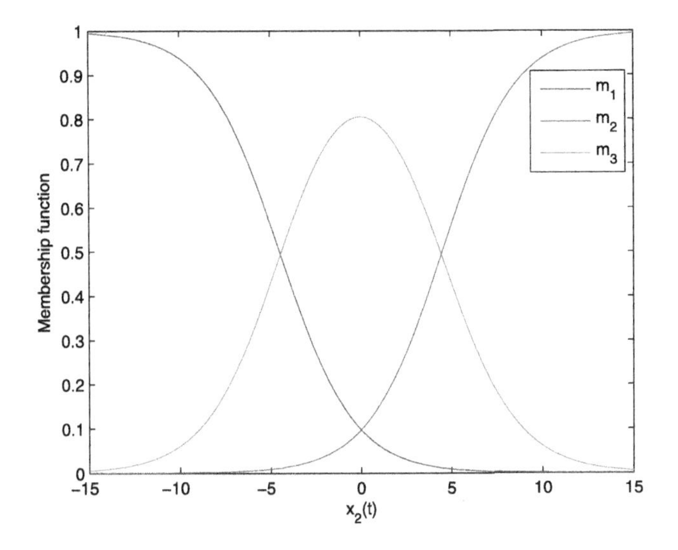

Figure 8.3 Membership functions of the IT2 fuzzy controller.

are given in Table 8.3. The membership functions of the IT2 fuzzy controller are depicted in Fig. 8.3. Based on Theorem 8.10, by setting $\rho_1 = 0.1$, $\rho_2 = 0.9$, $\rho_3 = 0.1$, and solving the LMIs (8.38)–(8.41), some decisions matrices are obtained:

$$H_1 = \begin{bmatrix} 0.0116 & -0.7576 \end{bmatrix},\ H_2 = \begin{bmatrix} 0.0177 & -0.7084, \end{bmatrix}$$

$$H_3 = \begin{bmatrix} 0.0097 & -0.7615 \end{bmatrix},\ \bar{P} = \begin{bmatrix} 0.8272 & -0.7172 \\ -0.0008 & -0.1092 \end{bmatrix}.$$

Then, the state feedback controller gain matrices are obtained as follows:

$$K_1 = \begin{bmatrix} 0.0202 & 6.8054 \end{bmatrix},\ K_2 = \begin{bmatrix} 0.0271 & 6.3098 \end{bmatrix}$$

$$K_3 = \begin{bmatrix} 0.0180 & 6.8561 \end{bmatrix}.$$

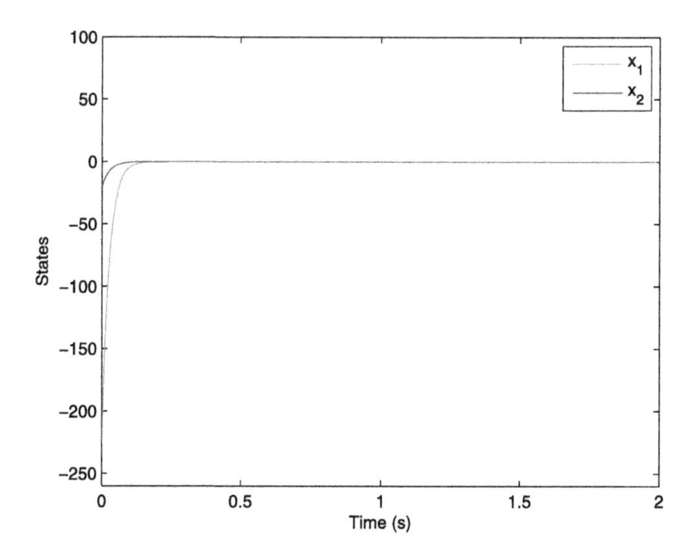

Figure 8.4 State trajectories of closed-loop singular IT2 fuzzy system.

The state response of closed-loop system with initial condition $x(0) = \begin{bmatrix} -226.0493 \\ -20.0000 \end{bmatrix}$ is given in Fig. 8.4, which demonstrates the effectiveness of the state feedback controller design method.

Example 8.3. The effectiveness of SOF controller design method provided in Theorem 8.13 is demonstrated in this example. Consider the singular IT2 T-S fuzzy system in (8.17) with $r = 2$ and the following system parameters:

$$E = \begin{bmatrix} 1 & 1 & 0 \\ 1 & 1 & 0 \\ 0 & 0 & 0 \end{bmatrix}, \ A_1 = \begin{bmatrix} -3 & -4 & 0 \\ -1 & -2 & 0 \\ 0 & 0 & 2 \end{bmatrix},$$

$$A_2 = \begin{bmatrix} -2 & -2 & 0 \\ -1 & 2 & 0 \\ 0 & 0 & 3 \end{bmatrix}, \ B_1 = \begin{bmatrix} -1 \\ 10 \\ 1 \end{bmatrix}, \ B_2 = \begin{bmatrix} -2 \\ 5 \\ 1 \end{bmatrix},$$

$$C_1 = \begin{bmatrix} 2 & -1 & 2 \\ -0.5 & 0 & 1 \end{bmatrix}, \ C_2 = \begin{bmatrix} 1 & -0.5 & 0.4 \\ 1 & 2 & 0 \end{bmatrix}, \ \bar{E}_L = \begin{bmatrix} 1 \\ 1 \\ 0 \\ 0 \end{bmatrix},$$

$$\bar{\Lambda} = \bar{U}^T = \begin{bmatrix} 1 & 1 & 0 \\ -1 & -1 & 0 \\ 0 & 1 & 0 \\ 0 & 0 & 1 \end{bmatrix}.$$

By using the same membership functions as that in Example 8.2, setting $\rho_1 = 0.1$, $\rho_2 = 0.9$, and solving the LMIs in Theorem 8.13, we get

$$G_3 = 1.1898, \ T_1 = \begin{bmatrix} -0.5278 & 0.1645 \end{bmatrix},$$
$$T_2 = \begin{bmatrix} -0.5570 & 0.1603 \end{bmatrix},$$

and the SOF controller is obtained as follows:

$$S_1 = \begin{bmatrix} -0.4436 & 0.1383 \end{bmatrix}, \ S_2 = \begin{bmatrix} -0.4681 & 0.1347 \end{bmatrix}.$$

Example 8.4. A two-rule IT2 T-S fuzzy model in the form of (8.17) is applied to describe an nonlinear plant with following system parameters:

$$E = \begin{bmatrix} 1 & 0 \\ 0 & 0 \end{bmatrix}, \ A_1 = \begin{bmatrix} 0.3 & -1 \\ 0.1 & 0.1 \end{bmatrix}, \ A_2 = \begin{bmatrix} -0.4 & 0.7 \\ 0 & 0.2 \end{bmatrix},$$
$$B_1 = \begin{bmatrix} -1 & -0.2 \end{bmatrix}, \ B_2 = \begin{bmatrix} 0.1 & -1 \end{bmatrix}.$$

The membership functions with uncertain parameter $\delta(t) \in [-1, 1]$, which may be caused by the uncertainty parameters of plant are given as follows:

$$w_1(x_1) = \mu_{W_{11}} = 1 - \frac{1}{1 + e^{-(x_1 + 4 + \delta(t))}}, \ w_2(x_1) = \mu_{W_{21}} = 1 - \mu_{W_{11}}.$$

This will lead to uncertain grades of membership, and the existing type-1 stability results under parallel distributed compensation (PDC) method cannot be applied.

Then the lower and upper membership functions for the IT2 T-S fuzzy system are chosen as

$$\underline{w}_1(x_1) = \underline{\mu}_{W_{11}} = 1 - \frac{1}{1 + e^{-(x_1 + 5)}},$$
$$\bar{w}_1(x_1) = \bar{\mu}_{W_{11}} = 1 - \frac{1}{1 + e^{-(x_1 + 3)}},$$
$$\underline{w}_2(x_1) = \underline{\mu}_{W_{21}} = 1 - \bar{w}_1(x_1), \ \bar{w}_2(x_1) = \bar{\mu}_{W_{21}} = 1 - \underline{w}_1(x_1).$$

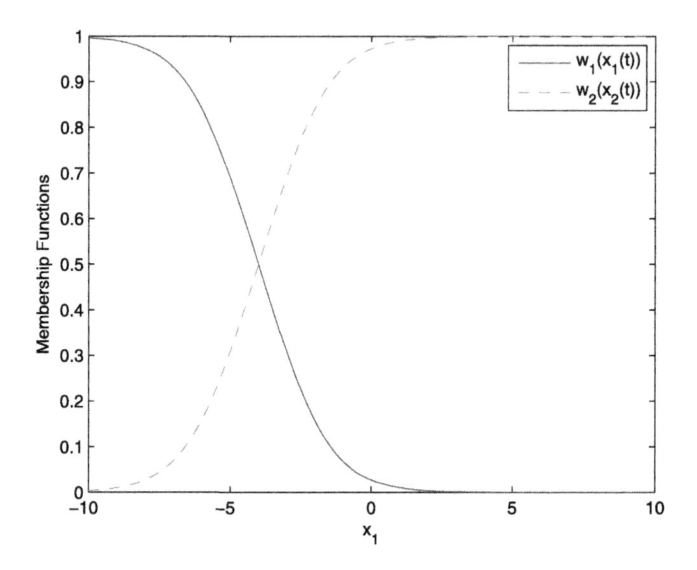

Figure 8.5 Membership functions of singular IT2 fuzzy system.

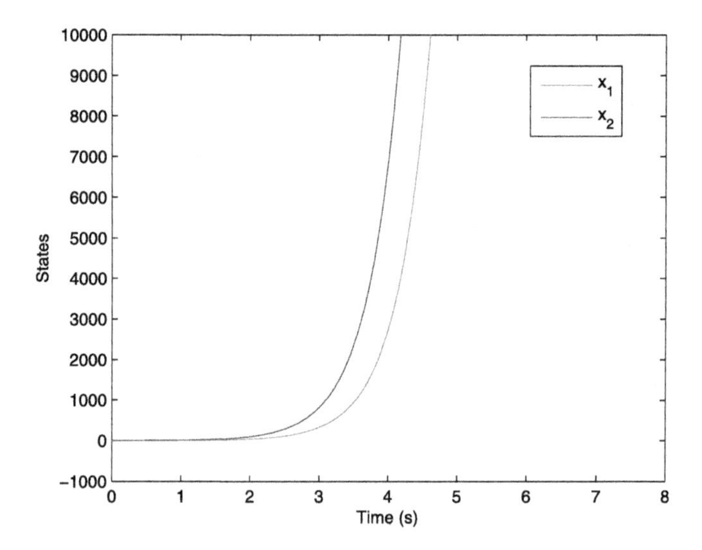

Figure 8.6 State trajectories of open-loop singular IT2 fuzzy system.

The membership functions of IT2 T-S fuzzy system are depicted in Fig. 8.5, and the state trajectories of the open-loop system are plotted in Fig. 8.6, respectively. From the simulation result, we can see that the open-loop singular IT2 T-S fuzzy system is not stable. To stabilize the system,

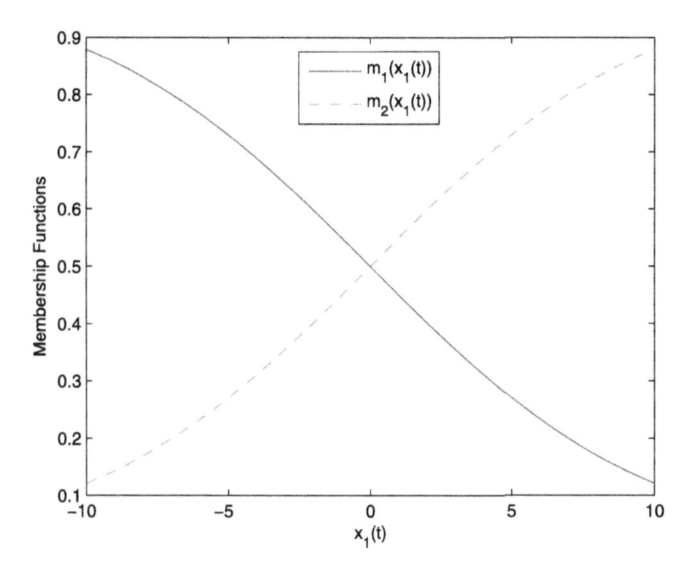

Figure 8.7 Membership functions of IT2 fuzzy controller.

state-feedback control method in Theorem 8.10 will be utilized. To this end, the membership functions of fuzzy controller are given as follows:

$$\begin{aligned}
\underline{m}_1(x_1) =& \underline{\mu}_{M_{11}} = 1 - \frac{1}{1 + e^{-\frac{x_1 - 1}{5}}}, \\
\bar{m}_1(x_1) =& \bar{\mu}_{M_{11}} = 1 - \frac{1}{1 + e^{-\frac{x_1 - 3}{5}}}, \\
\underline{m}_2(x_1) =& \underline{\mu}_{M_{21}} = 1 - \bar{m}_1(x_1), \ \bar{m}_2(x_1) = \bar{\mu}_{M_{21}} = 1 - \underline{m}_1(x_1).
\end{aligned}$$

The membership functions of fuzzy controller are illustrated in Fig. 8.7. By choosing $\rho_1 = 0.1$, $\rho_2 = 0.4$, and solving the LMIs in (8.42)–(8.45), the state-feedback controller is obtained:

$$K_1 = \begin{bmatrix} -0.4313 & -0.7700 \end{bmatrix}, \ K_2 = \begin{bmatrix} -0.5014 & -0.8541 \end{bmatrix}.$$

With this controller, the state trajectories of the closed-loop singular IT2 fuzzy system are depicted in Fig. 8.8, which implies the effectiveness of the state feedback control method.

Remark 8.16. Compared with existing results, the superiorities are given as follows:

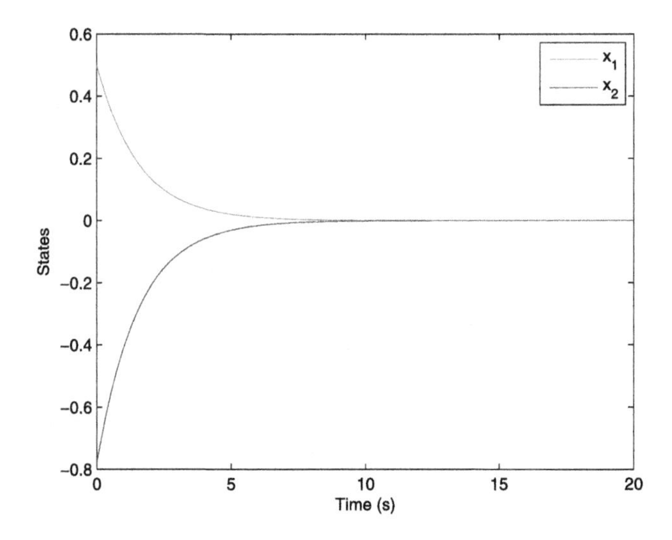

Figure 8.8 State trajectories of closed-loop singular IT2 fuzzy system.

- For singular fuzzy systems, when the membership functions subject to parameter uncertainties are uncertain and the fuzzy systems and fuzzy controllers do not share the same premise membership functions, many existing state-feedback control method cannot be applied, such as [103], [58], [76].
- When system matrix $E = I$ and matrix $Q = 0$ in this section, Theorem 8.10 will reduce to the stabilization condition for standard IT2 fuzzy systems considered in [96]. Although some works about IT2 fuzzy systems have been reported, such as [87], [94], [116], all these results in those papers will not apply when the considered system is singular IT2 fuzzy system.

8.2.5 Conclusion

The state feedback and SOF feedback admissibilization problems of singular IT2 fuzzy systems have been investigated in this section. Firstly, a sufficient admissibilization condition for existence of an IT2 fuzzy state feedback controller is proposed. Then, an SOF controller is designed such that the closed-loop system is admissible. Benefiting from two equivalent sets and the system augmentation approach, conditions are given in terms of strict LMIs that can be computed by standard software. Moreover, the output matrices are not necessarily the same matrices. Numerical examples are given to show the effectiveness of the proposed results.

References

[1] C.K.An. Linear, Matrix inequality approach to passive filtering for delayed neural networks, Proc. Inst. Mech. Eng., Part I, J. Syst. Control Eng. 224 (10) (2010) 1040–1047.

[2] B.D.O. Anderson, J.B. Moore, Optimal Filtering, Prentice-Hall, Englewood Cliffs, NJ, 1979.

[3] D.V. Balandin, M.M. Kogan, LMI-based H_∞-optimal control with transients, Int. J. Control 83 (8) (2010) 1664–1673.

[4] L.J. Banu, P. Balasubramaniam, Admissibility analysis for discrete-time singular systems with randomly occurring uncertainties via delay-divisioning approach, ISA Trans. 59 (2015) 354–362.

[5] E.K. Boukas, Stabilization of stochastic singular nonlinear hybrid systems, Nonlinear Anal. Theory Methods Appl. 65 (2) (2006) 217–228.

[6] E.K. Boukas, Singular linear systems with delay: H_∞ stabilization, Optim. Control Appl. Methods 28 (4) (2007) 259–274.

[7] E.K. Boukas, Control of Singular Systems with Random Abrupt Changes, Springer, Berlin, 2008.

[8] A. Boulkroune, A fuzzy adaptive control approach for nonlinear systems with unknown control gain sign, Neurocomputing 179 (2016) 318–325.

[9] S. Boyd, L. El Ghaoui, E. Feron, V. Balakrishnan, Linear Matrix Inequalities in System and Control Theory, SIAM, Philadelphia, PA, 1994.

[10] H. Bustince, E. Barrenechea, M. Pagola, J. Fernandez, Z. Xu, B. Bedregal, J. Montero, H. Hagras, F. Herrera, B. De Baets, A historical account of types of fuzzy sets and their relationships, IEEE Trans. Fuzzy Syst. 24 (1) (2016) 179–194.

[11] H. Bustince, J. Fernandez, H. Hagras, F. Herrera, M. Pagola, E. Barrenechea, Interval type-2 fuzzy sets are generalization of interval-valued fuzzy sets: toward a wider view on their relationship, IEEE Trans. Fuzzy Syst. 23 (5) (2015) 1876–1882.

[12] C.I. Byrnes, A. Isidori, J.C. Willems, Passivity, feedback equivalence and the global stabilization of minimum phase nonlinear systems, IEEE Trans. Autom. Control 36 (1999) 1228–1240.

[13] J. Cao, P. Li, H. Liu, An interval fuzzy controller for vehicle active suspension systems, IEEE Trans. Intell. Transp. Syst. 11 (4) (2010) 885–895.

[14] M. Chaabane, O. Bachelier, M. Souissi, D. Mehdi, Stability and stabilization of continuous descriptor systems: an LMI approach, Math. Probl. Eng. 2006 (2006) 1–15.

[15] M. Chadli, D. Darouach, Novel bounded real lemma for discrete-time descriptor systems: application to H_∞ control design, Automatica 48 (2012) 449–453.

[16] B. Chen, Y. Niu, Y. Zou, Sliding mode control for stochastic Markovian jumping systems with incomplete transition rate, IET Control Theory Appl. 7 (10) (2013) 1330–1338.

[17] B. Chen, Y. Niu, Y. Zou, Adaptive sliding mode control for stochastic Markovian jumping systems with actuator degradation, Automatica 49 (6) (2013) 1748–1754.

[18] J. Chen, Y. Sun, H. Min, F. Sun, Y. Zhang, New results on static output feedback H_∞ control for fuzzy singularly perturbed systems: a linear matrix inequality approach, Int. J. Robust Nonlinear Control 23 (6) (2013) 681–694.

[19] J.R. Chen, F. Liu, Robust reliable H_∞ control for discrete-time Markov jump linear systems with actuator failures, J. Syst. Eng. Electron. 19 (2008) 965–973.

[20] S.J. Chen, J.H. Chou, Stability robustness of linear discrete singular time-delay systems with structured parameter uncertainties, IEE Proc., Control Theory Appl. 150 (2003) 296–302.

[21] V. Chellaboina, W. Haddad, A. Kamath, A dissipative dynamical systems approach to stability analysis of time delay systems, Int. J. Robust Nonlinear Control 15 (2005) 25–33.
[22] Y.Y. Chen, S.L. Sun, C.T. Chen, Static output feedback stabilization for discrete singular large-scale control systems, in: 2010 International Conference on Machine Learning and Cybernetics (ICMLC), Qingdao, China, 2010, pp. 2317–2321.
[23] D. Cobb, Feedback and pole placement in descriptor variable systems, Int. J. Control 33 (1981) 1135–1146.
[24] D. Cobb, Controllability, observability and duality in singular systems, IEEE Trans. Autom. Control 29 (1984) 1076–1082.
[25] L. Dai, Singular Control Systems, Springer-Verlag, Berlin, 1989.
[26] D.P. De Farias, J.C. Geromel, J.B.R. Do Val, O.L.V. Costa, Output feedback control of Markov jump linear systems in continuous-time, IEEE Trans. Autom. Control 45 (2000) 944–949.
[27] Y. Ding, H. Liu, J. Cheng, H_∞ filtering for a class of discrete-time singular Markovian jump systems with time-varying delays, ISA Trans. 53 (4) (2014) 1054–1060.
[28] X. Dong, Robust strictly dissipative control for singular systems with time-delay and parameter uncertainties, in: Proceedings of Chinese Control Conference, Zhangjiajie, China, 2007, pp. 578–582.
[29] X. Dong, Robust strictly dissipative control for discrete singular systems, IET Control Theory Appl. 1 (2007) 1060–1067.
[30] B. Du, J. Lam, Stability analysis of static recurrent neural networks using delay-partitioning and projection, Neural Netw. 22 (2009) 343–347.
[31] B. Du, J. Lam, Z. Shu, A delay-partitioning projection approach to stability analysis of neutral systems, in: IFAC World Congress 2008, Seoul, 2008.
[32] B. Du, J. Lam, Z. Shu, Z. Wang, A delay-partitioning projection approach to stability analysis of continuous systems with multiple delay components, IET Control Theory Appl. 3 (2009) 383–390.
[33] Z. Du, Q. Zhang, L. Liu, New delay-dependent robust stability of discrete singular systems with time-varying delay, Asian J. Control 13 (1) (2011) 136–147.
[34] G. Duan, Analysis and Design of Descriptor Linear Systems, Springer, London, UK, 2010.
[35] G. Duan, Y. Li, Robust passive filtering for time-delay T-S fuzzy systems, in: Proceedings of 8th IEEE International Conference on Control and Automation, Xiamen, China, 2010, pp. 405–410.
[36] M. Fang, Delay-dependent stability analysis for discrete singular systems with time-varying delays, Acta Autom. Sin. 36 (5) (2010) 751–755.
[37] Y. Fang, K.A. Loparo, Stabilization of continuous-time jump linear systems, IEEE Trans. Autom. Control 47 (2002) 1590–1603.
[38] Z. Fei, H. Gao, P. Shi, New results on stabilization of Markovian jump systems with time delay, Automatica 45 (2009) 2300–2306.
[39] Y. Feng, M. Yagoubi, P. Chevrel, On dissipativity of continuous-time singular systems, in: Proceedings of 18th Mediterranean Conference on Control & Automation Congress, Marrakech, Morocco, 2010, pp. 839–844.
[40] Z. Feng, J. Lam, Stability and dissipativity analysis of distributed delay cellular neural networks, IEEE Trans. Neural Netw. 22 (2011) 976–981.
[41] Z. Feng, J. Lam, Robust reliable dissipative filtering for discrete delay singular systems, Signal Process. 92 (12) (2012) 3010–3025.
[42] Z. Feng, J. Lam, Reduced-order dissipative filtering for discrete-time singular systems, in: Proceedings of IEEE International Symposium on Industrial Electronics, Taipei, Taiwan, 2013.
[43] Z. Feng, J. Lam, Reliable dissipative control for singular Markovian systems, Asian J. Control 15 (3) (2013) 901–910.

[44] Z. Feng, J. Lam, H. Gao, B. Du, Improved stability and stabilization results for discrete singular delay systems via delay partitioning, in: Proceedings of 48th IEEE Conference on Decision and Control and 28th Chinese Control Conference, Shanghai, China, 2009, pp. 7210–7215.
[45] Z. Feng, J. Lam, H. Gao, B. Du, Delay-dependent robust H_∞ controller synthesis for discrete singular delay systems, Int. J. Robust Nonlinear Control 21 (2011) 1880–1902.
[46] Z. Feng, W. Li, J. Lam, New admissibility analysis for discrete singular systems with time-varying delay, Appl. Math. Comput. 265 (2015) 1058–1066.
[47] Z. Feng, P. Shi, Two equivalent sets: application to singular systems, Automatica 77 (2017) 198–205.
[48] Z. Feng, W.X. Zheng, L.G. Wu, Reachable set estimation of T-S fuzzy systems with time-varying delay, IEEE Trans. Fuzzy Syst. 25 (4) (2017) 878–891.
[49] L.R. Fletcher, A. Aasaraai, On disturbance decoupling in descriptor systems, SIAM J. Control Optim. 27 (1989) 1319–1332.
[50] E. Fridman, U. Shaked, H_∞ control of linear state-delay descriptor systems: an LMI approach, Linear Algebra Appl. 315–352 (2002) 271–302.
[51] E. Fridman, U. Shaked, Delay-dependent stability and H_∞ control: constant and time-varying delays, Int. J. Control 76 (2003) 48–60.
[52] E. Fridman, G. Tsodik, H_∞ control of distributed and discrete delay systems via discretized Lyapunov functional, Eur. J. Control 11 (2009) 1–11.
[53] H. Gao, T. Chen, New results on stability of discrete-time systems with time-varying state delay, IEEE Trans. Autom. Control 52 (2007) 328–334.
[54] H. Gao, T. Chen, J. Lam, A new delay system approach to network based control, Automatica 44 (2008) 39–52.
[55] H. Gao, J. Lam, C. Wang, Y. Wang, Delay-dependent output-feedback stabilisation of discrete-time systems with time-varying state delay, IEE Proc., Control Theory Appl. 151 (2004) 691–698.
[56] H. Gao, C. Wang, A delay-dependent approach to robust H_∞ filtering for uncertain discrete-time state-delayed systems, IEEE Trans. Signal Process. 6 (2004) 1631–1640.
[57] Y. Gao, H. Li, M. Chadli, H.K. Lam, Static output-feedback control for interval type-2 discrete-time fuzzy systems, Complexity 2 (2014) 74–88.
[58] H. Gassara, A. El Hajjaji, M. Kchaou, M. Chaabane, Observer based (Q, V, R)-α-dissipative control for TS fuzzy descriptor systems with time delay, J. Franklin Inst. 351 (1) (2014) 187–206.
[59] M.B. Gorzalczany, An interval-valued fuzzy inference method basic properties, Fuzzy Sets Syst. 31 (2) (1989) 243–251.
[60] F. Gouaisbaut, D. Peaucelle, Delay-dependent stability analysis of linear time delay systems, in: IFAC Workshop on Time Delay System, L'Aquila, Italy, 2006.
[61] J. Grimm, Realization and canonicity for implicit systems, SIAM J. Control Optim. 26 (1988) 1331–1347.
[62] K. Gu, Y. Liu, Lyapunov–Krasovskii functional for uniform stability of coupled differential-functional equations, Automatica 45 (2009) 798–804.
[63] H.A. Hagras, A hierarchical type-2 fuzzy logic control architecture for autonomous mobile robots, IEEE Trans. Fuzzy Syst. 12 (4) (2004) 524–539.
[64] C. Han, L. Wu, P. Shi, Q. Zeng, On dissipativity of Takagi–Sugeno fuzzy descriptor systems with time-delay, J. Franklin Inst. 349 (10) (2012) 3170–3184.
[65] P.S. He, F. Liu, Exponential passive filtering for a class of nonlinear jump systems, J. Syst. Eng. Electron. 20 (2009) 829–837.
[66] Y. He, Q.G. Wang, C. Lin, M. Wu, Delay-range-dependent stability for systems with time-varying delay, Automatica 43 (2007) 371–376.

[67] H. Hemami, B.F. Wyman, Modeling and control of constrained dynamic system with application to biped locomotion in the frontal plane, IEEE Trans. Autom. Control 24 (1979) 526–535.
[68] D. Hinrichsen, W. Manthey, U. Helmke, Minimal partial realization by descriptor systems, Linear Algebra Appl. 326 (2001) 45–84.
[69] L. Huang, X. Mao, Stability of singular stochastic systems with Markovian switching, IEEE Trans. Autom. Control 56 (2) (2011) 424–429.
[70] J.Y. Ishihara, M.H. Terra, A new Lyapunov equation for discrete-time descriptor systems, in: Proceedings of American Control Conference, Denver, Colorado, 2003, pp. 5078–5082.
[71] J.Y. Ishihara, M.H. Terra, R.M. Sales, The full information and state feedback H_2 optimal controllers for descriptor systems, Automatica 39 (2003) 391–402.
[72] X. Ji, H. Shu, J. Chu, Delay-dependent robust stability of uncertain discrete singular time-delay systems, in: Proceedings of American Control Conference, Minneapolis, Minnesota, USA, 2006, pp. 3843–3848.
[73] J. Jiao, Robust stability and stabilization of discrete singular systems with interval time-varying delay and linear fractional uncertainty, Int. J. Autom. Comput. 9 (1) (2012) 8–15.
[74] N. Kablar, Dissipativity theory for singular systems. Part I: continuous-time case, in: Proceedings of 44th IEEE Conference on Decision and Control, and the European Control Conference, Seville, Spain, 2005, pp. 5639–5644.
[75] Y. Kao, J. Xie, C. Wang, H.R. Karimi, A sliding mode approach to H_∞ non-fragile observer-based control design for uncertain Markovian neutral-type stochastic systems, Automatica 52 (2015) 218–226.
[76] M. Kchaou, M. Souissi, A. Toumi, Delay-dependent stability and robust $L_2 - L_\infty$ control for a class of fuzzy descriptor systems with time-varying delay, Int. J. Robust Nonlinear Control 23 (3) (2013) 284–304.
[77] M.A. Khanesar, E. Kayacan, M. Teshnehlab, O. Kaynak, Extended Kalman filter based learning algorithm for type-2 fuzzy logic systems and its experimental evaluation, IEEE Trans. Ind. Electron. 59 (11) (2012) 4443–4455.
[78] J.H. Kim, Delay-dependent robust H_∞ filtering for uncertain discrete-time singular systems with interval time-varying delay, Automatica 46 (2010) 591–597.
[79] J.H. Kim, Reduced-order delay-dependent robust H_∞ filtering for uncertain discrete-time singular systems with time-varying delay, Automatica 47 (2011) 2801–2804.
[80] J.H. Kim, Discrete passive filtering of singular systems with time-varying delay, Appl. Math. Sci. 6 (2012) 4047–4055.
[81] J.H. Kim, J.H. Lee, H.B. Park, Robust H_∞ control of singular systems with time delays and uncertainties via LMI approach, in: Proceedings of American Control Conference, Anchorage, USA, 2006, pp. 620–621.
[82] A. Kumar, P. Daoutidis, Feedback control of nonlinear differential-algebraic-equation systems, AIChE J. 41 (1995) 619–636.
[83] G.A. Kurina, R. März, On linear-quadratic optimal control problems for time-varying descriptor systems, SIAM J. Control Optim. 42 (2004) 2062–2077.
[84] O. Kwon, M. Park, J. Park, S. Lee, E. Cha, New criteria on delay-dependent stability for discrete-time neural networks with time-varying delays, Neurocomputing 121 (2013) 185–194.
[85] H.K. Lam, H. Li, C. Deters, E.L. Secco, H.A. Wurdemann, K. Althoefer, Control design for interval type-2 fuzzy systems under imperfect premise matching, IEEE Trans. Ind. Electron. 61 (2) (2014) 956–968.
[86] H.K. Lam, M. Narimani, L.D. Seneviratne, LMI-based stability conditions for interval type-2 fuzzy-model-based control systems, in: 2011 IEEE International Conference on Fuzzy Systems, Taipei, Taiwan, 2011, pp. 298–303.

[87] H.K. Lam, L.D. Seneviratne, Stability analysis of interval type-2 fuzzy-model-based control systems, IEEE Trans. Syst. Man Cybern., Part B, Cybern. 38 (3) (2008) 617–628.

[88] J. Lam, Z. Shu, S. Xu, E.K. Boukas, Robust H_∞ control of descriptor discrete-time Markovian jump systems, Int. J. Control 80 (2007) 374–385.

[89] J. Lam, B. Zhang, Reachable set estimation for discrete-time linear systems with time delays, Int. J. Robust Nonlinear Control 52 (2015) 146–153.

[90] F.L. Lewis, A survey of linear singular systems, Circuits Syst. Signal Process. 5 (1986) 3–36.

[91] F.L. Lewis, A tutorial on the geometric analysis of linear time-invariant implicit systems, Automatica 28 (1992) 119–137.

[92] C.J. Li, H.Y. An, Y.F. Feng, Dissipative filtering for linear discrete-time systems via LMI, in: 2009 Chinese Control and Decision Conference, Guilin, China, 2009, pp. 3866–3870.

[93] H. Li, Y. Pan, P. Shi, Y. Shi, Switched fuzzy output feedback control and its application to mass-spring-damping system, IEEE Trans. Fuzzy Syst. 24 (6) (2016) 1259–1269.

[94] H. Li, Y. Pan, Q. Zhou, Filter design for interval type-2 fuzzy systems with D stability constraints under a unified frame, IEEE Trans. Fuzzy Syst. 23 (3) (2015) 719–725.

[95] H. Li, P. Shi, D. Yao, L. Wu, Observer-based adaptive sliding mode control of nonlinear Markovian jump systems, Automatica 64 (2016) 133–142.

[96] H. Li, X. Sun, L. Wu, H.K. Lam, State and output feedback control of interval type-2 fuzzy systems with mismatched membership functions, IEEE Trans. Fuzzy Syst. 23 (6) (2015) 1943–1957.

[97] H. Li, C. Wu, P. Shi, Y. Gao, Control of nonlinear networked systems with packet dropouts: interval type-2 fuzzy model-based approach, IEEE Trans. Cybern. 45 (11) (2015) 2378–2389.

[98] H. Li, C. Wu, S. Yin, H.K. Lam, Observer-based fuzzy control for nonlinear networked systems under unmeasurable premise variables, IEEE Trans. Fuzzy Syst. 24 (5) (2016) 1233–1245.

[99] H. Li, C. Wu, S. Yin, L. Wu, H.K. Lam, Y. Gao, Filtering of interval type-2 fuzzy systems with intermittent measurements, IEEE Trans. Cybern. 46 (3) (2015) 668–678.

[100] H. Li, S. Yin, Y. Pan, H.K. Lam, Model reduction for interval type-2 Takagi–Sugeno fuzzy systems, Automatica 61 (2015) 308–314.

[101] Z. Li, J. Wang, H. Shao, Delay-dependent dissipative control for linear time-delay systems, J. Franklin Inst. 339 (2002) 529–542.

[102] Q. Liang, J.M. Mendel, Equalization of nonlinear time-varying channels using type-2 fuzzy adaptive filters, IEEE Trans. Fuzzy Syst. 8 (5) (2000) 551–563.

[103] C. Lin, Q.G. Wang, T.H. Lee, Delay-dependent LMI conditions for stability and stabilization of T-S fuzzy systems with bounded time-delay, Fuzzy Sets Syst. 157 (9) (2006) 1229–1247.

[104] J. Liu, J. Zhang, Note on stability of discrete-time time-varying delay systems, IET Control Theory Appl. 6 (2) (2012) 335–339.

[105] L. Liu, Z. Han, W. Li, H_∞ non-fragile observer-based sliding mode control for uncertain time-delay systems, J. Franklin Inst. 347 (2) (2010) 567–576.

[106] M. Liu, G. Sun, Observer-based sliding mode control for Itô stochastic time-delay systems with limited capacity channel, J. Franklin Inst. 39 (4) (2012) 1602–1616.

[107] W.Q. Liu, W.Y. Yan, K.L. Teo, On initial instantaneous jumps of singular systems, IEEE Trans. Autom. Control 40 (1995) 1650–1655.

[108] W.Q. Liu, W.Y. Yan, K.L. Teo, Initial and transient response improvement for singular systems, Automatica 32 (1996) 461–464.

[109] Y. Liu, Y. Niu, Y. Zou, Non-fragile observer-based sliding mode control for a class of uncertain switched systems, J. Franklin Inst. 351 (2) (2014) 952–963.

[110] Y. Liu, Z. Wang, W. Wang, Reliable H_∞ filtering for discrete time-delay systems with randomly occurred nonlinearities via delay-partitioning method, Signal Process. 91 (2011) 713–727.

[111] Y. Liu, Z. Wang, W. Wang, Reliable H_∞ filtering for discrete time-delay Markovian jump systems with partly unknown transition probabilities, Int. J. Adapt. Control Signal Process. 25 (2011) 554–570.

[112] J.C. Lo, D.L. Wu, Dissipative filtering for discrete fuzzy systems, in: 2008 IEEE International Conference on Fuzzy Systems, Hong Kong, China, 2008, pp. 361–365.

[113] J. Löfberg, Modeling and solving uncertain optimization problems in YALMIP, in: Proceedings of the 17th IFAC World Congress, Seoul, South Korea, 2008, pp. 1337–1341.

[114] R. Lozano, B. Maschke, B. Brogliato, O. Egeland, Dissipative Systems Analysis and Control: Theory and Applications, Springer-Verlag, New York, 2000.

[115] K. Lu, J. Qiu, M.S. Mahmoud, Robust passive filter design for uncertain singular stochastic Markov jump systems with mode-dependent time delays, in: Proceedings of International Conference on Intelligent Control and Information Processing, Dalian, China, 2011, pp. 385–390.

[116] Q. Lu, P. Shi, H.K. Lam, Y. Zhao, Interval type-2 fuzzy model predictive control of nonlinear networked control systems, IEEE Trans. Fuzzy Syst. 23 (6) (2015) 2317–2328.

[117] R. Lu, F. Han, A. Xue, Dissipative control for stochastic descriptor systems with time-delays, in: 2009 Chinese Control and Decision Conference, Shenyang, China, 2009, pp. 2825–2829.

[118] R. Lu, H. Su, J. Chu, S. Zhou, M. Fu, Reduced-order H_∞ filtering for discrete-time singular systems with lossy measurements, IET Control Theory Appl. 4 (1) (2010) 151–163.

[119] R. Lu, Y. Xu, A. Xue, J. Zheng, Networked control with state reset and quantized measurements: observer-based case, IEEE Trans. Ind. Electron. 60 (11) (2013) 5206–5213.

[120] D.G. Luenberger, A. Arbel, Singular dynamic Leontief systems, Econometrica 45 (1977) 991–995.

[121] S. Ma, Z. Cheng, C. Zhang, Delay-dependent robust stability and stabilisation for uncertain discrete singular systems with time-varying delays, IET Control Theory Appl. 1 (4) (2007) 1086–1095.

[122] S. Ma, C. Zhang, Z. Cheng, Delay-dependent robust H_∞ control for uncertain discrete-time singular systems with time-delays, J. Comput. Appl. Math. 217 (2008) 194–211.

[123] M.S. Mahmoud, Delay-dependent dissipativity of singular time-delay systems, IMA J. Math. Control Inf. 26 (2009) 45–58.

[124] M.S. Mahmoud, Y. Xia, Design of reduced-order l_2-l_∞ for singular discrete-time systems using strict linear matrix inequalities, IEE Proc., Control Theory Appl. 4 (2010) 509–519.

[125] Z. Mao, B. Jiang, P. Shi, H_∞ fault detection filter design for networked control systems modelled by discrete Markovian jump systems, IET Control Theory Appl. 1 (2007) 1336–1343.

[126] F. Martinelli, Optimality of a two-threshold feedback control for a manufacturing system with a production dependent failure rate, IEEE Trans. Autom. Control 52 (2007) 1937–1942.

[127] I. Masubuchi, Dissipativity inequalities for continuous-time descriptor systems with applications to synthesis of control gains, Syst. Control Lett. 55 (2006) 158–164.

[128] I. Masubuchi, Output feedback controller synthesis for descriptor systems satisfying closed-loop dissipativity, Automatica 43 (2007) 339–345.

[129] I. Masubuchi, Y. Kamitane, A. Ohara, N. Suda, H_∞ control for descriptor systems: a matrix inequalities approach, Automatica 33 (1997) 669–673.
[130] M. Meisami-Azad, J. Mohammadpour, K.M. Grigoriadis, Dissipative analysis and control of state-space symmetric systems, Automatica 45 (2009) 1574–1579.
[131] J.M. Mendel, R.I. John, F. Liu, Interval type-2 fuzzy logic systems made simple, IEEE Trans. Fuzzy Syst. 14 (6) (2006) 808–821.
[132] X. Meng, J. Lam, H. Gao, B. Du, A delay-partitioning approach to stability analysis of discrete-time systems, Automatica 46 (2010) 610–614.
[133] Y.S. Moon, P. Park, W.H. Kwon, Y.S. Lee, Delay-dependent robust stabilization of uncertain state-delayed systems, IEE Proc., Control Theory Appl. 74 (2001) 1447–1455.
[134] S. Mou, H. Gao, T. Chen, New delay-range-dependent stability condition for linear system, in: Proceedings of 7th World Congress on Intelligent Control and Automation, Chongqing, China, 2008, pp. 313–316.
[135] K. Mourad, T. Fernando, C. Mohamed, A partitioning approach for H_∞ control of singular time-delay systems, Optim. Control Appl. Methods 34 (2013) 472–486.
[136] P.T. Nam, P.N. Pathirana, H. Trinh, Discrete Wirtinger-based inequality and its application, J. Franklin Inst. 352 (5) (2015) 1893–1905.
[137] R. Newcomb, Some circuits and systems applications of semistate theory, Circuits Syst. Signal Process. 8 (1989) 235–260.
[138] R. Nikoukhah, S.L. Campbell, F. Delebecque, Kalman filtering for general discrete-time linear systems, IEEE Trans. Autom. Control 44 (1999) 1829–1839.
[139] P.N. Paraskevopoulos, F.N. Koumboulis, Unifying approach to observers for regular and singular systems, IEE Proc. Part D, Control Theory Appl. 138 (1991) 561–572.
[140] P. Park, J. Ko, C. Jeong, Reciprocally convex approach to stability of systems with time-varying delays, Automatica 47 (1) (2011) 235–238.
[141] P.G. Park, A delay-dependent stability criterion for systems with uncertain time-invariant delays, IEEE Trans. Autom. Control 44 (1999) 876–877.
[142] D. Peaucelle, D. Arzelier, O. Bachelier, J. Bernussou, A new robust D-stability condition for real convex polytopic uncertainty, Syst. Control Lett. 40 (2000) 21–30.
[143] L. Qin, G. Duan, Robust dissipative control for uncertain descriptor linear system with time delay, in: Proceedings of the 6th World Congress on Intelligent Control, Dalian, China, 2006, pp. 2327–2333.
[144] A. Rehm, F. Allgöwer, An LMI approach towards stabilizaiton of discrete-time descriptor systems, in: Proceedings of the IFAC 15th Triennial Word Congress, Barcelona, Spain, 2002.
[145] H. Rotstein, M. Sznaier, M. Idan, H_2/H_∞ filtering theory and an aerospace application, Int. J. Robust Nonlinear Control 6 (1996) 347–366.
[146] R. Sakthivel, M. Joby, K. Mathiyalagan, S. Santra, Mixed H_∞ and passive control for singular Markovian jump systems with time delays, J. Franklin Inst. 352 (2015) 4446–4466.
[147] A.V. Savkin, I.R. Petersen, Structured dissipativeness and absolute stability of nonlinear systems, Int. J. Control 62 (2) (1995) 443–460.
[148] R. Sepúlveda, O. Castillo, P. Melin, A. Rodríguez-Díaz, O. Montiel, Experimental study of intelligent controllers under uncertainty using type-1 and type-2 fuzzy logic, Inf. Sci. 177 (10) (2007) 2023–2048.
[149] A. Seuret, F. Gouaisbaut, Wirtinger-based integral inequality: application to time-delay systems, Automatica 49 (2013) 2860–2866.
[150] H. Shao, Q.L. Han, New stability criteria for linear discrete-time systems with interval-like time-varying delays, IEEE Trans. Autom. Control 56 (2011) 619–625.
[151] S. Shao, New delay-dependent criteria for systems with interval delay, Automatica 45 (2009) 744–749.

[152] H. Shen, J. Park, L. Zhang, Z. Wu, Robust extended dissipative control for sampled-data Markov jump systems, Int. J. Control 87 (8) (2014) 1549–1564.
[153] H. Shen, Z. Wu, J. Park, Reliable mixed passive and H_∞ filtering for semi-Markov jump systems with randomly occurring uncertainties and sensor failures, Int. J. Robust Nonlinear Control 25 (2015) 3231–3251.
[154] H. Shen, S. Xu, X. Song, G. Shi, Passivity-based control for Markovian jump systems via retarded output feedback, Circuits Syst. Signal Process. 31 (2012) 189–202.
[155] P. Shi, E.K. Boukas, H_∞-control for Markovian jumping linear systems with parametric uncertainty, Automatica 95 (1997) 75–99.
[156] P. Shi, E.K. Boukas, On H_∞ control design for singular continuous-time delay systems with parametric uncertainties, Nonlinear Dyn. Syst. Theory 4 (2004) 59–71.
[157] S. Shi, Q. Zhang, Z. Yuan, W.Q. Liu, Hybrid impulsive control for switched singular systems, IET Control Theory Appl. 5 (2011) 103–111.
[158] Z. Shu, J. Lam, An augmented system approach to static output-feedback stabilization with H_∞ performance for continuous-time plants, Int. J. Robust Nonlinear Control 19 (2009) 768–785.
[159] B.L. Stevens, F.L. Lewis, Aircraft Modeling, Dynamics and Control, Wiley, New York, 1991.
[160] J. Sun, G.P. Liu, J. Chen, D. Rees, Improved delay-range-dependent stability criteria for linear systems with time-varying delays, Automatica 46 (2) (2010) 466–470.
[161] K. Takaba, N. Morihira, T. Katayama, A generalized Lyapunov theorem for descriptor system, Syst. Control Lett. 24 (1995) 49–51.
[162] Z. Tan, Y.C. Soh, L. Xie, Dissipative control of linear discrete-time systems, Automatica 35 (1999) 1557–1564.
[163] F. Tao, Q. Zhao, Synthesis of active fault-tolerant control based on Markovian jump system models, IET Control Theory Appl. 1 (2007) 1160–1168.
[164] E. Tian, D. Yue, C. Peng, Reliable control for networked control systems with probabilistic sensors and actuators faults, IET Control Theory Appl. 4 (2010) 1478–1488.
[165] E. Uezato, M. Ikeda, Strict LMI conditions for stability, robust stabilization, and H_∞ control of descriptor systems, in: Proceedings of the 38th IEEE Conference on Decision and Control, Phoenix, Arizona, USA, 1999, pp. 4092–4097.
[166] A. Varga, On stabilization methods of descriptor systems, Syst. Control Lett. 24 (1995) 133–138.
[167] R.J. Veillette, J.B. Medanic, W.R. Perkins, Design of reliable control systems, IEEE Trans. Autom. Control 37 (1992) 290–304.
[168] C. Verghese, C. Bernard, K. Thomas, A generalized state-space for singular systems, IEEE Trans. Autom. Control 24 (1981) 811–831.
[169] C. Wang, Strictly passive static output feedback control for discrete singular system, in: 8th World Congress on Intelligent Control and Automation (WCICA), Jinan, China, 2010, pp. 3801–3804.
[170] G. Wang, H. Bo, Stabilization of singular Markovian jump systems by generally observer-based controllers, Asian J. Control 18 (2016) 328–339.
[171] G. Wang, Q. Zhang, X. Yan, Analysis and Design of Singular Markovian Jump Systems, Springer, Switzerland, 2015.
[172] H. Wang, X. Zhao, A. Xue, R. Lu, Delay-dependent robust control for uncertain discrete singular systems with time-varying delay, J. Zhejiang Univ. Sci. A 5 (2009) 1655–1664.
[173] H.S. Wang, C.F. Yung, F.R. Chang, Bounded real lemma and H_∞ control for descriptor systems, IEE Proc., Control Theory Appl. 145 (1998) 316–322.
[174] J. Wang, H. Wang, A. Xue, R. Lu, Delay-dependent H_∞ control for singular Markovian jump systems with time delay, Nonlinear Anal. Hybrid Syst. 8 (2013) 1–12.

[175] R. Wang, Y. Liu, Asymptotic stability and robustness for discrete-time singular systems with multiple time-delays, in: Proceedings of 3th World Congress on Intelligent Control and Automation, Hefei, China, 2000.
[176] T. Wang, H. Gao, J. Qiu, A combined adaptive neural network and nonlinear model predictive control for multirate networked industrial process control, IEEE Trans. Neural Netw. Learn. Syst. 27 (2) (2016) 416–425.
[177] Y. Wang, Z. Wang, J. Liang, A delay fractioning approach to global synchronization of delayed complex networks with stochastic disturbances, Phys. Lett. A 372 (2008) 6066–6073.
[178] Z. Wang, Daniel W.C. Ho, Y. Liu, X. Liu, Robust H_∞ control for a class of nonlinear discrete time-delay stochastic systems with missing measurements, Automatica 45 (2009) 684–691.
[179] Z. Wang, Y. Liu, X. Liu, H_∞ filtering for uncertain stochastic time-delay systems with sector-bounded nonlinearities, Automatica 44 (2008) 1268–1277.
[180] J.C. Willems, Dissipative dynamical systems part I: general theory, Arch. Ration. Mech. Anal. 45 (5) (1972) 321–351.
[181] L. Wu, D.W.C. Ho, Sliding mode control of singular stochastic hybrid systems, Automatica 46 (4) (2010) 779–783.
[182] L. Wu, P. Shi, H. Gao, State estimation and sliding-mode control of Markovian jump singular systems, IEEE Trans. Autom. Control 55 (2010) 1213–1219.
[183] L. Wu, P. Shi, H. Gao, C. Wang, H_∞ filtering for 2D Markovian jump systems, Automatica 44 (2008) 1849–1858.
[184] L. Wu, X. Su, P. Shi, Sliding mode control with bounded L_2 gain performance of Markovian jump singular time-delay systems, Automatica 48 (2012) 1929–1933.
[185] L. Wu, W. Zheng, Passivity-based sliding mode control of uncertain singular time-delay systems, Automatica 45 (2009) 2120–2127.
[186] W. Wu, X. Ma, X. Xu, Application research on data segment by robust H_∞ filter estimation, in: Chinese Control and Decision Conference, Guilin, China, 2009, pp. 680–683.
[187] Z. Wu, J.H. Park, H. Shu, B. Song, J. Chu, Mixed H_∞ and passive filtering for singular systems with time delays, IEE Proc., Control Theory Appl. 93 (2013) 1705–1711.
[188] Z. Wu, J.H. Park, H. Su, J. Chu, Admissibility and dissipativity analysis for discrete-time singular systems with mixed time-varying delays, Appl. Math. Comput. 218 (13) (2012) 7128–7138.
[189] Z. Wu, J.H. Park, H. Su, J. Chu, Dissipativity analysis for singular systems with time-varying delays, Appl. Math. Comput. 218 (2011) 4605–4613.
[190] Z. Wu, P. Shi, H. Su, J. Chu, Dissipativity analysis for discrete-time stochastic neural networks with time-varying delays, IEEE Trans. Neural Netw. Learn. Syst. 24 (3) (2013) 345–355.
[191] Z. Wu, H. Su, J. Chu, Robust stabilization for uncertain discrete singular systems with state delay, Int. J. Robust Nonlinear Control 18 (16) (2008) 1532–1550.
[192] Z. Wu, H. Su, J. Chu, H_∞ filtering for singular systems with time-varying delay, Int. J. Robust Nonlinear Control 20 (2009) 1269–1284.
[193] Z. Wu, H. Su, J. Chu, Delay-dependent H_∞ control for singular Markovian jump systems with time delay, Optim. Control Appl. Methods 30 (2009) 443–461.
[194] Z. Wu, H. Su, J. Chu, Improved results on delay-dependent H_∞ control for singular time-delay systems, Acta Autom. Sin. 35 (8) (2009) 1101–1106.
[195] Z. Wu, H. Su, J. Chu, H_∞ filtering for singular Markovian jump systems with time delay, Int. J. Robust Nonlinear Control 20 (2010) 939–957.
[196] Z. Wu, W. Zhou, Delay-dependent robust H_∞ control for uncertain singular time-delay systems, IEE Proc., Control Theory Appl. 1 (2007) 1234–1241.
[197] Z. Wu, W. Zhou, Delay-dependent robust stabilization for uncertain singular systems with discrete and distributed delays, J. Control Theory Appl. 6 (2) (2008) 171–176.

[198] X. Xia, R. Li, J. An, On delay-fractional-dependent stability criteria for Takagi-Sugeno fuzzy systems with interval delay, Math. Probl. Eng. 1 (2014) 1–13.
[199] Y. Xia, E. Boukas, P. Shi, J. Zhang, Stability and stabilization of continuous-time singular hybrid systems, Automatica 45 (2009) 1504–1509.
[200] L. Xie, C.E. de Souza, Robust H_∞ control for linear systems with norm-bounded time-varying uncertainty, IEEE Trans. Autom. Control 37 (1992) 1188–1191.
[201] S. Xie, L. Xie, C.E. De Souza, Robust dissipative control for linear systems with dissipative uncertainty, Int. J. Control 70 (1998) 169–192.
[202] S. Xie, L. Xie, S. Ge, Dissipative control for linear systems with time-varying uncertainty, in: Proceedings of American Control Conference, Albuquerque, New Mexico, 1997, pp. 2531–2535.
[203] X. Xie, D. Yue, H. Zhang, Y. Xue, Control synthesis of discrete-time T-S fuzzy systems via a multi-instant homogenous polynomial approach, IEEE Trans. Cybern. 46 (3) (2016) 630–640.
[204] J. Xiong, J. Lam, H. Gao, D.W.C. Ho, On robust stabilization of Markovian jump systems with uncertain switching probabilities, Automatica 41 (2005) 897–903.
[205] B. Xu, W. Ji, Y. Qian, Application of robust H_∞ filtering in TV tracking system for maneuvering targets, in: IEEE International Conference on Control and Automation, Guangzhou, China, 2007, pp. 757–760.
[206] S. Xu, J. Lam, H_∞ filtering for singular systems, IEEE Trans. Autom. Control 48 (2003) 2217–2222.
[207] S. Xu, J. Lam, H_∞ model reduction for discrete-time singular systems, Syst. Control Lett. 48 (2003) 121–133.
[208] S. Xu, J. Lam, Robust stability and stabilization of discrete singular systems: an equivalent characterization, IEEE Trans. Autom. Control 49 (2004) 568–574.
[209] S. Xu, J. Lam, Robust Control and Filtering of Singular Systems, Springer, Berlin, 2006.
[210] S. Xu, J. Lam, Reduced-order H_∞ filtering for singular systems, Syst. Control Lett. 56 (2007) 48–57.
[211] S. Xu, J. Lam, A survey of linear matrix inequality techniques in stability analysis of delay systems, Int. J. Syst. Sci. 39 (2008) 1095–1113.
[212] S. Xu, J. Lam, C. Yang, Robust H_∞ control for discrete singular systems with state delay and parameter uncertainty, Dyn. Contin. Discrete Impuls. Syst., Ser. B, Appl. Algorithms 9 (2002) 539–554.
[213] S. Xu, J. Lam, C. Yang, Robust H_∞ control for uncertain singular systems with state delay, Int. J. Robust Nonlinear Control 13 (2003) 1213–1223.
[214] S. Xu, J. Lam, Y. Zou, An improved characterization of bounded realness for singular delay systems and its applications, Int. J. Robust Nonlinear Control 18 (2008) 263–277.
[215] S. Xu, B. Song, J. Lu, J. Lam, Delay-dependent H_∞ filtering for singular Markovian jump time-delay systems, Signal Process. 90 (6) (2010) 1815–1824.
[216] S. Xu, P. Van Dooren, R. Stefan, J. Lam, Robust stability and stabilization for singular systems with state delay and parameter uncertainty, IEEE Trans. Autom. Control 47 (2002) 1122–1128.
[217] S. Xu, C. Yang, Stabilization of discrete-time singular systems: a matrix inequalities approach, Automatica 35 (1999) 1613–1617.
[218] Y. Xu, R. Lu, H. Peng, K. Xie, A. Xue, Asynchronous dissipative state estimation for stochastic complex networks with quantized jumping coupling and uncertain measurements, IEEE Trans. Neural Netw. Learn. Syst. 28 (2) (2017) 268–277.
[219] G.H. Yang, D. Ye, Adaptive reliable H_∞ filtering against sensor failures, IEEE Trans. Signal Process. 55 (2007) 3161–3171.

[220] X.F. Yang, W. Yang, J.C. Liu, Non-fragile strictly dissipative filter for discrete-time nonlinear systems with sector-bounded nonlinearities, in: 2010 Chinese Control and Decision Conference, Xuzhou, China, 2010, pp. 4483–4488.
[221] J. Yoneyama, K. Hoshino, Static output feedback control design for Takagi-Sugeno descriptor fuzzy systems, in: 2015 International Conference on Informatics, Electronics & Vision (ICIEV 2015), Fukuoka, Japan, 2015.
[222] D. Yue, Q.L. Han, Robust H_∞ filter design of uncertain descriptor systems with discrete and distributed delays, IEEE Trans. Signal Process. 52 (2004) 3200–3212.
[223] D. Yue, Q.L. Han, Delay-dependent robust H_∞ controller design for uncertain descriptor systems with time-varying discrete and distributed delays, IEE Proc., Control Theory Appl. 152 (2005) 628–638.
[224] D. Yue, J. Lam, D.W.C. Ho, Reliable H_∞ controller design for linear systems, Automatica 37 (2001) 717–725.
[225] D. Yue, J. Lam, D.W.C. Ho, Reliable H_∞ control of uncertain descriptor systems with multiple time delays, IET Control Theory Appl. 150 (2003) 557–564.
[226] D. Yue, J. Lam, D.W.C. Ho, Delay-dependent robust exponential stability of uncertain descriptor systems with time-varying delays, Dyn. Contin. Discrete Impuls. Syst., Ser. B, Appl. Algorithms 12 (2005) 129–150.
[227] L.A. Zadeh, Quantitative fuzzy semantics, Inf. Sci. 3 (2) (1971) 159–176.
[228] H. Zeng, Y. He, M. Wu, J. She, New results on stability analysis for systems with discrete distributed delay, Automatica 60 (2015) 189–192.
[229] B. Zhang, S. Xu, Y. Zou, Improved stability criterion and its applications in delayed controller design for discrete-time systems, Automatica 44 (2008) 2963–2967.
[230] B. Zhang, S. Zhou, D. Du, Robust H_∞ filtering of delayed singular systems with linear fractional parametric uncertainties, Circuits Syst. Signal Process. 25 (2006) 627–647.
[231] C.Z. Zhang, G. Feng, J.B. Qiu, Y.Y. Shen, Control synthesis for a class of linear network-based systems with communication constraints, IEEE Trans. Ind. Electron. 60 (8) (2013) 3339–3348.
[232] C.Z. Zhang, J.F. Hu, J.B. Qiu, Q.J. Chen, Event-triggered nonsynchronized H_∞ filtering for discrete-time T-S fuzzy systems based on piecewise Lyapunov functions, IEEE Trans. Syst. Man Cybern. Syst. 47 (8) (2017) 2330–2341.
[233] G. Zhang, Y. Xia, P. Shi, New bounded real lemma for discrete-time singular systems, Automatica 44 (2008) 886–890.
[234] H. Zhang, Z.H. Guan, G. Feng, Reliable dissipative control for stochastic impulsive systems, Automatica 44 (2008) 1004–1010.
[235] J. Zhang, Y. Xia, E.K. Boukas, New approach to H_∞ control for Markovian jump singular system, IET Control Theory Appl. 4 (11) (2010) 2273–2284.
[236] L. Zhang, J. Lam, Necessary and sufficient conditions for analysis and synthesis of Markov jump linear systems with incomplete transition descriptions, IEEE Trans. Autom. Control 55 (2010) 1695–1701.
[237] L. Zhang, J. Lam, S. Xu, On positive realness of descriptor systems, IEEE Trans. Circuits Syst. I, Fundam. Theory Appl. 49 (3) (2002) 401–407.
[238] P. Zhang, J. Cao, G. Wang, Mode-independent guaranteed cost control of singular Markovian delay jump systems with switching probability rate design, Int. J. Innov. Comput. Inf. Control 10 (4) (2014) 1291–1303.
[239] W. Zhang, J. Wang, Y. Liang, Z. Han, Improved delay-range-dependent stability criterion for discrete-time systems with interval time-varying delay, J. Inf. Comput. Sci. 8 (2011) 3321–3328.
[240] W. Zhang, Y. Zhao, L. Sheng, Some remarks on stability of stochastic singular systems with state-dependent noise, Automatica 51 (2015) 273–277.

[241] X.M. Zhang, Q.L. Han, Delay-dependent robust H_∞ filtering for uncertain discrete-time systems with time-varying delay based on a finite sum inequality, IEEE Trans. Circuits Syst. II, Express Briefs 53 (2006) 1446–1470.
[242] J. Zhao, D.J. Hill, Dissipativity theory for switched systems, IEEE Trans. Autom. Control 53 (4) (2008) 941–953.
[243] T. Zhao, J. Xiao, L. Han, C. Qiu, J. Huang, Static output feedback control for interval type-2 T-S fuzzy systems based on fuzzy Lyapunov functions, Asian J. Control 16 (6) (2014) 1702–1712.
[244] Y. Zhao, R. Zhang, C. Shao, Estimating delay bounds of time-delay systems with fuzzy model-based approach, Int. J. Innov. Comput. Inf. Control 9 (11) (2013) 4343–4357.
[245] Y. Zhao, W. Zhang, New results on stability of singular stochastic Markov jump systems with state-dependent noise, Int. J. Robust Nonlinear Control 26 (2016) 2169–2186.
[246] R.X. Zhong, Z. Yang, Delay-dependent robust control of descriptor systems with time delay, Asian J. Control 8 (1) (2008) 36–44.
[247] B. Zhou, J. Hu, G.R. Duan, Strict linear matrix inequality characterisation of positive realness for linear discrete-time descriptor systems, IEE Proc., Control Theory Appl. 4 (2010) 1277–1281.
[248] S. Zhou, W. Zheng, Delay-dependent dissipative control for a class of non-linear system via Takagi-Sugeno fuzzy descriptor model with time delay, IET Control Theory Appl. 8 (7) (2014) 451–461.
[249] W. Zhou, H.Q. Lu, Delay-dependent robust H_∞ stability for uncertain discrete-time singular system, in: Proceedings of 6th International Conference on Machine Learning Cybernetics, 2007, pp. 655–659.
[250] Y. Zhou, Z. Li, Energy-to-peak filtering for singular systems: the discrete-time case, IEE Proc., Control Theory Appl. 2 (2008) 773–781.
[251] Y. Zhou, Z. Li, Reduced-order L_2-L_∞ filtering for singular systems: a linear matrix inequality approach, IEE Proc., Control Theory Appl. 2 (2008) 228–238.
[252] H. Zhu, X. Zhang, S. Cui, Further results on H_∞ control for discrete-time uncertain singular systems with interval time-varying delays in state and input, Optim. Control Appl. Methods 34 (3) (2013) 328–347.

Notations

T	matrix transposition
$\mathbb{R}^n$	n-dimensional Euclidean space
$\mathbb{R}_+$	set of nonnegative real numbers
$\mathbb{Z}_+$	set of nonnegative integers
$\mathbb{Z}_{\geq s_1}$	$\{k \in \mathbb{Z}_+ \mid k \geq s_1\}$
$\mathbb{Z}_{[s_1, s_2]}$	$\{k \in \mathbb{Z}_+ \mid s_2 \geq k \geq s_1\}$
$\Vert\cdot\Vert$	Euclidean vector norm
$\Vert w\Vert_2$	$\sqrt{\sum_{k=0}^{\infty} w^T(k)w(k)}$ for $\{w(k)\} \in l_2[0, \infty)$
$\Vert e\Vert_\infty$	$\sqrt{\sup_k\{e^T(k)e(k)\}}$ for $\{e(k)\} \in l_\infty[0, \infty)$
$\Vert x\Vert_{\mathcal{S}}$	distance of a vector x to set $\mathcal{S}$, $x \in \mathbb{R}^n$, $\mathcal{S} \subset \mathbb{R}^n$, $\Vert x\Vert_{\mathcal{S}} := \inf_{y \in \mathcal{S}} \Vert x - y\Vert$
$l_2\,[0, \infty)$	space of square-summable infinite sequences
$l_\infty[0, \infty)$	space of all essentially bounded functions
$\lceil a \rceil$	nearest integer greater than or equal to a
$\star$	an ellipsis for the terms that are introduced by symmetry
$\mathcal{S}^n_{\succ 0}$	set of $n \times n$ symmetric positive definite matrices
$\mathcal{S}_1 \ominus \mathcal{S}_2$	$\{s \in \mathbb{R}^n \vert s + s_2 \in \mathcal{S}_1, \forall s_2 \in \mathcal{S}_2\}$
$\mathcal{S}_1 \oplus \mathcal{S}_2$	$\{s_1 + s_2 \in \mathbb{R}^n \vert s_1 \in \mathcal{S}_1, s_2 \in \mathcal{S}_2\}$
$co\{\mathcal{S}\}$	convex hull of $\mathcal{S}$
$\mathcal{B}^n$	$\{x \in \mathbb{R}^n : \Vert x\Vert_2 \leq 1\}$
$\mathcal{C}^1$	space of continuously differentiable functions
$A \Leftrightarrow B$	A is equivalent to B
$diag\{\cdots\}$	block-diagonal matrix
I	identity matrix
0	zero matrix
$P > 0\ (\geq 0)$	P is a real symmetric and positive definite (semipositive definite) matrix
$eig(A)$	set of eigenvalues of a matrix A
sup	supremum or least upper bound
inf	infimum or greatest lower bound
$d(a, S)$	distance between an element a and a set S
$d_H(S_1, S_2)$	Hausdorff distance between sets S_1 and S_2
$\triangleq$	equal by definition or is defined by

Index

T

U

W

Printed in the United States
By Bookmasters